Biochemical Biomarkers and Neurodegenerative Diseases

Biochemical Biomarkers and Neurodegenerative Diseases

Editor

Marcello Ciaccio

MDPI • Basel • Beijing • Wuhan • Barcelona • Belgrade • Manchester • Tokyo • Cluj • Tianjin

Editor
Marcello Ciaccio
Department of Biomedicine,
Neurosciences and Advanced
Diagnostics, Institute of Clinical
Biochemistry, Clinical Molecular
Medicine and Laboratory
Medicine
University of Palermo
Palermo
Italy

Editorial Office
MDPI
St. Alban-Anlage 66
4052 Basel, Switzerland

This is a reprint of articles from the Special Issue published online in the open access journal *Brain Sciences* (ISSN 2076-3425) (available at: www.mdpi.com/journal/brainsci/special_issues/ND_Biomarkers).

For citation purposes, cite each article independently as indicated on the article page online and as indicated below:

LastName, A.A.; LastName, B.B.; LastName, C.C. Article Title. *Journal Name* **Year**, *Volume Number*, Page Range.

ISBN 978-3-0365-1722-3 (Hbk)
ISBN 978-3-0365-1721-6 (PDF)

Contents

About the Editor . . . . **vii**

Preface to "Biochemical Biomarkers and Neurodegenerative Diseases" . . . . **ix**

Marcello Ciaccio
Biochemical Biomarkers and Neurodegenerative Diseases
Reprinted from: *Brain Sciences* **2021**, *11*, 940, doi:10.3390/brainsci11070940 . . . . **1**

**Paola Feraco, Cesare Gagliardo, Giuseppe La Tona, Eleonora Bruno, Costanza D'angelo,
Maurizio Marrale, Anna Del Poggio, Maria Chiara Malaguti, Laura Geraci, Roberta Baschi,
Benedetto Petralia, Massimo Midiri and Roberto Monastero**
Imaging of Substantia Nigra in Parkinson's Disease: A Narrative Review
Reprinted from: *Brain Sciences* **2021**, *11*, 769, doi:10.3390/brainsci11060769 . . . . **3**

**Concetta Scazzone, Luisa Agnello, Bruna Lo Sasso, Giuseppe Salemi, Caterina Maria
Gambino, Paolo Ragonese, Giuseppina Candore, Anna Maria Ciaccio, Rosaria Vincenza
Giglio, Giulia Bivona, Matteo Vidali and Marcello Ciaccio**
FOXP3 and GATA3 Polymorphisms, Vitamin D3 and Multiple Sclerosis
Reprinted from: *Brain Sciences* **2021**, *11*, 415, doi:10.3390/brainsci11040415 . . . . **17**

**Antonino Lupica, Vincenzo Di Stefano, Andrea Gagliardo, Salvatore Iacono, Antonia
Pignolo, Salvatore Ferlisi, Angelo Torrente, Sonia Pagano, Massimo Gangitano and Filippo
Brighina**
Inherited Neuromuscular Disorders: Which Role for Serum Biomarkers?
Reprinted from: *Brain Sciences* **2021**, *11*, 398, doi:10.3390/brainsci11030398 . . . . **27**

**Giulia Bivona, Bruna Lo Sasso, Caterina Maria Gambino, Rosaria Vincenza Giglio, Concetta
Scazzone, Luisa Agnello and Marcello Ciaccio**
The Role of Vitamin D as a Biomarker in Alzheimer's Disease
Reprinted from: *Brain Sciences* **2021**, *11*, 334, doi:10.3390/brainsci11030334 . . . . **47**

**Tiziana Colletti, Luisa Agnello, Rossella Spataro, Lavinia Guccione, Antonietta Notaro,
Bruna Lo Sasso, Valeria Blandino, Fabiola Graziano, Caterina Maria Gambino, Rosaria
Vincenza Giglio, Giulia Bivona, Vincenzo La Bella, Marcello Ciaccio and Tommaso Piccoli**
Prognostic Role of CSF -amyloid 1–42/1–40 Ratio in Patients Affected by Amyotrophic Lateral
Sclerosis
Reprinted from: *Brain Sciences* **2021**, *11*, 302, doi:10.3390/brainsci11030302 . . . . **55**

**Marcello Ciaccio, Bruna Lo Sasso, Concetta Scazzone, Caterina Maria Gambino, Anna Maria
Ciaccio, Giulia Bivona, Tommaso Piccoli, Rosaria Vincenza Giglio and Luisa Agnello**
COVID-19 and Alzheimer's Disease
Reprinted from: *Brain Sciences* **2021**, *11*, 305, doi:10.3390/brainsci11030305 . . . . **65**

Filomena Iannuzzi, Vincenza Frisardi, Lucio Annunziato and Carmela Matrone
Might Fibroblasts from Patients with Alzheimer's Disease Reflect the Brain Pathology? A Focus
on the Increased Phosphorylation of Amyloid Precursor Protein Tyr_{682} Residue
Reprinted from: *Brain Sciences* **2021**, *11*, 103, doi:10.3390/brainsci11010103 . . . . **75**

Cesare Gagliardo, Roberto Cannella, Costanza D'Angelo, Patrizia Toia, Giuseppe Salvaggio, Paola Feraco, Maurizio Marrale, Domenico Gerardo Iacopino, Marco D'Amelio, Giuseppe La Tona, Ludovico La Grutta and Massimo Midiri
Transcranial Magnetic Resonance Imaging-Guided Focused Ultrasound with a 1.5 Tesla Scanner: A Prospective Intraindividual Comparison Study of Intraoperative Imaging
Reprinted from: *Brain Sciences* 2021, *11*, 46, doi:10.3390/brainsci11010046 85

Efthalia Angelopoulou, Yam Nath Paudel, Chiara Villa and Christina Piperi
Arylsulfatase A (ASA) in Parkinson's Disease: From Pathogenesis to Biomarker Potential
Reprinted from: *Brain Sciences* 2020, *10*, 713, doi:10.3390/brainsci10100713 97

Hyun-Jun Choi, Sun Joo Cha, Jang-Won Lee, Hyung-Jun Kim and Kiyoung Kim
Recent Advances on the Role of GSK3 in the Pathogenesis of Amyotrophic Lateral Sclerosis
Reprinted from: *Brain Sciences* 2020, *10*, 675, doi:10.3390/brainsci10100675 111

Germán Fernando Gutiérrez Aguilar, Iván Alquisiras-Burgos, Javier Franco-Pérez, Narayana Pineda-Ramírez, Alma Ortiz-Plata, Ismael Torres, José Pedraza-Chaverri and Penélope Aguilera
Resveratrol Prevents GLUT3 Up-Regulation Induced by Middle Cerebral Artery Occlusion
Reprinted from: *Brain Sciences* 2020, *10*, 651, doi:10.3390/brainsci10090651 127

About the Editor

Marcello Ciaccio

Marcello Ciaccio is a full professor of Clinical Biochemistry and Molecular Medicine and is the dean of the School of Medicine at the University of Palermo. He is a director of the Institute of Clinical Biochemistry, Clinical Molecular Medicine, and Laboratory Medicine within the Department of Biomedicine, Neurosciences, and Advanced Diagnostics at the University of Palermo. He is also the director of the Department of Laboratory Medicine—A.O.U.P. Palermo "P. Giaccone". He has been the chair of the Italian Society of Clinical Biochemistry and Clinical Molecular Biology (SIBioC). He is part of the editorial board of several peer-reviewed journals, and he serves as a reviewer for many prestigious journals. He is the author of more than 400 scientific publications in national and international journals.

Preface to "Biochemical Biomarkers and Neurodegenerative Diseases"

Neurodegenerative diseases represent an important health burden, and their early identification is crucial. However, today, this remains challenging. Thus, intense research is being conducted and remains ongoing in order to identify biomarkers for assisting clinicians in the management of neurodegenerative diseases, from screening to diagnosis, prognosis, and treatment.

Marcello Ciaccio
Editor

Editorial

Biochemical Biomarkers and Neurodegenerative Diseases

Marcello Ciaccio [1,2] 

[1] Institute of Clinical Biochemistry, Clinical Molecular Medicine and Laboratory Medicine, Department of Biomedicine, Neurosciences, and Advanced Diagnostics, University of Palermo, 90127 Palermo, Italy; marcello.ciaccio@unipa.it

[2] Department of Laboratory Medicine, AOUP "P. Giaccone", 90127 Palermo, Italy

Citation: Ciaccio, M. Biochemical Biomarkers and Neurodegenerative Diseases. *Brain Sci.* **2021**, *11*, 940. https://doi.org/10.3390/brainsci11070940

Received: 28 June 2021
Accepted: 12 July 2021
Published: 16 July 2021

Publisher's Note: MDPI stays neutral with regard to jurisdictional claims in published maps and institutional affiliations.

Neurodegenerative diseases (ND) are a heterogeneous group of disorders characterized by progressive dysfunction and loss of neurons in different areas of the central nervous system or peripheral nervous system. NDs, including Alzheimer's disease (AD), Parkinson's disease (PD), and motor neuron disease (MND), represent a big challenge for scientific research due to their prevalence, cost, basic pathophysiological mechanisms, and lack of mechanism-based treatments. The diagnosis, prognosis, and monitoring of such disorders are complex and rely mainly on clinical criteria. In the last decades, biochemical markers have emerged as promising tools in the field of ND. The articles belonging to this Special Issue of "Biochemical Biomarkers and Neurodegenerative Disorders" encompass the last literature evidence on the importance of biomarkers in the management of ND, from screening to diagnosis, prognosis, and treatment.

Scazzone et al. explored the association among Vitamin D3, single nucleotide polymorphisms (SNPs), and Multiple Sclerosis (MS) in a retrospective case-control study [1]. They showed that MS patients had significantly lower levels of Vitamin D3 than controls, but no association among SNPs, Vitamin D3, and MS risk was found. The role of hypovitaminosis D in MS risk has been widely investigated in the last decades, and some literature evidence supports the hypothesis that Vitamin D3 could be involved in MS pathogenesis. Noteworthy, Vitamin D3 status is influenced by both genetic and environmental factors. Thus, many Authors investigated the possible influence of genetic variants in Vitamin D3 related genes on MS risk, achieving contrasting results [2].

Beyond its well-known role in calcium homeostasis, Vitamin D3 has pleiotropic functions, including immune-regulation and neurological function [3]. Thus, its possible role as a biomarker or risk factor in several autoimmune and neurodegenerative diseases has been evaluated. Bivona et al. described the current knowledge on the role of Vitamin D3 in Alzheimer's Disease (AD), stating that a definite conclusion cannot be drawn because controversial findings have been found across the studies [4].

Another interesting area of research is the role of circulating biomarkers in Inherited Neuromuscular Disorders (INMD), defined as a heterogeneous group of genetic diseases characterized by progressive muscle degeneration and weakness and associated with long-term disability. They represent rare disorders whose diagnosis is based on an extensive clinical evaluation with complementary genetic analysis. Due to the presence of genetic heterogeneity and lack of segregation in sporadic cases, a definite diagnosis is challenging. Thus, serum biomarkers are strongly sought after. Lupica et al. described several promising biomarkers that could help clinicians in the diagnostic workup of INMD [5].

Another rare disease with important clinical consequences is Amyotrophic Lateral Sclerosis (ALS). Many efforts are ongoing to find prognostic biomarkers of this devastating disease. Colletti et al. found that beta-amyloid 1–42 (Aβ 1–42) could be involved in the pathogenesis of ALS, and the Aβ 1–42/Aβ 1–40 ratio could represent a biomarker of prognosis [6].

Finally, an interesting article was focused on the current COVID-19 pandemic, raising the question if SARS-CoV-2 infection could induce long-term neurological consequences [7]. Notable, SARS-CoV-2 is a neurotropic virus and, consequently, it could predispose and accelerate the development of neurological disorders, such as AD. However, we may have the answer to such an interesting question in the next few years.

Funding: This research received no external funding.

Conflicts of Interest: The author declares no conflict of interests.

References

1. Scazzone, C.; Agnello, L.; Lo Sasso, B.; Salemi, G.; Gambino, C.M.; Ragonese, P.; Candore, G.; Ciaccio, A.M.; Giglio, R.V.; Bivona, G.; et al. FOXP3 and GATA3 Polymorphisms, Vitamin D3 and Multiple Sclerosis. *Brain Sci.* **2021**, *11*, 415. [CrossRef] [PubMed]
2. Scazzone, C.; Agnello, L.; Bivona, G.; Lo Sasso, B.; Ciaccio, M. Vitamin D and Genetic Susceptibility to Multiple Sclerosis. *Biochem. Gen.* **2021**, *59*, 1–30. [CrossRef] [PubMed]
3. Bivona, G.; Agnello, L.; Bellia, C.; Iacolino, G.; Scazzone, C.; Lo Sasso, B.; Ciaccio, M. Non-Skeletal Activities of Vitamin D: From Physiology to Brain Pathology. *Medicina* **2019**, *55*, 341. [CrossRef] [PubMed]
4. Bivona, G.; Lo Sasso, B.; Gambino, C.M.; Giglio, R.V.; Scazzone, C.; Agnello, L.; Ciaccio, M. The Role of Vitamin D as a Biomarker in Alzheimer's Disease. *Brain Sci.* **2021**, *11*, 334. [CrossRef] [PubMed]
5. Lupica, A.; Di Stefano, V.; Gagliardo, A.; Iacono, S.; Pignolo, A.; Ferlisi, S.; Torrente, A.; Pagano, S.; Gangitano, M.; Brighina, F. Inherited Neuromuscular Disorders: Which Role for Serum Biomarkers? *Brain Sci.* **2021**, *11*, 398. [CrossRef] [PubMed]
6. Colletti, T.; Agnello, L.; Spataro, R.; Guccione, L.; Notaro, A.; Lo Sasso, B.; Blandino, V.; Graziano, F.; Gambino, C.M.; Giglio, R.V.; et al. Prognostic Role of CSF β-amyloid 1-42/1-40 Ratio in Patients Affected by Amyotrophic Lateral Sclerosis. *Brain Sci.* **2021**, *11*, 302. [CrossRef] [PubMed]
7. Ciaccio, M.; Lo Sasso, B.; Scazzone, C.; Gambino, C.M.; Ciaccio, A.M.; Bivona, G.; Piccoli, T.; Giglio, R.V.; Agnello, L. COVID-19 and Alzheimer's Disease. *Brain Sci.* **2021**, *11*, 305. [CrossRef] [PubMed]

Review

Imaging of Substantia Nigra in Parkinson's Disease: A Narrative Review

Paola Feraco [1,2], Cesare Gagliardo [3,*], Giuseppe La Tona [3], Eleonora Bruno [3], Costanza D'angelo [3], Maurizio Marrale [4], Anna Del Poggio [5], Maria Chiara Malaguti [6], Laura Geraci [7], Roberta Baschi [8], Benedetto Petralia [2], Massimo Midiri [3] and Roberto Monastero [8]

1. Department of Experimental, Diagnostic and Specialty Medicine (DIMES), University of Bologna, Via S. Giacomo 14, 40138 Bologna, Italy; paola.feraco@apss.tn.it
2. Neuroradiology Unit, S. Chiara Hospital, 38122 Trento, Italy; benedetto.petralia@apss.tn.it
3. Section of Radiological Sciences, Department of Biomedicine, Neurosciences & Advanced Diagnostics, School of Medicine, University of Palermo, 90127 Palermo, Italy; giuseppe.latona@unipa.it (G.L.T.); elebru.91@gmail.com (E.B.); costanza.dangelo@gmail.com (C.D.); massimo.midiri@unipa.it (M.M.)
4. Department of Physics and Chemistry, University of Palermo, 90128 Palermo, Italy; maurizio.marrale@unipa.it
5. Department of Neuroradiology and CERMAC, San Raffaele Scientific Institute, San Raffaele Vita-Salute University, 20132 Milan, Italy; delpoggio.anna@hsr.it
6. Neurology Unit, S. Chiara Hospital, 38122 Trento, Italy; mariachiara.malaguti@apss.tn.it
7. Diagnostic and Interventional Neuroradiology Unit, A.R.N.A.S. Civico-Di Cristina-Benfratelli, 90127 Palermo, Italy; laura.geraci@inwind.it
8. Section of Neurology, Department of Biomedicine, Neurosciences & Advanced Diagnostics, School of Medicine, University of Palermo, 90127 Palermo, Italy; roberta.baschi@gmail.com (R.B.); roberto.monastero@unipa.it (R.M.)
* Correspondence: cesare.gagliardo@unipa.it

Citation: Feraco, P.; Gagliardo, C.; La Tona, G.; Bruno, E.; D'angelo, C.; Marrale, M.; Del Poggio, A.; Malaguti, M.C.; Geraci, L.; Baschi, R.; et al. Imaging of Substantia Nigra in Parkinson's Disease: A Narrative Review. *Brain Sci.* **2021**, *11*, 769. https://doi.org/10.3390/brainsci 11060769

Academic Editor: Emma Lane

Received: 5 March 2021
Accepted: 5 June 2021
Published: 9 June 2021

Publisher's Note: MDPI stays neutral with regard to jurisdictional claims in published maps and institutional affiliations.

Abstract: Parkinson's disease (PD) is a progressive neurodegenerative disorder, characterized by motor and non-motor symptoms due to the degeneration of the pars compacta of the substantia nigra (SNc) with dopaminergic denervation of the striatum. Although the diagnosis of PD is principally based on a clinical assessment, great efforts have been expended over the past two decades to evaluate reliable biomarkers for PD. Among these biomarkers, magnetic resonance imaging (MRI)-based biomarkers may play a key role. Conventional MRI sequences are considered by many in the field to have low sensitivity, while advanced pulse sequences and ultra-high-field MRI techniques have brought many advantages, particularly regarding the study of brainstem and subcortical structures. Nowadays, nigrosome imaging, neuromelanine-sensitive sequences, iron-sensitive sequences, and advanced diffusion weighted imaging techniques afford new insights to the non-invasive study of the SNc. The use of these imaging methods, alone or in combination, may also help to discriminate PD patients from control patients, in addition to discriminating atypical parkinsonian syndromes (PS). A total of 92 articles were identified from an extensive review of the literature on PubMed in order to ascertain the-state-of-the-art of MRI techniques, as applied to the study of SNc in PD patients, as well as their potential future applications as imaging biomarkers of disease. Whilst none of these MRI-Imaging biomarkers could be successfully validated for routine clinical practice, in achieving high levels of accuracy and reproducibility in the diagnosis of PD, a multimodal MRI-PD protocol may assist neuroradiologists and clinicians in the early and differential diagnosis of a wide spectrum of neurodegenerative disorders.

Keywords: magnetic resonance imaging; neuromelanin; nigrosome-1; iron; biomarkers; radiomics; neurodegenerative diseases; Parkinson's disease; parkinsonian disorders

1. Introduction

Parkinson's disease (PD) is a progressive neurodegenerative disease that is characterized by motor and non-motor symptoms. The disease is mostly sporadic, and it is caused by the interplay between genetic and environmental factors [1]. The neuropathology of PD is characterized by neuronal degeneration in the pars compacta of the substantia nigra (SNc) with dopaminergic denervation of the striatum. Subsequently, this neuronal loss is also seen in other brain regions and non-dopaminergic neurons, with multisite involvement of the central, peripheral, and autonomic nervous system [2]. The histological hallmark of PD are Lewy bodies, which are cytoplasmic inclusions resulting from an abnormal deposition of α-synuclein aggregates. The latter are not specific to PD, but they characterize other Parkinsonisms, such as Lewy body dementia and multiple system atrophy. It is currently unknown how Lewy bodies are related to the progression of PD, and current knowledge suggests that neuronal degeneration occurs due to several processes, including neuroinflammation, oxidative stress abnormalities, mitochondrial dysfunction, and abnormalities of protein quality control [3].

According to the *gut-to-brain* transmission model of PD pathology proposed by Braak et al., changes in brainstem and subcortical structures are more evident in early disease stages, while cortical structures are principally involved in advanced-stage PD [4]. Clinical manifestations of PD primarily include bradykinesia plus at least one of resting tremor and rigidity. Supportive criteria for a PD diagnosis are a beneficial response to dopamine therapy, the presence of medication-induced dyskinesia, and (early) olfactory dysfunction [5]. Motor symptoms progressively worsen with age, leading to near total immobility in advanced-stage PD. Although PD has been primarily identified as a movement disorder, non-motor symptoms (such as hyposmia, autonomic dysfunction, mood and sleep disorders, and cognitive impairment) are very common features of the disease, and they have been associated with a poor quality of life [6]. Specifically, cognitive impairment—which encompasses a spectrum varying from mild cognitive impairment to dementia [7,8]—has been associated with poor outcomes and mortality [9].

The diagnosis of PD is to date based on clinical features, with motor symptoms constituting the core criteria [1,5]. Over the past two decades, great efforts have been invested in evaluating reliable biomarkers for PD; none of these parameters, however, have been successfully validated for routine clinical practice [10]. Among these biomarkers, magnetic resonance imaging (MRI)-based biomarkers have undoubtedly contributed to the differential diagnosis between degenerative from secondary Parkinsonism [11].

Although the sensitivity of conventional MRI sequences (i.e., T2 or T1 weighted) has been considered as *poor*, particularly in early PD, the advent of high- and ultra-high-field MRI techniques has brought many advantages to the study of brainstem and subcortical structures [11]. The distinction between PD and atypical parkinsonian syndromes (PS) (including multiple system atrophy (MSA), progressive supranuclear palsy (PSP), corticobasal syndrome (CBS), and dementia with Lewy bodies (DLB)) is challenging to establish, particularly in the early stages. However, diagnostic accuracy is important in predicting a response to levodopa or anticholinergic therapy. In addition, whilst many studies have described the MRI features of PS (especially regarding PD and MSA subtype P), it is not easy to distinguish these diseases with routine MRI. Recently, improvements in MRI technology have made possible the study of changes within the SNc, which is particularly vulnerable to degeneration in PD [12]. The SNc, which is subdivided into nigrosomes and the nigral matrix, plays an essential role in regulating movements, with classic PD motor symptoms appearing when 30% or more of its dopaminergic neurons have vanished [13,14]. Recent efforts have been focused on the development of MRI sequences in order to enhance the characterization of SNc damage in PD. These efforts regard nigrosome imaging, neuromelanin-sensitive sequences, iron-sensitive sequences, and advanced diffusion imaging [11,13,15,16]. The use of these imaging methods, alone or in combination, is emerging as an encouraging early diagnostic biomarker of PD [17]. These techniques may help to discriminate PD patients from control patients or to discriminate PD patients from atypical PS. Whilst these imaging methods are not in common use and they require specific training to achieve high levels of accuracy and reproducibility [18], their inclusion in a multimodal MRI-PD protocol may assist clinicians and neuroradiologists an arriving at a differential diagnosis. The purpose of this narrative review is to evaluate the state-of-the-art of MRI techniques, as applied to the study of SN, and their potential future applications regarding the diagnosis and treatment of PD.

2. Materials and Methods

Extensive research in English was performed in January 2021 on the literature contained in PubMed (https://pubmed.ncbi.nlm.nih.gov, accessed on 10 January 2021), using the following keywords and their combinations: *Parkinsonisms, Parkinson's disease, magnetic resonance imaging, sustantia nigra, neuromelanin imaging, iron imaging, nigrosome imaging*, and *diffusion weighted and/or diffusion tensor imaging*. Preclinical and clinical studies from the last six years (January 2015–December 2020) were meticulously reviewed, focusing on new MRI sequences applied to PD. The publication date was restricted to the last six years to facilitate detailed comprehension regarding future perspectives of MRI in the study of SN in PD. Various relevant articles with this time range were included in order to maximize the topic coverage of this review. Where available, full texts in English were included, together with the most significant corresponding references. The exclusion criteria were unavailability of full text; non-English publications; case reports; reviews; and publications unrelated to PD, PS, or SN. Thereafter, the results were assessed according to the PRISMA statement (Figure 1). Recent MRI applications in PD were described from the included studies, and they were systematically organized and grouped according to a particular field of study and perspectives.

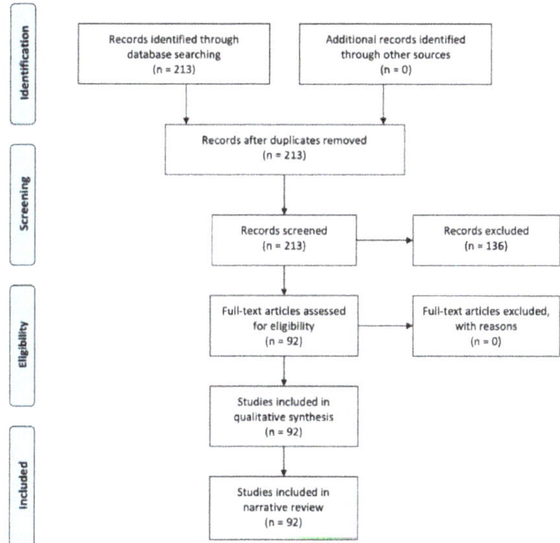

Figure 1. PRISMA flow diagram of studies selection.

3. Results

Two-hundred and thirteen articles were identified from the PubMed literature search. These were subsequently screened for relevance: 136 studies were excluded according to the exclusion criteria, while 92 were included. The full text was available for all of the 92 included studies, which were included in the qualitative analysis. Considering the different study techniques, we identified the following: 20 articles relating to neuromelanin, 23 regarding nigrosome-1 imaging, 22 discussing iron imaging, 16 relating to diffusion-weighted imaging, and 11 articles referring to radiomics.

4. Discussion

Various neuroimaging techniques (structural and functional) have been applied to Parkinsonism over the past two decades, each providing specific information regarding underlying brain disorders [11]. Specifically, MRI has been used as a tool with which to improve diagnostic accuracy in characterizing patients with extrapyramidal symptoms. Recent efforts have focused on the development of more precise and performing MRI sequences in order to obtain an enhanced characterization of the SNc damage in Parkinsonism. These efforts include nigrosome imaging, neuromelanin-sensitive sequences, iron-sensitive sequences, and advanced diffusion imaging [11,13,15,16]. The use of these imaging methods, alone or in combination, is emerging as an encouraging early diagnostic biomarker of PD. Recent and forthcoming applications of MRI have been summarized from the available literature and grouped by *field/s of application* for this review.

4.1. Neuromelanin Imaging

Neuromelanin (NM) is a black pigment that is composed of melanin, proteins, lipids, and metal ions, and it is found in the SNc (in the nigral matrix and the nigrosomes). NM plays a protective role against the accumulation of toxic catecholamine derivatives and oxidative stress [19]. NM normally accumulates during aging but is strongly reduced in patients with PD as a result of the selective loss of dopaminergic neurons containing NM. The latter has a paramagnetic T1 reduction effect on MRI due to the presence of melanin-iron complexes [20]. With high-resolution turbo spin echo (TSE) T1W images with a magnetization transfer (MT) pulse, it is possible to suppress brain tissue signals due to the prolongation of the T1 relaxation time [21]. Hence, nuclei-containing NM can be visualized as a separate hyperintense area relative to the surrounding hypointense brain tissue. Although the use of TSE T1W images has been consistently applied to visualizing NM, the gradient recalled echo (GRE) sequence with MT pulse has recently been demonstrated to achieve the sharpest contrast and lowest variability when compared with a T1W TSE-MT sequence [22].

NM-MRI is a validated technique with which to quantify the loss of dopaminergic neurons in the SN of patients with PD. The loss of SN hyperintensity in the T1W-MT sequence is associated with the loss of neuromelanin-containing neurons in PD and DLB, as confirmed in post-mortem studies [23]. Indeed, patients suffering from PD have significantly reduced NM signal in the SN (Figure 2), which invariably decreases on follow-up [24–29].

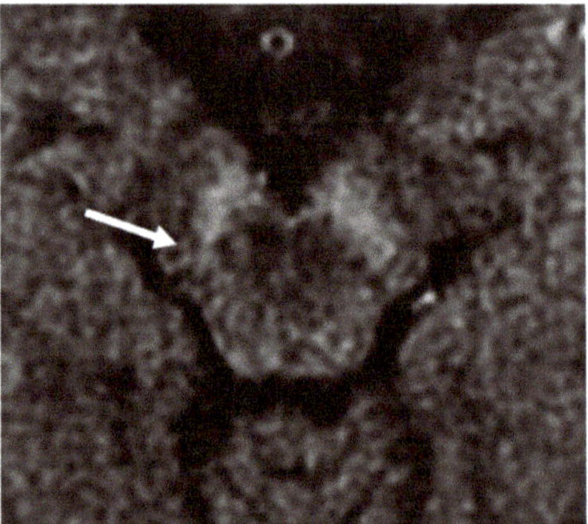

Figure 2. NM-MRI sequence with an explicit MT preparation pulse, scanned with a 1.5T MR scanner at the level of the SN in a PD patient with asymmetrical motor symptoms onset: the loss of hyperintensity in the posterolateral aspect of the right SN (arrow) correlated well with the clinical presentation.

Measuring NM-sensitive images correlates with elevated diagnostic accuracy for PD: the sensitivity and specificity of this technique to distinguish between PD and control patients are 88% and 80%, respectively [30]. The NM signal changes commence in the posterolateral motor areas of the SN, and then proceed to the medial areas [31]. Hence, the evaluation of longitudinal changes in the NM signal in PD patients could be used as a marker indicating disease progression. A reduction in NM signal has been reported to be not specific for motor or non-motor PD subtypes [32]. On the other hand, a potential diagnostic value of NM-MRI in discriminating PD motor phenotypes has been proposed [33]. Indeed, patients with postural instability gait difficulty phenotype display increased severe signal attenuation in the medial part of the SNc, in comparison with tremor-dominant PD patients [33]. Furthermore, the use of NM-MRI-based imaging is capable of differentiating between untreated essential tremor (ET) and de novo PD with a tremor-dominant phenotype [34]. Finally, a NM signal decrease has been observed in patients suffering from idiopathic rapid eye movement sleep behavior disorder, which is considered a prodromal phase of Parkinsonism and PD [35,36].

4.2. Nigrosome-1 Imaging

Nigrosomes are dopaminergic neurons within the SNc that are characterized by high NM levels and a paucity of iron. They can be subdivided into five different regions (nigrosome 1 to 5), the largest of which, nigrosome-1 (located in the dorsolateral part of SNc [12]), has been shown to play a key role in the neuropathology of PD. Indeed, the greatest loss of dopaminergic neurons in PD patients occurs in the nigrosome-1. It was first detected in vivo by 7.0-Tesla (7T) MRI as a hyperintense, ovoid area on T2*-weighted images, within the dorsolateral border of the hypointense SN pars compacta [37,38]. Similar findings can be found with the more commonly used 3-Tesla (3T) MRI [39]. By using T2* or susceptibility-weighted imaging (SWI), researchers have also termed this region *dorsolateral nigral hyperintensity* or a *swallow-tail sign* (STS) (Figure 3).

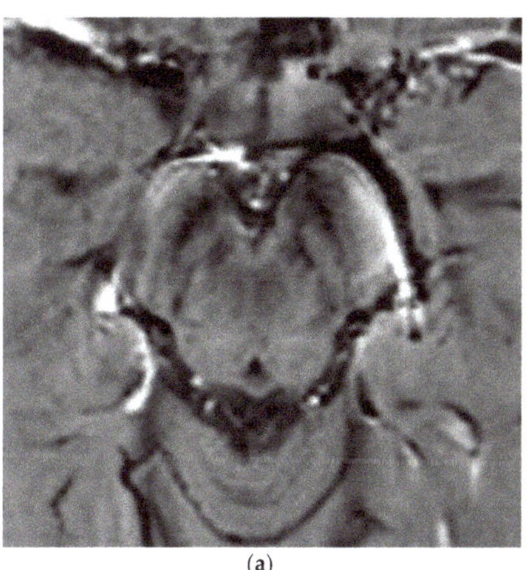

(a)

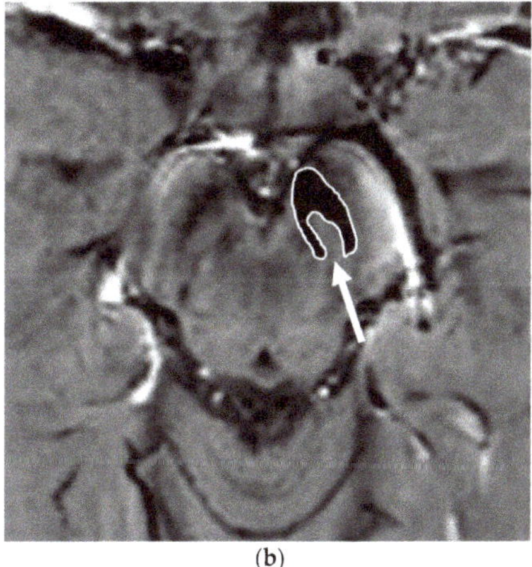

(b)

Figure 3. Susceptibility-weighted imaging (SWI) scan performed with a 3T MR scanner in a normally aging brain of a 65-year-old male who underwent a brain MRI examination for persistent headaches. A raw slice passing through the mesencephalon (a) and the same slice with superimposed highlighted SNc (white surrounded black ROI), thereby demonstrating the normal appearance of the nigrosome-1 (hyperintense area pointed by the white arrow) or swallow-tail sign (b).

Normal nigrosome-1 and the surrounding structure of the dorsolateral SN appear as a swallow tail [40], and they can be visualized in 95% of healthy subjects [41,42]. Iron deposits and microvessels have been reported as contributing to the hyposignal surrounding nigrosome-1 in the SWI of normal aged midbrains [43]. Nigrosome-1 in PD patients displays a significant loss of STS on T2* weighted images, probably due to a reduction in NM within dopaminergic neurons, an increase in free iron (which induces local inhomogeneity in the magnetic field resulting in signal loss), or a loss of paramagnetic NM–iron complexes [44,45]. As the disease advances, a loss of T2* hyperintensity in PD has been demonstrated to progress from nigrosome-1 to nigrosome-4 [46]. The absence of STS may assist in the differential diagnosis for PD if compared with controls and ET, ultimately reaching high sensitivity and specificity [17,40,47,48] (Figure 4).

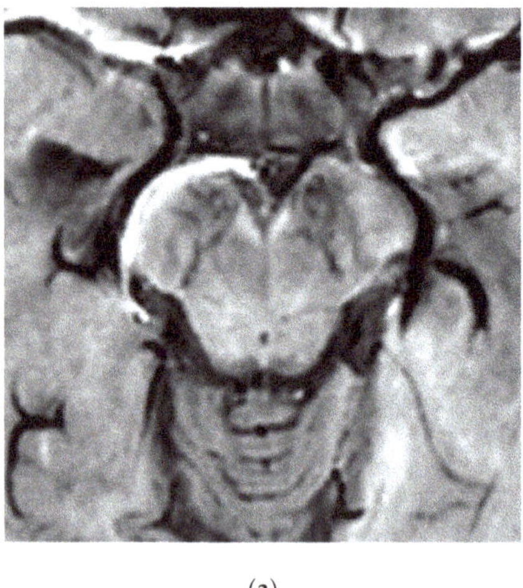

(**a**)

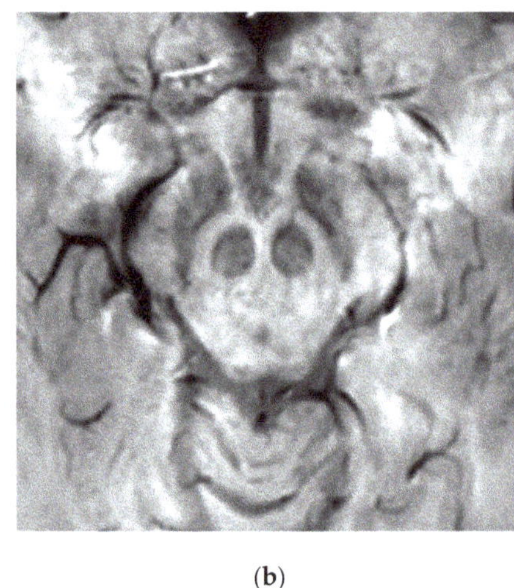

(**b**)

Figure 4. Susceptibility-weighted imaging (SWI) scan performed with a 3T MR. (**a**) Presence of regular swallow tail sign in a healthy patient; (**b**) loss of swallow tail sign in a patient with Parkinson's disease.

Moreover, the imaging of nigrosome-1 with 3T MR has been demonstrated to differentiate drug-induced Parkinsonism from idiopathic PD with elevated accuracy, thereby being of assistance in screening patients who required dopamine transporter imaging [49]. Furthermore, a loss of STS has also been observed in patients suffering from idiopathic rapid eye movement sleep behavior disorder and DLB [50,51]. Whilst the loss of nigrosome-1 on SWI sequences may assist as a potential imaging biomarker in the diagnosis of degenerative parkinsonian syndromes, it cannot differentiate between idiopathic PD and PS [52,53]. Nevertheless, it has been reported that anatomical changes of SN, detected via the SWI sequence at 7T, may distinguish MSA and PSP from CBD [54], thereby confirming the pathological heterogeneity of these diseases. Of note, nigrosome-1 has also been visualized on 3D FLAIR images as an hyperintense structure within otherwise surrounding hypointense dorsolateral SN. Its loss can be used to predict presynaptic dopaminergic function and to diagnose PD with a high degree of accuracy [55].

Recently, it has been reported that the combined visual analysis of SN (by using NM-MRI and nigrosome-1 imaging, displaying normal NM in SNc and nigrosome-1 loss) has enabled the distinction of MSA-P from PD and healthy controls [56]. Moreover, it has also been described that a stratification of the swallow tail sign, using a scale on SWI-map imaging, may serve as a useful imaging biomarker regarding the differential diagnosis of Parkinsonism [57]. However, the veracity of these results must be confirmed by larger cohort studies.

4.3. Iron Imaging

Together with a degeneration of dopaminergic neurons, iron overload has been implicated in the pathology and pathogenesis of PD and PS. Iron deposition initially occurs in

SN; however, abnormal iron levels have also been detected in the basal ganglia, thalamus, and cortex of PD patients [58].

With the introduction of MRI, the in vivo characterization of brain iron content has become possible. The possibility of quantifying regional brain iron overload may provide more knowledge regarding the correlation between iron accumulation and parkinsonian symptoms. Indeed, extensive data have emphasized the importance of SN iron increase in PD patients compared to controls [30,35,59]. From a technical perspective, the iron content can be assessed by evaluating T2 and T2* relaxation rates, using either magnitude (R2*) or phase (quantitative susceptibility mapping, QSM) imaging. Among these methods, R2 and R2* relaxometry (i.e.,: 1/T2*, proton transverse relaxation rate which reflects increased tissue iron content) considers heterogeneities from local and adjacent tissue as being more susceptible to influence from disturbances due to calcification, micro bleeds, and myelinated fibers [11]. The R2* values in the SNc have been reported to be significantly higher in de novo PD patients with a gradual increase, which is related to disease progression [60,61] (Figure 5).

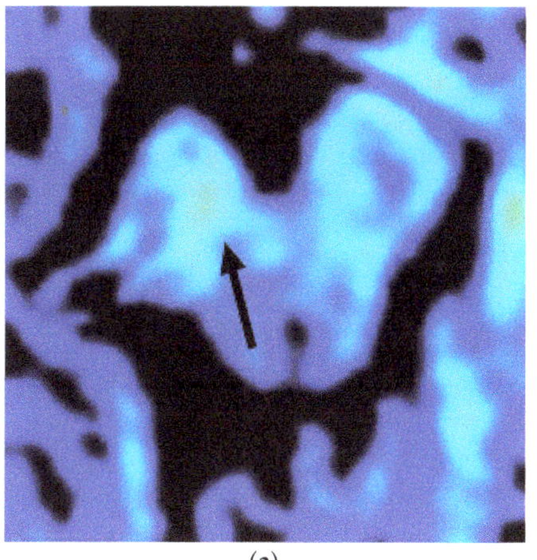
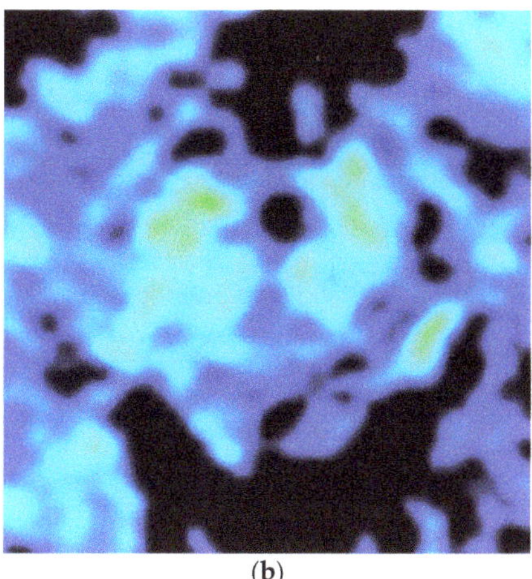

(a)   (b)

Figure 5. T2* map study (color scale) of two patients with PD in an evaluation of iron deposition within the SN (blue: less iron deposition; green: more iron deposition). The patient in (**a**) has a more evident asymmetrical iron deposition when compared to the patient in (**b**).

Since correlations between motor symptoms and high levels of R2* values within the SNc have been reported in PD, and R2* changes rapidly with disease progression, these methods can also be used in the prospective evaluation of PD patients [60,61]. Moreover, it has been reported that PD patients with early gait freezing pattern will have higher iron content, as evaluated by means of R2* relaxometry in the SNc, in comparison to those who do not [62].

Furthermore, QSM provides a direct measure of the local heterogeneities of the magnetic field by using a deconvolution method, which assists in eliminating the susceptibilities of surrounding structures [63]. It has been demonstrated that QSM is more sensitive than R2* in identifying iron overload in PD [63–65], even in the prodromal stage of the disease [66]. Values from QSM correlate with disease condition and duration [64,65,67,68], and they distinguish PD and PS [69]. Moreover, QSM can address iron variation within the SN [70] and lateral asymmetry of iron deposition, which is related to a manifestation of asymmetric signs and symptoms in PD [24]. When QSM is used in early- and advanced-stage PD patients, it is of note that it has been demonstrated that iron deposition affected SNc exclusively in the early stages of the disease, while in the late PD stage, iron deposition involved other regions, concomitant with SNc [71]. This latter finding indicates that QSM is a tool with which to monitor iron deposition and disease progression in PD. Specifically, changes in iron seem to be limited to the ventral aspect of SN [70], which has been reported to degenerate early in the course of the disease [72]. According to the distribution of the pathological involvement distinguishing the various forms of Parkinsonism, red and subtalamic nuclei are involved in PSP, together with SN, while iron deposition in MSA is significantly higher in the putamen [73]. Finally, all Parkinsonisms have been demonstrated to display increased susceptibility in the subcortical structures, thereby reflecting distinct topographical patterns of abnormal brain iron accumulation [74].

Both QSM and R2* may be effective tools in the differential diagnoses of degenerative PS, a fact that permits the tracking of dynamic changes that are associated with the pathological progression of these disorders. In addition, while QSM is more sensitive to the iron content of SN, R2* can be said to reflect pathological features, such as α-synuclein, in addition to iron deposits [75].

4.4. Advanced Diffusion Weighted Imaging Techniques

The loss of dopaminergic neurons in the SNc in the midbrain of patients with PD, as well as related nigral changes, are useful in differentiating neurodegenerative Parkinsonism from ET and other non-degenerative PS [64]. Routine conventional brain MRI, with an assessment of T1, T2, FLAIR, and proton density weighted sequences, is usually normal in early PD, while several studies have shown that advanced diffusion weighted imaging (diffusion tensor imaging, DTI) can assist in the early diagnosis of the disease [76]. The SN can be most clearly depicted when the diffusion gradient is applied in a left–right direction, thereby providing sharp contrast between the SN and the surrounding white matter. By depicting the white matter around SN as an area of high signal intensity, diffusion weighted imaging (DWI) reveals SN as a crescent-shaped area of low signal intensity between the tegmentum of the midbrain and the cerebral peduncle [77]. Several DTI studies have described early within-SN changes of PD patients, as compared to controls, and characterized by low fractional anisotropy (FA) values [78–81]. High resolution DTI in the SN can be useful in the diagnosis of PD, distinguish early-stage disease from controls, and has the potential to be a non-invasive early biomarker for PD diagnosis [76]. Moreover, higher SN-DTI changes have been reported to correlate with increasing dopaminergic deficits and declining α-synuclein and total tau protein concentrations in cerebrospinal fluid [80]. Furthermore, a nigral diffusion measure has been proposed as a measure of disease progression [81].

Whilst several authors have evaluated the application of DTI to studying SN in PD in the last 10 years, the results of these studies are conflicting [78–83]. For example, in their systematic review and meta-analysis, Hirata et al. estimated the mean change in SN-FA induced by PD and related diagnostic accuracy, and they concluded that SN-FA cannot be used as an isolated measure with which to diagnose PD since it displayed low sensitivity and specificity [83]. These discordant results have been hypothesized to be due to variable approaches used to demarcate the SN or unpredictable contamination of DTI evaluations from extracellular water compound or free water (FW). Hence, a FW mapping was developed, permitting the separation of the contribution of FW to DTI assessments (FW-corrected DTI) [84]. Using this approach, FW levels were observed to have increased in the posterior SN of PD patients, if compared to healthy controls, with a progressive increase during the progression of the disease. Moreover, the FW predicted the future changes in bradykinesia and cognitive status in a 1-year period, thereby providing a potential non-invasive progression marker of SN [84–87].

In addition to early PD, the FW in the posterior part of the SN has been reported to also have been increased in early MSA and PSP, as demonstrated by Arribarat et al. [88]. It has also been correlated with striatal dopaminergic denervation, thus reflecting motor and cognitive deficits. Compared PD and control patients, Planetta et al. observed an FW increase in the SN, in addition to the subthalamic nucleus, red nucleus, pedunculopontine nuclei, cerebellum, and basal ganglia in patients with PSP and MSA [86]. Several studies have demonstrated changes in water diffusivity in the SN (measured as a reduction in FA) in patients with MSA and a predominance of parkinsonian symptoms; this permits the differentiation with PD even in its early stages, when a volumetric reduction or signal change on conventional MRI are still absent [89].

There are other anatomical regions in PS (PSP, CBS, and MSA-C) that reveal microstructural anomalies, as detected by reduced FA and an increased MD. Studying changes in SN is, therefore, not indicated regarding a differential diagnosis of atypical Parkinsonism. For example, abnormal DTI in the cerebellum and MCP seems to be mainly involved in MSA-C; the DTI of SCP is mostly vulnerable in PSP. Abnormal DTI in supratentorial white matter regions appears to be mainly involved in CBS [90].

4.5. Radiomics, Artificial Intelligence, and Future Perspectives

Nowadays, an interest in NM-dedicated imaging and iron content quantification by means of artificial intelligence tools has only increased. A radiomic approach can be adopted to extract and analyze quantitative imaging features from medical images in garnering information to lend support to clinical decision making. These features are commonly correlated with patients' clinical data by advanced computational methods, including machine and deep learning algorithms; the latter are ever more frequently used to aid in the early or differential diagnosis of PD [91–94].

Machine learning techniques are typically based on an analysis of training data (i.e., features extracted from images) and the transformation of these features into class labels. The aim of this is to develop a model that is capable of classification, prediction, and the estimation of a situation from selected known data (e.g., images) [95]. Also known as *deep neural learning* or *deep neural network*, deep learning is a subfield of machine learning, the aim of which is to "imitate" the human brain in processing data and decision making. Deep learning permits the differential interpretation of data by means of the use of different layers in the network: each network defines specific features of the data in a hierarchical system [95], and data representation is performed in conjunction with prediction (obtained via classification or regression).

Various reports have described PD diagnoses by means of machine learning techniques, such as a support vector machine algorithm, as applied to DTI, and a voxel-based morphometry of the whole brain [96–98]. Recently, deep neural networks have shown great promise when creating markers for PD prognosis and diagnosis by adopting convolutional neural network (CNN) regarding NM-MRI acquisitions. This algorithm automatically segments the SN region; computes class activation maps for patient classification; and, therefore, acts as a computer-aided PD diagnostic framework, using the NM signal [92]. Using CNNs to create prognostic and diagnostic biomarkers of PD from NM-MRI, Shinde et al. demonstrated the higher performance of this method when compared to a radiomics classifier, discriminating PD from PS with an accuracy of 85.7%. They also demonstrated that the left SNc plays a key role in this classification, as compared to the right SNc [92].

Another application of SN segmentation via CNN has been reported by Krupička et al. [99]. Artificial neural networks were also used to validate a dynamic, atlas-based segmentation process of the SN and to quantify NM-rich brainstem structures in PD [100]. Moreover, the application of texture analysis, by means of QSM, has been reported to successfully distinguish PD from healthy control patients, with higher performance indices, compared to R2* texture analysis [93]. This combination of radiomics and CNN features from QSM could enhance the diagnostic accuracy of PD [94]. Finally, applications of artificial intelligence tools appear to promise much since they may support the identification of radiological biomarkers in PD, and they may also reveal deeper understanding of the pathophysiological alterations in SNc.

5. Conclusions

MRI-based biomarkers have undoubtedly contributed to the diagnosis of PD and a differential diagnosis of PD and atypical PS over the past two decades. Improvements in

MRI technology have made the study of SN microstructural changes and metal deposits possible, with both being of major importance to PD patients. An increasing number of MRI sequences and methods have been developed, resulting in more precise imaging findings that characterize SNc damage in PD. These images comprise nigrosome imaging, neuromelanin-sensitive sequences, iron-sensitive sequences, and DWI. The use of these imaging methods, alone or in combination, is emerging as an encouraging early diagnostic biomarker of PD. These techniques may also permit the discrimination of PD from control patients or PD patients from atypical PS. However, the diagnosis of PD is still based on clinical features, and these imaging methods are not yet in widespread use. Accordingly, multi-center studies deploying large cohorts are required. Results from these studies may result in the identification of new imaging biomarker of PD, thereby enabling the neuroradiologist to support clinicians in the final diagnosis of the disease.

Finally, the application of artificial intelligence tools has only increased in assisting the early or differential diagnosis of PD. A radiomic approach has also been increasingly adopted to extract and analyze quantitative imaging features from medical images, which are beyond those identifiable by an expert eye. The next step will be the inclusion of these radiomic features into the clinical decision making workflow. Such a process may also lead to extending our knowledge relating to the pathophysiological alterations of impaired brain areas, nuclei, and networks.

Author Contributions: Conceptualization, P.F., C.G. and R.M.; methodology, P.F.; formal analysis, P.F. and C.G.; investigation, all authors; data curation, P.F. and C.G.; writing—original draft preparation, P.F., C.G., E.B., C.D. and R.M.; writing—review and editing, P.F., C.G. and R.M.; manuscript final version validation, P.F., C.G., G.L.T., E.B., C.D., M.M. (Massimo Midiri), M.M. (Maurizio Marrale), A.D.P., M.C.M., L.G., R.B. and R.M.; supervision, P.F., C.G., M.M. (Massimo Midiri), R.M., P.F., C.G., G.L.T., E.B., C.D., M.M. (Massimo Midiri), M.M. (Maurizio Marrale), A.D.P., M.C.M., L.G., R.B. and R.M. All authors have read and agreed to the published version of the manuscript.

Funding: This research has received no external funding.

Institutional Review Board Statement: No additional ethical approval was sought for this narrative review because all MRI examinations were performed within and covered by Italian National Health System.

Informed Consent Statement: All the images included in this review are original and not previously published. All patients signed an informed consent form after having been fully informed about the MRI examination and its risks.

Conflicts of Interest: The authors declare no conflict of interest.

References

1. Kalia, L.V.; Lang, A.E. Parkinson's disease. *Lancet* **2015**, *386*, 896–912. [CrossRef]
2. Dickson, D.W. Neuropathology of Parkinson disease. *Park. Relat. Disord.* **2018**, *46*, S30–S33. [CrossRef] [PubMed]
3. Cacabelos, R. Parkinson's Disease: From Pathogenesis to Pharmacogenomics. *Int. J. Mol. Sci.* **2017**, *18*, 551. [CrossRef] [PubMed]
4. Braak, H.; Ghebremedhin, E.; Rüb, U.; Bratzke, H.; Del Tredici, K. Stages in the development of Parkinson's disease-related pathology. *Cell Tissue Res.* **2004**, *318*, 121–134. [CrossRef]
5. Postuma, R.B.; Berg, D.; Stern, M.B.; Poewe, W.; Olanow, C.W.; Oertel, W.H.; Obeso, J.; Marek, K.; Litvan, I.; Lang, A.E.; et al. MDS clinical diagnostic criteria for Parkinson's disease. *Mov. Disord.* **2015**, *30*, 1591–1601. [CrossRef]
6. Schapira, A.H.; Chaudhuri, K.R.; Jenner, P. Non-motor features of Parkinson disease. *Nat. Rev. Neurosci.* **2017**, *18*, 435–450. [CrossRef]
7. Monastero, R.; Cicero, C.E.; Baschi, R.; Davì, M.; Luca, A.; Restivo, V.; Zangara, C.; Fierro, B.; Zappia, M.; Nicoletti, A. Mild cognitive impairment in Parkinson's disease: The Parkinson's disease cognitive study (PACOS). *J. Neurol.* **2018**, *265*, 1050–1058. [CrossRef]
8. Nicoletti, A.; Luca, A.; Baschi, R.; Cicero, C.E.; Mostile, G.; Davì, M.; Pilati, L.; Restivo, V.; Zappia, M.; Monastero, R. Incidence of Mild Cognitive Impairment and Dementia in Parkinson's Disease: The Parkinson's Disease Cognitive Impairment Study. *Front. Aging Neurosci.* **2019**, *11*, 21. [CrossRef]
9. Aarsland, D.; Creese, B.; Politis, M.; Chaudhuri, K.R.; Ffytche, D.H.; Weintraub, D.; Ballard, C. Cognitive decline in Parkinson disease. *Nat. Rev. Neurol.* **2017**, *13*, 217–231. [CrossRef]

10. Chen-Plotkin, A.S.; Albin, R.; Alcalay, R.; Babcock, D.; Bajaj, V.; Bowman, D.; Buko, A.; Cedarbaum, J.; Chelsky, D.; Cookson, M.R.; et al. Finding useful biomarkers for Parkinson's disease. *Sci. Transl. Med.* **2018**, *10*, eaam6003. [CrossRef]
11. Heim, B.; Krismer, F.; De Marzi, R.; Seppi, K. Magnetic resonance imaging for the diagnosis of Parkinson's disease. *J. Neural Transm.* **2017**, *124*, 915–964. [CrossRef] [PubMed]
12. Damier, P.; Hirsch, E.C.; Agid, Y.; Graybiel, A.M. The substantia nigra of the human brain. I. Nigrosomes and the nigral matrix, a compartmental organization based on calbindin D(28K) immunohistochemistry. *Brain* **1999**, *122*, 1421–1436. [CrossRef] [PubMed]
13. Arribarat, G.; De Barros, A.; Péran, P. Modern Brainstem MRI Techniques for the Diagnosis of Parkinson's Disease and Parkinsonisms. *Front. Neurol.* **2020**, *11*, 791. [CrossRef] [PubMed]
14. Cheng, H.-C.; Ulane, C.M.; Burke, R.E. Clinical progression in Parkinson disease and the neurobiology of axons. *Ann. Neurol.* **2010**, *67*, 715–725. [CrossRef]
15. Saeed, U.; Lang, A.E.; Masellis, M. Neuroimaging Advances in Parkinson's Disease and Atypical Parkinsonian Syndromes. *Front. Neurol.* **2020**, *11*, 572976. [CrossRef]
16. Castellanos, G.; Fernández-Seara, M.A.; Betancor, O.L.; Ortega-Cubero, S.; Puigvert, M.; Uranga, J.; Vidorreta, M.; Irigoyen, J.; Lorenzo, E.; Muñoz-Barrutia, A.; et al. Automated Neuromelanin Imaging as a Diagnostic Biomarker for Parkinson's Disease. *Mov. Disord.* **2015**, *30*, 945–952. [CrossRef] [PubMed]
17. Akly, M.S.P.; Stefani, C.V.; Ciancaglini, L.; Bestoso, J.S.; Funes, J.A.; Bauso, D.J.; Besada, C.H. Accuracy of nigrosome-1 detection to discriminate patients with Parkinson's disease and essential tremor. *Neuroradiol. J.* **2019**, *32*, 395–400. [CrossRef]
18. Zorzenon, C.D.P.F.; Bienes, G.H.A.A.; Alves, E.D.; Tibana, L.A.T.; Júnior, H.C.; Ferraz, H.B. Magnetic resonance imaging evaluation of nigrosome 1 and neuromelanin can assist Parkinson's disease diagnosis, but requires an expert neuroradiologist. *Park. Relat. Disord.* **2021**, *83*, 8–12. [CrossRef]
19. Zucca, F.A.; Basso, E.; Cupaioli, F.A.; Ferrari, E.; Sulzer, D.; Casella, L.; Zecca, L. Neuromelanin of the Human Substantia Nigra: An Update. *Neurotox. Res.* **2014**, *25*, 13–23. [CrossRef]
20. Trujillo, P.; Summers, P.; Ferrari, E.; Zucca, F.A.; Sturini, M.; Mainardi, L.T.; Cerutti, S.; Smith, A.K.; Smith, S.A.; Zecca, L.; et al. Contrast mechanisms associated with neuromelanin-MRI. *Magn. Reson. Med.* **2017**, *78*, 1790–1800. [CrossRef]
21. Sasaki, M.; Shibata, E.; Kudo, K.; Tohyama, K. Neuromelanin-sensitive MRI basics, technique, and clinical applications. *Clin. Neuroradiol.* **2008**, *18*, 147–153. [CrossRef]
22. Pluijm, M.; Cassidy, C.; Ms, M.Z.; Wallert, E.; Bruin, K.; Booij, J.; Haan, L.; Horga, G.; Giessen, E. Reliability and Reproducibility of Neuromelanin-Sensitive Imaging of the Substantia Nigra: A Comparison of Three Different Sequences. *J. Magn. Reson. Imaging* **2021**, *53*, 712–721. [CrossRef]
23. Kitao, S.; Matsusue, E.; Fujii, S.; Miyoshi, F.; Kaminou, T.; Kato, S.; Ito, H.; Ogawa, T. Correlation between pathology and neuromelanin MR imaging in Parkinson's disease and dementia with Lewy bodies. *Neuroradiology* **2013**, *55*, 947–953. [CrossRef] [PubMed]
24. Reimão, S.; Lobo, P.P.; Neutel, D.; Guedes, L.C.; Coelho, M.; Rosa, M.M.; Azevedo, P.; Ferreira, J.; Abreu, D.; Gonçalves, N.; et al. Substantia nigra neuromelanin-MR imaging differentiates essential tremor from Parkinson's disease. *Mov. Disord.* **2015**, *30*, 953–959. [CrossRef] [PubMed]
25. Sulzer, D.; Cassidy, C.; Horga, G.; Kang, U.J.; Fahn, S.; Casella, L.; Pezzoli, G.; Langley, J.; Hu, X.P.; Zucca, F.A.; et al. Neuromelanin detection by magnetic resonance imaging (MRI) and its promise as a biomarker for Parkinson's disease. *NPJ Park. Dis.* **2018**, *4*, 1–13. [CrossRef]
26. Matsuura, K.; Maeda, M.; Tabei, K.-I.; Umino, M.; Kajikawa, H.; Satoh, M.; Kida, H.; Tomimoto, H. A longitudinal study of neuromelanin-sensitive magnetic resonance imaging in Parkinson's disease. *Neurosci. Lett.* **2016**, *633*, 112–117. [CrossRef]
27. Takahashi, H.; Watanabe, Y.; Tanaka, H.; Mihara, M.; Mochizuki, H.; Yamamoto, K.; Liu, T.; Wang, Y.; Tomiyama, N. Comprehensive MRI quantification of the substantia nigra pars compacta in Parkinson's disease. *Eur. J. Radiol.* **2018**, *109*, 48–56. [CrossRef] [PubMed]
28. Takahashi, H.; Watanabe, Y.; Tanaka, H.; Mihara, M.; Mochizuki, H.; Liu, T.; Wang, Y.; Tomiyama, N. Quantifying changes in nigrosomes using quantitative susceptibility mapping and neuromelanin imaging for the diagnosis of early-stage Parkinson's disease. *Br. J. Radiol.* **2018**, *91*, 20180037. [CrossRef]
29. Prasad, S.; Stezin, A.; Lenka, A.; George, L.; Saini, J.; Yadav, R.; Pal, P.K. Three-dimensional neuromelanin-sensitive magnetic resonance imaging of the substantia nigra in Parkinson's disease. *Eur. J. Neurol.* **2018**, *25*, 680–686. [CrossRef]
30. Pyatigorskaya, N.; Magnin, B.; Mongin, M.; Yahia-Cherif, L.; Valabregue, R.; Arnaldi, D.; Ewenczyk, C.; Poupon, C.; Vidailhet, M.; Lehéricy, S. Comparative Study of MRI Biomarkers in the Substantia Nigra to Discriminate Idiopathic Parkinson Disease. *Am. J. Neuroradiol.* **2018**, *39*, 1460–1467. [CrossRef]
31. Biondetti, E.; Gaurav, R.; Yahia-Cherif, L.; Mangone, G.; Pyatigorskaya, N.; Valabrègue, R.; Ewenczyk, C.; Hutchison, M.; François, C.; Arnulf, I.; et al. Spatiotemporal changes in substantia nigra neuromelanin content in Parkinson's disease. *Brain* **2020**, *143*, 2757–2770. [CrossRef] [PubMed]
32. Wang, J.; Li, Y.; Huang, Z.; Wan, W.; Zhang, Y.; Wang, C.; Cheng, X.; Ye, F.; Liu, K.; Fei, G.; et al. Neuromelanin-sensitive magnetic resonance imaging fea-tures of the substantia nigra and locus coeruleus in de novo Parkinson's disease and its phenotypes. *Eur. J. Neurol.* **2018**, *25*, 949-e73. [CrossRef] [PubMed]

33. Xiang, Y.; Gong, T.; Wu, J.; Li, J.; Chen, Y.; Wang, Y.; Li, S.; Cong, L.; Lin, Y.; Han, Y.; et al. Subtypes evaluation of motor dysfunction in Parkinson's disease using neuromelanin-sensitive magnetic resonance imaging. *Neurosci. Lett.* **2017**, *638*, 145–150. [CrossRef] [PubMed]

34. Wang, J.; Huang, Z.; Li, Y.; Ye, F.; Wang, C.; Zhang, Y.; Cheng, X.; Fei, G.; Liu, K.; Zeng, M.; et al. Neuromelanin-sensitive MRI of the substantia nigra: An im-aging biomarker to differentiate essential tremor from tremor-dominant Parkinson's disease. *Parkinsonism Relat. Disord.* **2019**, *58*, 3–8. [CrossRef] [PubMed]

35. Chougar, L.; Pyatigorskaya, N.; Degos, B.; Grabli, D.; Lehéricy, S. The Role of Magnetic Resonance Imaging for the Diagnosis of Atypical Parkinsonism. *Front. Neurol.* **2020**, *11*, 665. [CrossRef]

36. Pyatigorskaya, N.; Gaurav, R.; Arnaldi, D.; Leu-Semenescu, S.; Yahia-Cherif, L.; Valabregue, R.; Vidailhet, M.; Arnulf, I.; Lehéricy, S. Magnetic Resonance Imaging Biomarkers to Assess Substantia Nigra Damage in Idiopathic Rapid Eye Movement Sleep Behavior Disorder. *Sleep* **2017**, *40*. [CrossRef]

37. Schwarz, S.T.; Mougin, O.; Xing, Y.; Blazejewska, A.; Bajaj, N.; Auer, D.P.; Gowland, P. Parkinson's disease related signal change in the nigrosomes 1–5 and the substantia nigra using T2* weighted 7T MRI. *NeuroImage Clin.* **2018**, *19*, 683–689. [CrossRef]

38. Lehéricy, S.; Bardinet, E.; Poupon, C.; Vidailhet, M.; François, C. 7 tesla magnetic resonance imaging: A closer look at sub-stantia nigra anatomy in Parkinson's disease. *Mov Disord* **2014**, *29*, 1574–1581. [CrossRef]

39. Schwarz, S.T.; Afzal, M.; Morgan, P.S.; Bajaj, N.; Gowland, P.A.; Auer, D.P. The "swallow tail" appearance of the healthy nigro-some—A new accurate test of Parkinson's disease: A case-control and retrospective cross-sectional MRI study at 3T. *PLoS ONE* **2014**, *9*, 93814. [CrossRef]

40. Gao, P.; Zhou, P.-Y.; Wang, P.-Q.; Zhang, G.-B.; Liu, J.-Z.; Xu, F.; Yang, F.; Wu, X.-X.; Li, G. Universality analysis of the existence of substantia nigra "swallow tail" appearance of non-Parkinson patients in 3T SWI. *Eur. Rev. Med. Pharmacol. Sci.* **2016**, *20*, 1307–1314.

41. Cheng, Z.; He, N.; Huang, P.; Li, Y.; Tang, R.; Sethi, S.K.; Ghassaban, K.; Yerramsetty, K.K.; Palutla, V.K.; Chen, S.; et al. Imaging the Nigrosome 1 in the substantia nigra using sus-ceptibility weighted imaging and quantitative susceptibility mapping: An application to Parkinson's disease. *Neuroimage Clin.* **2020**, *25*, 102–103.

42. Schmidt, M.A.; Engelhorn, T.; Marxreiter, F.; Winkler, J.; Lang, S.; Kloska, S.; Goelitz, P.; Doerfler, A. Ultra high-field SWI of the substantia nigra at 7T: Reliability and consistency of the swallow-tail sign. *BMC Neurol.* **2017**, *17*, 1–6. [CrossRef] [PubMed]

43. Kau, T.; Hametner, S.; Endmayr, V.; Deistung, A.; Prihoda, M.; Haimburger, E.; Menard, C.; Haider, T.; Höftberger, R.; Robinson, S.; et al. Microvessels may Confound the "Swallow Tail Sign" in Normal Aged Midbrains: A Postmortem 7 T SW-MRI Study. *J. Neuroimaging* **2018**, *29*, 65–69. [CrossRef] [PubMed]

44. Gao, P.; Zhou, P.-Y.; Li, G.; Zhang, G.-B.; Wang, P.-Q.; Liu, J.-Z.; Xu, F.; Yang, F.; Wu, X.-X. Visualization of nigrosomes-1 in 3T MR susceptibility weighted imaging and its absence in diagnosing Parkinson's disease. *Eur. Rev. Med. Pharmacol. Sci.* **2015**, *19*, 4603–4609.

45. Mahlknecht, P.; Krismer, F.; Poewe, W.; Seppi, K. Meta-analysis of dorsolateral nigral hyperintensity on magnetic reso-nance imaging as a marker for Parkinson's disease. *Mov. Disord.* **2017**, *32*, 619–623. [CrossRef] [PubMed]

46. Sung, Y.H.; Lee, J.; Nam, Y.; Shin, H.-G.; Noh, Y.; Shin, D.H.; Kim, E.Y. Differential involvement of nigral subregions in idiopathic parkinson's disease. *Hum. Brain Mapp.* **2018**, *39*, 542–553. [CrossRef] [PubMed]

47. Calloni, S.F.; Conte, G.; Sbaraini, S.; Cilia, R.; Contarino, V.E.; Avignone, S.; Sacilotto, G.; Pezzoli, G.; Triulzi, F.M.; Scola, E. Multiparametric MR imaging of Parkin-sonisms at 3 tesla: Its role in the differentiation of idiopathic Parkinson's disease versus atypical Parkinsonian disorders. *Eur. J. Radiol.* **2018**, *109*, 95–100. [CrossRef]

48. Stezin, A.; Naduthota, R.M.; Botta, R.; Varadharajan, S.; Lenka, A.; Saini, J.; Yadav, R.; Pal, P.K. Clinical utility of visualisation of nigro-some-1 in patients with Parkinson's disease. *Eur. Radiol.* **2018**, *28*, 718–726. [CrossRef]

49. Sung, Y.H.; Noh, Y.; Lee, J.; Kim, E.Y. Drug-induced Parkinsonism versus Idiopathic Parkinson Disease: Utility of Nigro-some 1 with 3-T Imaging. *Radiology* **2016**, *279*, 849–858. [CrossRef]

50. De Marzi, R.; Seppi, K.; Högl, B.; Müller, C.; Scherfler, C.; Stefani, A.; Iranzo, A.; Tolosa, E.; Santamarìa, J.; Gizewski, E.R.; et al. Loss of dorsolateral nigral hyperintensity on 3.0 tesla susceptibility-weighted imaging in idiopathic rapid eye movement sleep behavior disorder. *Ann. Neurol.* **2016**, *79*, 1026–1030. [CrossRef]

51. Rizzo, G.; De Blasi, R.; Capozzo, R.; Tortelli, R.; Barulli, M.R.; Liguori, R.; Grasso, D.; Logroscino, G. Loss of Swallow Tail Sign on Susceptibil-ity-Weighted Imaging in Dementia with Lewy Bodies. *J. Alzheimers Dis.* **2019**, *67*, 61–65. [CrossRef]

52. Kathuria, H.; Mehta, S.; Ahuja, C.K.; Chakravarty, K.; Ray, S.; Mittal, B.R.; Singh, P.; Lal, V. Utility of Imaging of Nigrosome-1 on 3T MRI and Its Comparison with 18F-DOPA PET in the Diagnosis of Idiopathic Parkinson Disease and Atypical Parkinsonism. *Mov. Disord. Clin. Pr.* **2021**, *8*, 224–230. [CrossRef] [PubMed]

53. Kim, J.-M.; Jeong, H.-J.; Bae, Y.J.; Park, S.-Y.; Kim, E.; Kang, S.Y.; Oh, E.S.; Kim, K.J.; Jeon, B.; Kim, S.E.; et al. Loss of substantia nigra hyperintensity on 7 Tesla MRI of Parkinson's disease, multiple system atrophy, and progressive supranuclear palsy. *Park. Relat. Disord.* **2016**, *26*, 47–54. [CrossRef]

54. Frosini, D.; Ceravolo, R.; Tosetti, M.; Bonuccelli, U.; Cosottini, M. Nigral involvement in atypical parkinsonisms: Evidence from a pilot study with ultra-high field MRI. *J. Neural Transm.* **2016**, *123*, 509–513. [CrossRef] [PubMed]

55. Oh, S.W.; Shin, N.Y.; Lee, J.J.; Lee, S.K.; Lee, P.H.; Lim, S.M.; Kim, J.W. Correlation of 3D FLAIR and Dopamine Transporter Im-aging in Patients With Parkinsonism. *AJR Am. J. Roentgenol.* **2016**, *207*, 1089–1094. [CrossRef]

56. Simões, R.M.; Caldas, A.C.; Grilo, J.; Correia, D.; Guerreiro, C.; Lobo, P.P.; Valadas, A.; Fabbri, M.; Guedes, L.C.; Coelho, M.; et al. A distinct neuromelanin magnetic resonance imaging pattern in parkinsonian multiple system atrophy. *BMC Neurol.* **2020**, *20*, 1–12. [CrossRef] [PubMed]

57. Vitali, P.; Pan, M.I.; Palesi, F.; Germani, G.; Faggioli, A.; Anzalone, N.; Francaviglia, P.; Minafra, B.; Zangaglia, R.; Pacchetti, C.; et al. Substantia Nigra Volumetry with 3-T MRI in De Novo and Advanced Parkinson Disease. *Radiology* **2020**, *296*, 401–410. [CrossRef] [PubMed]

58. Wang, J.Y.; Zhuang, Q.Q.; Zhu, L.B.; Zhu, H.; Li, T.; Li, R.; Chen, S.F.; Huang, C.P.; Zhang, X.; Zhu, J.H. Meta-analysis of brain iron levels of Parkinson's disease pa-tients determined by postmortem and MRI measurements. *Sci. Rep.* **2016**, *6*, 36669. [CrossRef]

59. Wang, Y.; Butros, S.R.; Shuai, X.; Dai, Y.; Chen, C.; Liu, M.; Haacke, E.M.; Hu, J.; Xu, H. Different iron-deposition patterns of multiple system atro-phy with predominant parkinsonism and idiopathetic Parkinson diseases demonstrated by phase-corrected susceptibil-ity-weighted imaging. *AJNR Am. J. Neuroradiol.* **2012**, *33*, 266–273. [CrossRef]

60. Hopes, L.; Grolez, G.; Moreau, C.; Lopes, R.; Ryckewaert, G.; Carrière, N.; Auger, F.; Laloux, C.; Petrault, M.; Devedjian, J.-C.; et al. Magnetic Resonance Imaging Features of the Nigrostriatal System: Biomarkers of Parkinson's Disease Stages? *PLoS ONE* **2016**, *11*, e0147947. [CrossRef]

61. Du, G.; Lewis, M.M.; Sica, C.; He, L.; Connor, J.R.; Kong, L.; Mailman, R.B.; Huang, X. Distinct progression pattern of susceptibility MRI in the substantia nigra of Parkinson's patients. *Mov. Disord.* **2018**, *33*, 1423–1431. [CrossRef]

62. Wieler, M.; Gee, M.; Camicioli, R.; Martin, W.W. Freezing of gait in early Parkinson's disease: Nigral iron content estimated from magnetic resonance imaging. *J. Neurol. Sci.* **2016**, *361*, 87–91. [CrossRef] [PubMed]

63. Du, G.; Liu, T.; Lewis, M.M.; Kong, L.; Wang, Y.; Connor, J.; Mailman, R.B.; Huang, X. Quantitative susceptibility mapping of the midbrain in Parkinson's disease. *Mov. Disord.* **2016**, *31*, 317–324. [CrossRef]

64. Azuma, M.; Hirai, T.; Yamada, K.; Yamashita, S.; Ando, Y.; Tateishi, M.; Iryo, Y.; Yoneda, T.; Kitajima, M.; Wang, Y.; et al. Lateral asymmetry and spatial difference of iron deposition in the substantia nigra of patients with Parkinson disease measured with quantitative susceptibility map-ping. *AJNR Am. J. Neuroradiol.* **2016**, *37*, 782–788. [CrossRef] [PubMed]

65. Langkammer, C.; Pirpamer, L.; Seiler, S.; Deistung, A.; Schweser, F.; Franthal, S.; Homayoon, N.; Katschnig-Winter, P.; Koegl-Wallner, M.; Pendl, T.; et al. Quantitative Susceptibility Mapping in Parkinson's Disease. *PLoS ONE* **2016**, *11*, e0162460. [CrossRef] [PubMed]

66. Sun, J.; Lai, Z.; Ma, J.; Gao, L.; Chen, M.; Chen, J.; Fang, J.; Fan, Y.; Bao, Y.; Zhang, D.; et al. Quantitative Evaluation of Iron Content in Idiopathic Rapid Eye Movement Sleep Behavior Disorder. *Mov. Disord.* **2020**, *35*, 478–485. [CrossRef]

67. Saikiran, P.; Priyanka. Effectiveness of QSM over R2* in assessment of parkinson's disease—A systematic review. *Neurol. India.* **2020**, *68*, 278–281. [CrossRef]

68. An, H.; Zeng, X.; Niu, T.; Li, G.; Yang, J.; Zheng, L.; Zhou, W.; Liu, H.; Zhang, M.; Huang, D.; et al. Quantifying iron deposition within the substantia nigra of Parkin-son's disease by quantitative susceptibility mapping. *J. Neurol. Sci.* **2018**, *386*, 46–52. [CrossRef]

69. Fedeli, M.P.; Contarino, V.E.; Siggillino, S.; Samoylova, N.; Calloni, S.; Melazzini, L.; Conte, G.; Sacilotto, G.; Pezzoli, G.; Triulzi, F.M.; et al. Iron deposition in Parkinsonisms: A Quantitative Susceptibility Mapping study in the deep grey matter. *Eur. J. Radiol.* **2020**, *133*, 109394. [CrossRef]

70. Bergsland, N.; Zivadinov, R.; Schweser, F.; Hagemeier, J.; Lichter, D.; Guttuso, T., Jr. Ventral posterior substantia nigra iron increases over 3 years in Parkinson's disease. *Mov. Disord.* **2019**, *34*, 1006–1013. [CrossRef]

71. Guan, X.; Xuan, M.; Gu, Q.; Huang, P.; Liu, C.; Wang, N.; Xu, X.; Luo, W.; Zhang, M. Regionally progressive accumulation of iron in Parkinson's disease as measured by quantitative susceptibility mapping. *NMR Biomed.* **2017**, *30*, e3489. [CrossRef] [PubMed]

72. Damier, P.; Hirsch, E.C.; Agid, Y.; Graybiel, A.M. The substantia nigra of the human brain. II. Patterns of loss of dopa-mine-containing neurons in Parkinson's disease. *Brain* **1999**, *122*, 1437–1448. [CrossRef] [PubMed]

73. Mazzucchi, S.; Frosini, D.; Costagli, M.; Del Prete, E.; Donatelli, G.; Cecchi, P.; Migaleddu, G.; Bonuccelli, U.; Ceravolo, R.; Cosottini, M. Quantitative susceptibility mapping in atypical Parkinsonisms. *NeuroImage Clin.* **2019**, *24*, 101999. [CrossRef]

74. Sjöström, H.; Granberg, T.; Westman, E.; Svenningsson, P. Quantitative susceptibility mapping differentiates between par-kinsonian disorders. *Parkinsonism Relat. Disord.* **2017**, *44*, 51–57. [CrossRef] [PubMed]

75. Lewis, M.M.; Du, G.; Baccon, J.; Snyder, A.M.; Murie, B.; Cooper, F.; Stetter, C.; Kong, L.; Sica, C.; Mailman, R.B.; et al. Susceptibility MRI captures nigral pathology in pa-tients with parkinsonian syndromes. *Mov. Disord.* **2018**, *33*, 1432–1439. [CrossRef] [PubMed]

76. Atkinson-Clement, C.; Pinto, S.; Eusebio, A.; Coulon, O. Diffusion tensor imaging in Parkinson's disease: Review and me-ta-analysis. *Neuroimage. Clin.* **2017**, *16*, 98–110. [CrossRef] [PubMed]

77. Moseley, M.E.; Cohen, Y.; Kucharczyk, J.; Mintorovitch, J.; Asgari, H.S.; Wendland, M.F.; Tsuruda, J.; Norman, D. Diffusion-weighted MR imaging of anisotropic water diffusion in cat central nervous system. *Radiology* **1990**, *176*, 439–445. [CrossRef]

78. Knossalla, F.; Kohl, Z.; Winkler, J.; Schwab, S.; Schenk, T.; Engelhorn, T.; Doerfler, A.; Gölitz, P. High-resolution diffusion tensor-imaging in-dicates asymmetric microstructural disorganization within substantia nigra in early Parkinson's disease. *J. Clin. Neurosci.* **2018**, *50*, 199–202. [CrossRef]

79. Zhang, Y.; Wu, I.-W.; Tosun, D.; Foster, E.; Schuff, N.; Initiative, T.P.P.M. Progression of Regional Microstructural Degeneration in Parkinson's Disease: A Multicenter Diffusion Tensor Imaging Study. *PLoS ONE* **2016**, *11*, e0165540. [CrossRef]

80. Langley, J.; Huddleston, D.E.; Merritt, M.; Chen, X.; McMurray, R.; Silver, M.; Factor, S.A.; Hu, X. Diffusion tensor imaging of the substan-tia nigra in Parkinson's disease revisited. *Hum. Brain Mapp.* **2016**, *37*, 2547–2556. [CrossRef]

81. Loane, C.; Politis, M.; Kefalopoulou, Z.; Valle-Guzman, N.; Paul, G.; Widner, H.; Foltynie, T.; Barker, R.A.; Piccini, P. Aberrant nigral diffusion in Parkin-son's disease: A longitudinal diffusion tensor imaging study. *Mov. Disord.* **2016**, *31*, 1020–1026. [CrossRef]

82. Guttuso, T., Jr.; Bergsland, N.; Hagemeier, J.; Lichter, D.G.; Pasternak, O.; Zivadinov, R. Substantia Nigra Free Water Increases Longitudinally in Parkinson Disease. *AJNR Am. J. Neuroradiol.* **2018**, *39*, 479–484. [CrossRef] [PubMed]

83. Hirata, F.C.C.; Sato, J.R.; Vieira, G.; Lucato, L.; Leite, C.C.; Bor-Seng-Shu, E.; Pastorello, B.F.; Otaduy, M.C.G.; Chaim, K.T.; Campanholo, K.R.; et al. Substantia nigra fractional anisotropy is not a diagnostic biomarker of Parkinson's disease: A diagnostic performance study and meta-analysis. *Eur. Radiol.* **2016**, *27*, 2640–2648. [CrossRef] [PubMed]

84. Ofori, E.; Pasternak, O.; Planetta, P.J.; Li, H.; Burciu, R.G.; Snyder, A.F.; Lai, S.; Okun, M.; Vaillancourt, D.E. Longitudinal changes in free-water within the substantia nigra of Parkinson's disease. *Brain* **2015**, *138*, 2322–2331. [CrossRef] [PubMed]

85. Ofori, E.; Pasternak, O.; Planetta, P.J.; Burciu, R.; Snyder, A.; Febo, M.; Golde, T.E.; Okun, M.; Vaillancourt, D.E. Increased free water in the substantia nigra of Parkinson's disease: A single-site and multi-site study. *Neurobiol. Aging* **2015**, *36*, 1097–1104. [CrossRef] [PubMed]

86. Planetta, P.J.; Ofori, E.; Pasternak, O.; Burciu, R.G.; Shukla, P.; DeSimone, J.C.; Okun, M.S.; McFarland, N.R.; Vaillancourt, D.E. Free-water imaging in Parkinson's disease and atypical parkinsonism. *Brain* **2016**, *139*, 495–508. [CrossRef]

87. Yang, J.; Archer, D.B.; Burciu, R.G.; Müller, M.L.T.M.; Roy, A.; Ofori, E.; Bohnen, N.I.; Albin, R.L.; Vaillancourt, D.E. Multimodal dopaminergic and free-water im-aging in Parkinson's disease. *Park. Relat. Disord.* **2019**, *62*, 10–15. [CrossRef]

88. Arribarat, G.; Pasternak, O.; De Barros, A.; Galitzky, M.; Rascol, O.; Péran, P. Substantia nigra locations of iron-content, free-water and mean diffusivity abnormalities in moderate stage Parkinson's disease. *Park. Relat. Disord.* **2019**, *65*, 146–152. [CrossRef]

89. Barbagallo, G.; Sierra-Peña, M.; Nemmi, F.; Traon, A.P.-L.; Meissner, W.G.; Rascol, O.; Péran, P. Multimodal MRI assessment of nigro-striatal pathway in multiple system atrophy and Parkinson disease. *Mov. Disord.* **2016**, *31*, 325–334. [CrossRef] [PubMed]

90. Zhang, Y.; Walter, R.; Ng, P.; Luong, P.N.; Dutt, S.; Heuer, H.; Rojas-Rodriguez, J.C.; Tsai, R.; Litvan, I.; Dickerson, B.C.; et al. Progressionof microstructural degeneration in progres-sive supranuclear palsy and corticobasal syndrome: A longitudinal diffusion tensor imaging study. *PLoS ONE* **2016**, *11*, e0157218.

91. Salvatore, C.; Castiglioni, I.; Cerasa, A. Radiomics approach in the neurodegenerative brain. *Aging Clin. Exp. Res.* **2019**, 1–3. [CrossRef] [PubMed]

92. Shinde, S.; Prasad, S.; Saboo, Y.; Kaushick, R.; Saini, J.; Pal, P.K.; Ingalhalikar, M. Predictive markers for Parkinson's disease using deep neural nets on neuromelanin sensitive MRI. *NeuroImage Clin.* **2019**, *22*, 101748. [CrossRef] [PubMed]

93. Li, G.; Zhai, G.; Zhao, X.; An, H.; Spincemaille, P.; Gillen, K.M.; Ku, Y.; Wang, Y.; Huang, D.; Li, J. 3D texture analyses within the substantia nigra of Par-kinson's disease patients on quantitative susceptibility maps and R2* maps. *Neuroimage* **2019**, *188*, 465–472. [CrossRef]

94. Xiao, B.; He, N.; Wang, Q.; Cheng, Z.; Jiao, Y.; Haacke, E.M.; Yan, F.; Shi, F. Quantitative susceptibility mapping based hybrid feature extraction for diagnosis of Parkinson's disease. *NeuroImage Clin.* **2019**, *24*, 102070. [CrossRef]

95. Avanzo, M.; Wei, L.; Stancanello, J.; Vallières, M.; Rao, A.; Morin, O.; Mattonen, S.A.; El Naqa, I. Machine and deep learning methods for radi-omics. *Med. Phys.* **2020**, *47*, e185–e202. [CrossRef]

96. Cherubini, A.; Morelli, M.; Nisticó, R.; Salsone, M.; Arabia, G.; Vasta, R.; Augimeri, A.; Caligiuri, M.E.; Quattrone, A. Magnetic resonance support vector machine discriminates between Parkinson disease and progressive supranuclear palsy. *Mov. Disord.* **2014**, *29*, 266–269. [CrossRef]

97. Cherubini, A.; Nistico, R.; Novellino, F.; Salsone, M.; Nigro, S.; Donzuso, G.; Quattrone, A. Magnetic resonance support vector ma-chine discriminates essential tremor with rest tremor from tremor-dominant Parkinson disease. *Mov. Disord.* **2014**, *29*, 1216–1219. [CrossRef] [PubMed]

98. Peran, P.; Barbagallo, G.; Nemmi, F.; Sierra, M.; Galitzky, M.; Traon, A.P.; Payoux, P.; Meissner, W.G.; Rascol, O. MRI supervised and unsupervised classifica-tion of Parkinson's disease and multiple system atrophy. *Mov. Disord.* **2018**, *33*, 600–608. [CrossRef]

99. Krupička, R.; Mareček, S.; Malá, C.; Lang, M.; Klempíř, O.; Duspivová, T.; Široká, R.; Jarošíková, T.; Keller, J.; Šonka, K.; et al. Automatic Substantia Nigra Segmentation in Neuromelanin-Sensitive MRI by Deep Neural Network in Patients With Prodromal and Manifest Synucleinopathy. *Physiol. Res.* **2019**, *68* (Suppl. 4), S453–S458. [CrossRef] [PubMed]

100. Ariz, M.; Abad, R.C.; Castellanos, G.; Martinez, M.; Munoz-Barrutia, A.; Fernandez-Seara, M.A.; Pastor, P.; Pastor, M.A.; Ortiz-De-Solorzano, C. Dynamic Atlas-Based Segmentation and Quantification of Neuromelanin-Rich Brainstem Structures in Parkinson Disease. *IEEE Trans. Med. Imaging* **2019**, *38*, 813–823. [CrossRef]

Article

FOXP3 and GATA3 Polymorphisms, Vitamin D3 and Multiple Sclerosis

Concetta Scazzone [1,†], Luisa Agnello [1,†], Bruna Lo Sasso [1], Giuseppe Salemi [2], Caterina Maria Gambino [1], Paolo Ragonese [2], Giuseppina Candore [3], Anna Maria Ciaccio [4], Rosaria Vincenza Giglio [1], Giulia Bivona [1], Matteo Vidali [5] and Marcello Ciaccio [1,6,*]

[1] Department of Biomedicine, Neurosciences and Advanced Diagnostics, Institute of Clinical Biochemistry, Clinical Molecular Medicine and Laboratory Medicine, University of Palermo, 90127 Palermo, Italy; concetta.scazzone@unipa.it (C.S.); luisa.agnello@unipa.it (L.A.); bruna.losasso@unipa.it (B.L.S.); cmgambino@libero.it (C.M.G.); rosaria.vincenza.giglio@alice.it (R.V.G.), giulia.bivona@unipa.it (G.B.)
[2] Unit of Neurology, Department of Biomedicine, Neurosciences and Advanced Diagnostics, University of Palermo, 90127 Palermo, Italy; giuseppe.salemi@unipa.it (G.S.); paolo.ragonese@unipa.it (P.R.)
[3] Laboratory of Immunopathology and Immunosenescence, Department of Biomedicine, Neurosciences and Advanced Diagnostics, University of Palermo, 90127 Palermo, Italy; giuseppina.candore@unipa.it
[4] Unit of Clinical Biochemistry, University of Palermo, 90127 Palermo, Italy; amciaccio21@gmail.com
[5] Foundation IRCCS Ca' Granda Ospedale Maggiore Policlinico, 20122 Milan, Italy; matteo.vidali@gmail.com
[6] Department of Laboratory Medicine, Azienda Ospedaliera Universitaria Policlinico "P. Giaccone", 90127 Palermo, Italy
*   Correspondence: marcello.ciaccio@unipa.it; Tel.: +39-091-655-3296
†   These Authors contributed equally to the study.

Citation: Scazzone, C.; Agnello, L.; Lo Sasso, B.; Salemi, G.; Gambino, C.M.; Ragonese, P.; Candore, G.; Ciaccio, A.M.; Giglio, R.V.; Bivona, G.; et al. FOXP3 and GATA3 Polymorphisms, Vitamin D3 and Multiple Sclerosis. *Brain Sci.* **2021**, *11*, 415. https://doi.org/10.3390/brainsci11040415

Academic Editors: Andrew Clarkson and Cris S. Constantinescu

Received: 8 February 2021
Accepted: 24 March 2021
Published: 25 March 2021

Publisher's Note: MDPI stays neutral with regard to jurisdictional claims in published maps and institutional affiliations.

Abstract: Background: Regulatory T cells (Tregs) alterations have been implicated in the pathogenesis of Multiple Sclerosis (MS). Recently, a crucial role of the X-Linked Forkhead Box P3 (FoxP3) for the development and the stability of Tregs has emerged, and FOXP3 gene polymorphisms have been associated with the susceptibility to autoimmune diseases. The expression of Foxp3 in Tregs is regulated by the transcription factor GATA binding-protein 3 (GATA3) and vitamin D_3. The aim of this retrospective case-control study was to investigate the potential association between FOXP3 and GATA3 genetic variants, Vitamin D_3, and MS risk. Methods: We analyzed two polymorphisms in the FOXP3 gene (rs3761547 and rs3761548) and a polymorphism in the GATA3 gene (rs3824662) in 106 MS patients and 113 healthy controls. Serum 25(OH)D3 was also measured in all participants. Results: No statistically significant genotypic and allelic differences were found in the distribution of FOXP3 rs3761547 and rs3761548, or GATA3 rs3824662 in the MS patients, compared with controls. Patients that were homozygous for rs3761547 had lower 25(OH)D3 levels. Conclusions: Our findings did not show any association among FOXP3 and GATA3 SNPs, vitamin D_3, and MS susceptibility.

Keywords: multiple sclerosis; genetic; polymorphisms; FOXP3; GATA3; vitamin D

1. Introduction

Multiple Sclerosis (MS) is a chronic autoimmune inflammatory disease of the central nervous system (CNS). Studies on Experimental Autoimmune Encephalomyelitis (EAE), which represents the best animal model of MS, made a significant contribution to the understanding of MS pathogenesis. It is now well documented that CD4 (+) and CD8 (+) T lymphocytes and their related cytokines, as well as B-lymphocytes, take part in the development of MS. Among CD4 (+) T cells, it is possible to distinguish different cell subsets according to their cytokine secretion pattern [1]. Specifically, Th1 and Th17 cells produce pro-inflammatory cytokines, such as IFN-γ and IL-17, respectively, whereas Th2 and regulatory T cells (Tregs) produce anti-inflammatory cytokines, such as IL-10 [2,3]. Tregs have an essential role in controlling the immune system by several mechanisms, including regulation of antigen-presenting cells (APC) function, induction of tolerance, cytolysis, and expression of inhibitory cytokines [4]. Overall, Tregs are fundamental in maintaining immune self-tolerance and immune homeostasis, limiting excessive inflammation. Alterations of Tregs, including numerical reduction and functional changes, have been implied in the immune-mediated damage of myelin and axons, leading to neuronal damage and neuroinflammation in MS [5]. Moreover, reduced migration of Tregs into CNS of MS patients has been described [6]. Noteworthy, master-regulators of transcription are essential for T lymphocytes function [1]. Among these, the X-Linked Forkhead Box P3 (FoxP3) has a crucial role in Tregs development and stability, as shown by in vivo and in vitro studies [7–9]. In particular, FOXP3-deficient Treg cells have been shown to reduce expression of Treg cell signature genes, such as TGF-β, IL-10, and CTLA4, which are critical for tolerance and immunosuppression, while gained the expression of cytokine genes, such as IFN-γ, TNF-α, and IL-17, which stimulate the immune response [7]. Many polymorphisms in the gene codifying for Foxp3 have been associated with reduced levels of Foxp3 and impaired suppressive function of Treg cells, resulting in the development of autoimmune diseases [10]. An association between single nucleotide polymorphisms (SNPs) of the FOXP3 gene and autoimmune diseases, such as allergy, Graves' disease, and systemic lupus erythematosus, has been described [11–13].

Additionally, the sustained trek expression of Foxp3 is regulated by several factors, including the transcription factor GATA binding-protein 3 (GATA3) and vitamin D_3. In vivo and in vitro studies showed that GATA3 expression has a fundamental role in maintaining high-levels of Foxp3 in Tregs [14]. GATA3 has been reported to prevent excessive polarization toward Th17 and inflammatory cytokine production of Treg cells. Indeed, GATA-3-null Treg cells have been shown to fail to maintain peripheral homeostasis and suppressive function, shifting toward Th17 cell phenotypes and expressing reduced amounts of Foxp3 [15].

Vitamin D_3 has a pivotal role in regulating the immune system [16–19]. An association between reduced levels of vitamin D_3 and increased risk of several autoimmune diseases, including MS, has been documented. Several hypotheses have been proposed to explain the potential role of vitamin D in the pathophysiology of MS. Among these, experimental studies revealed that 1,25-dihydroxivitamin D3 (1,25(OH)$_2$D$_3$) regulates FOXP3 expression in Tregs [20]. Thus, reduced vitamin D_3 levels could be associated with reduced FOXP3 expression and, consequently, could increase the risk of MS.

The aim of this study was to investigate the association among SNPs in FOXP3 and GATA3 genes, vitamin D_3, and MS susceptibility. Specifically, we selected two SNPs in FOXP3 gene, namely rs3761547 and rs3761548, and the rs3824662 in the GATA3 gene, which could influence the FOXP3 and GATA3 expression, respectively. Thus, they could predispose to the development of autoimmune diseases, such as MS.

2. Materials and Methods

2.1. Study Population

We performed a retrospective case-control study on a cohort consisting of 106 patients with MS and 113 healthy controls. Cases were enrolled from June 2013 to December 2014 at the Department of Neurology, University Hospital of Palermo. Healthy controls were blood donors, enrolled from April 2015 to July 2016, at the Unit of Transfusion Medicine of Villa Sofia-Cervello Hospital in Palermo. The study was performed in accordance with the Declaration of Helsinki, and the local medical ethics committee approved the protocol. All subjects provided informed consent. An experienced neurologist made the diagnosis of MS according to revised McDonald criteria [21]. The neurological status of patients was assessed using Kurtzke's Expanded Disability Status Scale (EDSS). The progression of disability was assessed using the Multiple Sclerosis Severity Score (MSSS) [22]. The annualized relapse rate (ARR) was calculated in the year before genotyping. The MS group consisted of 27 men and 79 women, median (IQR) age 39 (34–48) years. Eighty-four percent of patients were diagnosed with the relapsing-remitting form of the disease (RRMS), 15%

with the secondary-progressive form (SPMS), and 1% with the primary progressive form (PPMS). The overall median (IQR) of disease duration was 10.6 (2.4–20.2), EDSS score 2.3 (1.4–5.0), MSSS score 3.3 (1.5–5.5), and ARR scores 1 (1–2).

The control group consisted of 58 men and 55 women with a median (IQR) age of 40 years (28–49).

The study protocol was approved by the Ethics Committee of the University Hospital of Palermo (nr 07/2016) and was performed in accordance with the current revision of the Helsinki Declaration. Informed consent was obtained from all individual participants included in the study.

2.2. Molecular Analysis

Whole blood samples from patients and controls were collected in EDTA tubes and stored at 4 °C for subsequent DNA extraction. Genomic DNA was extracted from 200 μL of whole blood using a commercial kit (Qiagen, Valencia, CA, USA), according to the manufacturer's instructions. The DNA quality was evaluated by electrophoresis in a 0.8% agarose gel, quantified by using absorbance spectrophotometric analysis and stored at −20 °C for subsequent analysis.

We selected two SNPs in the FOXP3 gene, namely rs3761548 and rs3761547, and a SNP in the GATA3 gene, namely rs3824662, based on evidence in the literature [23,24]. Characteristics of all selected SNPs are shown in Table 1. We used the following primers (VIC/FAM): TGTCTGCAGGGCTTCAAGTTGACAA(T/C)TGCCCCTCTATCCAGGGGACTGGCT for rs3761547; GGTGCTGAGGGGTAAACTGAGGCCT(T/G)CAGTTGGGGGAGAGAGCC AGAACCAG for rs3761548; AGGAAGGCGCCTTTGGCATGCACTG(A/C)AGCGTG TTTGTGTTTAATCTCAGGG for rs3824662.

Table 1. Characteristics of FOXP3 and GATA3 single nucleotide polymorphisms (SNPs).

Gene	Chromosome	SNP	Ancestral Allele	Substitution Allele	SNP Location	Functional Effect	Ref.
FOXP3	X	rs3761547	T	C	Intron	Unknown	-
		rs3761548	G	T	Intron	Reduced expression	[25]
GATA3	10	rs3824662	C	A	Intron	Altered expression	[23]

All samples were genotyped using real-time allelic discrimination TaqMan assay (Applied Biosystems). The genotyping was performed by a 7500 real-time PCR system. All PCR reaction mixtures contained 1 μL DNA ($\approx$25 ng), 5 μL TaqMan Genotyping Master Mix, and 0.25 μL genotyping Assay mix containing primers and FAM- or VIC-labeled probes and distilled water for a final volume of 20 μL. The real-time PCR conditions were initially 60 °C for 30 s and then 95 °C for 10 min, and subsequently 40 cycles of amplification (95 °C for 15 s and 60 °C for 1 min), and finally 60 °C for 30 s (Applied Biosystems).

2.3. Biochemical Analysis

Serum 25(OH)D$_3$ levels were measured by high-performance liquid chromatography (HPLC) using a Chromsystem reagent kit (Chromsystems Instruments & Chemicals GmbH; Grafelfing, Munich, Germany).

According to the recommendation of the Institute of Medicine, vitamin D$_3$ deficiency was defined as serum 25(OH)D$_3$ < 20 ng/mL, vitamin D$_3$ insufficiency as serum 25(OH)D$_3$ levels 20–30 ng/mL, and vitamin D$_3$ sufficiency as serum 25(OH)D$_3$ > 30 ng/mL.

2.4. Statistical Analysis

Statistical analysis was performed by SPSS version 17.0 (SPSS Inc., Chicago, IL, USA) and R Language v.3.6.1 (R Foundation for Statistical Computing, Vienna, Austria). Quantitative variables were expressed by the median and interquartile range (IQR) while categorical variables by absolute and relative frequencies. All genotypes were tested for

Hardy–Weinberg equilibrium by using an exact test. Differences in age or vitamin D levels between MS patients and controls were evaluated by both parametric *t*-test and non-parametric Mann–Whitney test. The association between MS diagnosis (dependent dichotomous variable) or vitamin D levels (continuous dependent variable) and predictors was evaluated, respectively, by multivariate logistic regression and General Linear Model. Association with chrX SNP was evaluated assuming (0, 2) dosage for males and adjusting for sex.

3. Results

We enrolled 106 MS patients and 113 healthy controls. Table 2 shows the clinical characteristics of the study population. The polymorphisms were in Hardy–Weinberg equilibrium ($p > 0.05$). Genotype and allele frequencies of cases and controls are shown in Tables 3 and 4. No significant statistical association was found by logistic regression between FOXP3 rs3761547 and rs3761548 as well as GATA3 rs3824662 genotypes and MS disease. Moreover, no association was found between FOXP3 rs3761548 and rs3761547 or GATA3 rs3824662 genotypes on the age of disease onset (p ranging from 0.284 to 0.955), diseases duration (p ranging from 0.259 to 0.547) EDSS (p ranging from 0.631 to 0.985), MSSS (p ranging from 0.601 to 0.680) and ARR (p ranging from 0.203 to 0.900).

Table 2. Demographic and clinical characteristics of multiple sclerosis (MS) patients and controls.

	MS (n = 106)	Controls (n = 113)	p-Value
Age (years)	39 (34–48)	40 (28–49)	0.703
Sex, n (male/female)	27/79	58/55	<0.001
25(OH)$_3$, µg/L	20.0 (15.0–25.0)	39.0 (28.5–49.0)	<0.001
Disease duration (years)	10.6 (2.4–20.2)	-	
Age of MS onset (years)	28 (22–32)	-	
MS-type (%) RR/SP/PP	84/15/1	-	
EDSS	2.3 (1.4–5)	-	
MSSS	3.3 (1.5–5.5)	-	
ARR	1 (1–2)	-	

Data are shown as: median (interquartile range), RR, Relapsing Remitting; SP, Secodary Progressive; PP, Primary Progressive; EDSS = Expanded Disability Status Scale; MSSS = Multiple Sclerosis Severity Score; ARR = annualized relapse rate.

Table 3. Distribution of genotypic and allelic frequencies of the FOXP3 SNPs in MS patients and controls.

SNP	Patients (n/%)		Controls (n/%)		p-Value
rs3761548 (FOXP3)	Male (27)	Female (79)	Male (58)	Female (55)	
GG		12 (15)		11 (21)	
TG		41 (52)		30 (54)	
TT		26 (33)		14 (25)	0.997
G	15 (56)		26 (45)		
T	12 (44)		32 (55)		
rs3761547 (FOXP3)	Male (27)	Female (79)	Male (58)	Female (55)	
TT		63 (80)		44 (80)	
TC		14 (18)		9 (16)	
CC		2 (2)		2 (4)	0.460
T	24 (88)		56 (97)		
C	3 (12)		2 (3)		

Table 4. Distribution of genotypic and allelic frequencies of the GATA3 SNP in MS patients and controls.

SNP	Patients (n/%)	Controls (n/%)	*p*-Value
rs3824662 (GATA3)			
CC	61 (57)	64 (54)	
CA	39(37)	43 (38)	
AA	6 (6)	6 (8)	0.945
C	161 (76)	171 (75)	
A	51 (24)	55 (25)	

We found that serum 25(OH)D$_3$ levels were significantly lower in MS patients than in controls (20.0 (15.0–25.0)µg/L and 39.0 (28.5–49.0)µg/L, respectively; $p < 0.001$). In particular, vitamin D$_3$ insufficiency was prevalent in MS patients (59%); 26% had vitamin D$_3$ deficiency, and only 15% had optimal levels. Moreover, men in the whole sample displayed significantly higher levels of vitamin D$_3$ than women (median 33 (24–45) µg/L vs. 25 (19–35) µg/L; $p = 0.007$). Nevertheless, vitamin D$_3$ was not associated with age ($p = 0.683$).

Multivariate analysis was performed using vitamin D$_3$ levels as a dependent variable, while age, sex, MS diagnosis, and studied polymorphisms (three different models) as independent variables. The analysis showed that none of the three studied polymorphisms were associated with vitamin D levels (p ranging from 0.270 to 0.894). Interestingly, the only independent predictor that was found significantly associated with vitamin D$_3$ levels in all three models investigated was the presence of the MS, further supporting previous literature results on the role of vitamin D in MS.

4. Discussion

MS is a multifactorial disease that occurs in genetically susceptible individuals after exposure to environmental factors, which contribute to the loss of tolerance and activation of T cells to myelin antigens [26,27].

Genetic studies uncovered several gene variants, which are putatively associated with MS susceptibility, including those codifying molecules involved in vitamin D$_3$ metabolism [28–33]. However, an essential role for Foxp3 has recently emerged. Foxp3 is a transcription factor belonging to the forkhead/winged-helix transcription factor family, and it is fundamental for maintaining the suppressive activity of Tregs [34,35]. Genetic variants of FOXP3 have been associated with an impaired function and differentiation of Treg cells, resulting in autoimmune dysfunction. Some Authors reported altered expression of Foxp3 in patients with MS [36–38]. A transcriptional factor critical for Treg cell function and Foxp3 expression is GATA3 [39], as revealed by in vivo studies [15]. Wang et al. [15] showed that GATA-3-null Treg cells failed to maintain peripheral homeostasis and suppressive function, gained Th17 cell phenotypes, and expressed reduced Foxp3 levels. Finally, vitamin D$_3$ has an important role in regulating the expression of FOXP3.

Given this evidence, we performed a case-control study to evaluate the possible influence of two SNPs in the promoter region of FOXP3 and an SNP of GATA3 on genetic predisposition to MS. Moreover, we investigated the relationship between such polymorphisms and 25(OH)D$_3$.

Among the selected FOXP3 SNPs, a functional effect has been reported only for the rs3761548 [25]. In particular, the minor allele A has been associated with an impaired interaction of the promoter region of FOXP3 with transcription factors, such as E47 and C-Myb, which reduce FOXP3 transcription. Similarly, the rs3824662 GATA3 has been associated with altered GATA3 expression [23].

In this study, we did not find any association among the selected SNPs, MS susceptibility, and 25(OH)D$_3$. Although the influence of FOXP3 SNPs on disease risk has been established in several autoimmune diseases, such as systemic sclerosis [40] and asthma [41], contrasting results have been reported in MS patients. Jafarzadeh et al. [42] and Eftekharian et al. [43] found an association between rs3761548 FOXP3 gene and MS in an Iranian population. Recently, Wawrusiewicz-Kurylonek et al. [24] investigated the association of three SNPs of FOXP3, including the rs3761548 and the rs3761547, on MS risk in a Polish population. The authors found a gender-specific relation between rs3761547 FOXP3 gene polymorphism and MS susceptibility. In particular, such polymorphism was associated with increased risk only in male patients. On the other hand, Gajdošechová et al. failed to find any association between FOXP3 SNPs and MS risk in a Slovak population [44]. Similar results were found by Işik et al. [45]. Flauzino et al. found an association between the rs3761548 FOXP3 and MS in females in a Brazilian population [46]. Additionally, Zhang et al. [47] performed a meta-analysis on five studies investigating the correlation between FOXP3 polymorphisms and MS risk. Authors showed that the rs3761548 could be associated with MS susceptibility, especially in Asians. However, such meta-analysis has several biases, including the low number of studies included. Additionally, four out of five studies were performed on Asians. Thus, the results may do not apply to different ethnic populations, such as Europeans. Indeed, it is well recognized that ethnicity affects the genetic background, resulting in contrasting results among different populations. The inconsistency of the results among the different studies on the association of FOXP3 SNPs and MS could be due to several reasons, including the sample size, the genotyping methods (restriction fragment length polymorphism, real-time PCR), and the selection of controls (hospital-based controls, community-based controls, healthy blood donors) [48]. To the best of our knowledge, this is the first study evaluating the role of FOXP3 polymorphisms on MS risk in an Italian population and the first investigating the possible influence of GATA3 polymorphism in MS.

Limitations of this study are small sample size, case-control design, the lack of match between cases and controls, the lack of sample size estimation prior to starting the study, and the lack of assessment of cytokines in order to characterize better the complex relationship among FOXP3, chemical mediators and MS.

5. Conclusions

It is sound to investigate FOXP3, and GATA3 SNPs within the contest of immune-mediated diseases as MS since functional alterations of these proteins could be involved in the development of such diseases. FOXP3 and GATA3 exert immune suppressive activity; however, some of their gene variants have been shown to impair immune-suppressive activity, thus contributing to the development of autoimmune diseases. In this case-control study, we evaluated the possible role of two SNPs of the FOXP3 gene, the rs3761547 and rs3761548, and an SNP of the GATA3 gene, rs3824662, as a susceptibility risk factor for MS. We did not find an association between the selected SNPs and MS risk. However, it is possible that several factors, including the small sample size, could have influenced our results. Therefore, we cannot draw final conclusions on the role of GATA3 and FOXP3 polymorphisms on the MS risk. Accordingly, it is worth studying FOXP3 and GATA3 SNPs in larger populations to understand better whether they could have a role in MS susceptibility.

Author Contributions: C.S. and L.A. conceptualization, formal analysis, and writing—original draft preparation; B.L.S., C.M.G., G.C., A.M.C., R.V.G. and G.B. data analysis and writing—review and editing; G.S. and P.R. patients enrollment and data collection, writing—review and editing; M.V. statistical analysis, writing—review and editing; M.C. conceptualization, writing—review and editing, supervision. All authors have read and agreed to the published version of the manuscript.

Funding: This research received no external funding.

Institutional Review Board Statement: The study was conducted according to the guidelines of the Declaration of Helsinki and approved by the Institutional Review Board (or Ethics Committee) of Azienda Ospedaliera Universitaria Policlinico "Paolo Giaccone" of Palermo (protocol code n°7/2016 and 18/07/2016).

Informed Consent Statement: Informed consent was obtained from all subjects involved in the study.

Data Availability Statement: The data that support the findings of this study are available from the corresponding author upon reasonable request.

Conflicts of Interest: The authors declare no conflict of interest.

References

1. O'Connor, R.A.; Anderton, S.M. Inflammation-associated genes: Risks and benefits to Foxp3+ regulatory T-cell function. *Immunology* **2015**, *146*, 194–205. [CrossRef]
2. Hedegaard, C.J.; Krakauer, M.; Bendtzen, K.; Lund, H.; Sellebjerg, F.; Nielsen, C.H. T helper cell type 1 (Th1), Th2 and Th17responses to myelin basic protein and disease activity in multiplesclerosis. *Immunology* **2008**, *125*, 161–169. [CrossRef] [PubMed]
3. Peelen, E.; Damoiseaux, J.; Smolders, J.; Knippenberg, S.; Menheere, P.; Tervaert, J.W.; Hupperts, R.; Thewissen, M. Th17 expansion in MS patients iscounterbalanced by an expanded CD39+regulatory T cell population during remission but not during relapse. *J. Neuroimmunol.* **2011**, *240–241*, 97–103. [CrossRef] [PubMed]
4. Kleinewietfeld, M.; Hafler, D.A. Regulatory T cells in autoimmune neuroinflammation. *Immunol. Rev.* **2014**, *259*, 231–244. [CrossRef]
5. Grant, C.R.; Liberal, R.; Mieli-Vergani, G.; Vergani, D.; Longhi, M.S. Regulatory T-cells in autoimmune diseases: Challenges, controversies and–yet–unanswered questions. *Autoimmun. Rev.* **2015**, *14*, 105–116. [CrossRef]
6. Schneider-Hohendorf, T.; Stenner, M.P.; Weidenfeller, C.; Zozulya, A.L.; Simon, O.J.; Schwab, N.; Wiendl, H. Regulatory T cells exhibit enhanced migratory characteristics, a feature impaired in patients with multiple sclerosis. *Eur. J. Immunol.* **2010**, *40*, 3581–3590. [CrossRef] [PubMed]
7. Williams, L.M.; Rudensky, A.Y. Maintenance of the Foxp3-dependent developmental program in mature regulatory T cells requires continued expression of Foxp3. *Nat. Immunol.* **2007**, *8*, 277–284. [CrossRef]
8. Lin, W.; Haribhai, D.; Relland, L.M.; Truong, N.; Carlson, M.R.; Williams, C.B.; Chatila, T.A. Regulatory T cell development in the absence of functional Foxp3. *Nat. Immunol.* **2007**, *8*, 359–368. [CrossRef]
9. Wan, Y.Y.; Flavell, R.A. Regulatory T-cell functions are subverted and converted owing to attenuated Foxp3 expression. *Nature* **2007**, *445*, 766–770. [CrossRef]
10. Nie, J.; Li, Y.Y.; Zheng, S.G.; Tsun, A.; Li, B. FOXP3(+) Treg Cells and Gender Bias in Autoimmune Diseases. *Front. Immunol.* **2015**, *6*, 493. [CrossRef]
11. Owen, C.J.; Eden, J.A.; Jennings, C.E.; Wilson, V.; Cheetham, T.D.; Pearce, S.H. Genetic association studies of the FOXP3 gene in Graves' disease and autoimmune Addison's disease in the United Kingdom population. *J. Mol. Endocrinol.* **2006**, *37*, 97–104. [CrossRef]
12. Lin, Y.C.; Lee, J.H.; Wu, A.S.; Tsai, C.Y.; Yu, H.H.; Wang, L.C.; Yang, Y.H.; Chiang, B.L. Association of single-nucleotide polymorphisms in FOXP3 gene with systemic lupus erythematosus susceptibility: A case-control study. *Lupus* **2011**, *20*, 137–143. [CrossRef]
13. Ruan, Y.; Zhang, Y.; Zhang, L. Association between single-nucleotide polymorphisms of key genes in T regulatory cells signaling pathways and the efficacy of allergic rhinitis immune therapy. *Zhonghua Er Bi Yan Hou Tou Jing Wai Ke Za Zhi* **2016**, *51*, 34–42. [CrossRef] [PubMed]
14. Wohlfert, E.A.; Grainger, J.R.; Bouladoux, N.; Konkel, J.E.; Oldenhove, G.; Ribeiro, C.H.; Hall, J.A.; Yagi, R.; Naik, S.; Bhairavabhotla, R.; et al. GATA3 controls Foxp3$^+$ regulatory T cell fate during inflammation in mice. *J. Clin. Investig.* **2011**, *121*, 4503–4515. [CrossRef] [PubMed]
15. Wang, Y.; Su, M.A.; Wan, Y.Y. An essential role of the transcription factor GATA-3 for the function of regulatory T cells. *Immunity* **2011**, *35*, 337–348. [CrossRef] [PubMed]
16. Bivona, G.; Agnello, L.; Bellia, C.; Iacolino, G.; Scazzone, C.; Lo Sasso, B.; Ciaccio, M. Non-Skeletal Activities of Vitamin D: From Physiology to Brain Pathology. *Medicina* **2019**, *55*, 341. [CrossRef]
17. Bivona, G.; Agnello, L.; Ciaccio, M. The immunological implication of the new vitamin D metabolism. *Cent. Eur. J. Immunol.* **2018**, *13*, 331 334. [CrossRef]
18. Bivona, G.; Agnello, L.; Ciaccio, M. Vitamin D and Immunomodulation: Is It Time to Change the Reference Values? *Ann. Clin. Lab. Sci.* **2017**, *47*, 508–510.
19. Bivona, G.; Lo Sasso, B.; Iacolino, G.; Gambino, C.M.; Scazzone, C.; Agnello, L.; Ciaccio, M. Standardized measurement of circulating vitamin D [25(OH)D] and its putative role as a serum biomarker in Alzheimer's disease and Parkinson's disease. *Clin. Chim. Acta* **2019**, *497*, 82–87. [CrossRef]
20. Kang, S.W.; Kim, S.H.; Lee, N.; Lee, W.W.; Hwang, K.A.; Shin, M.S.; Lee, S.H.; Kim, W.U.; Kang, I. 1,25-Dihyroxyvitamin D3 promotes FOXP3 expression via binding to vitamin D response elements in its conserved noncoding sequence region. *J. Immunol.* **2012**, *188*, 5276–5282. [CrossRef]
21. Polman, C.H.; Reingold, S.C.; Banwell, B.; Clanet, M.; Cohen, J.A.; Filippi, M.; Fujihara, K.; Havrdova, E.; Hutchinson, M.; Kappos, L.; et al. Diagnostic criteria for multiple sclerosis: 2010 revisions to the McDonald criteria. *Ann. Neurol.* **2011**, *69*, 292–302. [CrossRef] [PubMed]

22. Roxburgh, R.H.; Seaman, S.R.; Masterman, T.; Hensiek, A.E.; Sawcer, S.J.; Vukusic, S.; Achiti, I.; Confavreux, C.; Coustans, M.; le Page, E.; et al. Multiple Sclerosis Severity Score: Using disability and disease duration to rate disease severity. *Neurology* **2005**, *64*, 1144–1151. [CrossRef]

23. Perez-Andreu, V.; Roberts, K.G.; Harvey, R.C.; Yang, W.; Cheng, C.; Pei, D.; Xu, H.; Gastier-Foster, J.E.S.; Lim, J.Y.; Chen, I.M.; et al. Inherited GATA3 variants are associated with Ph-like childhood acute lymphoblastic leukemia and risk of relapse. *Nat. Genet.* **2013**, *45*, 1494–1498. [CrossRef] [PubMed]

24. Wawrusiewicz-Kurylonek, N.; Chorąży, M.; Posmyk, R.; Zajkowska, O.; Zajkowska, A.; Krętowski, A.J.; Tarasiuk, J.; Kochanowicz, J.; Kułakowska, A. The FOXP3 rs3761547 Gene Polymorphism in Multiple Sclerosis as a Male-Specific Risk Factor. *Neuromolecular. Med.* **2018**, *20*, 537–543. [CrossRef]

25. Hanel, S.A.; Velavan, T.P.; Kremsner, P.G.; Kun, J.F. Novel and functional regulatory SNPs in the promoter region of FOXP3 gene in a Gabonese population. *Immunogenetics* **2011**, *63*, 409–415. [CrossRef] [PubMed]

26. Goverman, J.M. Immune tolerance in multiple sclerosis. *Immunol. Rev.* **2011**, *241*, 228–240. [CrossRef]

27. Scazzone, C.; Agnello, L.; Bivona, G.; Lo Sasso, B.; Ciaccio, M. Vitamin D and Genetic Susceptibility to Multiple Sclerosis. *Biochem. Genet.* **2021**, *59*, 1–30. [CrossRef]

28. Elkama, A.; Karahalil, B. Role of gene polymorphisms in vitamin D metabolism and in multiple sclerosis. *Arh. Hig. Rada. Toksikol.* **2018**, *69*, 25–31. [CrossRef]

29. Agnello, L.; Scazzone, C.; Lo Sasso, B.; Bellia, C.; Bivona, G.; Realmuto, S.; Brighina, F.; Schillaci, R.; Ragonese, P.; Salemi, G.; et al. VDBP, CYP27B1, and 25-Hydroxyvitamin D Gene Polymorphism Analyses in a Group of Sicilian Multiple Sclerosis Patients. *Biochem. Gene* **2017**, *55*, 183–192. [CrossRef] [PubMed]

30. Agnello, L.; Scazzone, C.; Lo Sasso, B.; Ragonese, P.; Milano, S.; Salemi, G.; Ciaccio, M. CYP27A1, CYP24A1, and RXR-α Polymorphisms, Vitamin D, and Multiple Sclerosis: A Pilot Study. *J. Mol. Neurosci.* **2018**, *66*, 77–84. [CrossRef]

31. Agnello, L.; Scazzone, C.; Ragonese, P.; Salemi, G.; Lo Sasso, B.; Schillaci, R.; Musso, G.; Bellia, C.; Ciaccio, M. Vitamin D receptor polymorphisms and 25-hydroxyvitamin D in a group of Sicilian multiple sclerosis patients. *Neurol. Sci.* **2016**, *37*, 261–267. [CrossRef] [PubMed]

32. Di Resta, C.; Ferrari, M. Next Generation Sequencing: From Research Area to Clinical Practice. *Electron. J. Int. Fed. Clin. Chem. Lab. Med.* **2018**, *29*, 215–220.

33. Scazzone, C.; Agnello, L.; Ragonese, P.; Lo Sasso, B.; Bellia, C.; Bivona, G.; Schillaci, R.; Salemi, G.; Ciaccio, M. Association of CYP2R1 rs10766197 with MS risk and disease progression. *J. Neurosci. Res.* **2018**, *96*, 297–304. [CrossRef] [PubMed]

34. Fontenot, J.D.; Gavin, M.A.; Rudensky, A.Y. Foxp3 programs the development and function of CD4+CD25+ regulatory T cells. *Nat. Immunol.* **2003**, *4*, 330–336. [CrossRef] [PubMed]

35. Hori, S.; Nomura, T.; Sakaguchi, S. Control of regulatory T cell development by the transcription factor Foxp3. *Science* **2003**, *299*, 1057–1061. [CrossRef] [PubMed]

36. Huan, J.; Culbertson, N.; Spencer, L.; Bartholomew, R.; Burrows, G.G.; Chou, Y.K.; Bourdette, D.; Ziegler, S.F.; Offner, H.; Vandenbark, A.A. Decreased FOXP3 levels in multiple sclerosis patients. *J. Neurosci. Res.* **2005**, *81*, 45–52. [CrossRef]

37. Venken, K.; Hellings, N.; Hensen, K.; Rummens, J.L.; Medaer, R.; D'hooghe, M.B.; Dubois, B.; Raus, J.; Stinissen, P. Secondary progressive in contrast to relapsing-remitting multiple sclerosis patients show a normal CD4+CD25+ regulatory T-cell function and FOXP3 expression. *J. Neurosci. Res.* **2006**, *83*, 1432–1446. [CrossRef] [PubMed]

38. Kumar, M.; Putzki, N.; Limmroth, V.; Remus, R.; Lindemann, M.; Knop, D.; Mueller, N.; Hardt, C.; Kreuzfelder, E.; Grosse-Wilde, H. CD4+CD25+FoxP3+ T lymphocytes fail to suppress myelin basic protein-induced proliferation in patients with multiple sclerosis. *J. Neuroimmunol.* **2006**, *180*, 178–184. [CrossRef]

39. Chen, T.; Hou, X.; Ni, Y.; Du, W.; Han, H.; Yu, Y.; Shi, G. The Imbalance of FOXP3/GATA3 in Regulatory T Cells from the Peripheral Blood of Asthmatic Patients. *J. Immunol. Res.* **2018**, *2018*, 3096183. [CrossRef]

40. D'Amico, F.; Skarmoutsou, E.; Marchini, M.; Malaponte, G.; Caronni, M.; Scorza, R.; Mazzarino, M.C. Genetic polymorphisms of FOXP3 in Italian patients with systemic sclerosis. *Immunol. Lett.* **2013**, *152*, 109–113. [CrossRef]

41. Marques, C.R.; Costa, R.S.; Costa, G.N.O.; da Silva, T.M.; Teixeira, T.O.; de Andrade, E.M.M.; Galvão, A.A.; Carneiro, V.L.; Figueiredo, C.A. Genetic and epigenetic studies of FOXP3 in asthma and allergy. *Asthma Res. Pract.* **2015**, *1*, 10. [CrossRef]

42. Jafarzadeh, A.; Jamali, M.; Mahdavi, R.; Ebrahimi, H.A.; Hajghani, H.; Khosravimashizi, A.; Nemati, M.; Najafipour, H.; Sheikhi, A.; Mohammadi, M.M.; et al. Circulating levels of interleukin-35 in patients with multiple sclerosis: Evaluation of the influences of FOXP3 gene polymorphism and treatment program. *J. Mol. Neurosci.* **2015**, *55*, 891–897. [CrossRef]

43. Eftekharian, M.M.; Sayad, A.; Omrani, M.D.; Ghannad, M.S.; Noroozi, R.; Mazdeh, M.; Mirfakhraie, R.; Movafagh, A.; Roshanaei, G.; Azimi, T.; et al. Single nucleotide polymorphisms in the FOXP3 gene are associated with increased risk of relapsing-remitting multiple sclerosis. *Hum. Antibodies* **2016**, *24*, 85–90. [CrossRef]

44. Gajdošechová, B.; Javor, J.; Čierny, D.; Michalik, J.; Ďurmanová, V.; Shawkatová, I.; Párnická, Z.; Cudráková-Čopíková, D.; Lisá, I.; Peterajová, L.; et al. Association of FOXP3 polymorphisms rs3761547 and rs3761548 with multiple sclerosis in the Slovak population. *Activitas Nervosa Superior Rediviva* **2017**, *59*, 9–15.

45. Işik, N.; Yildiz, M.N.; Aydin, C.İ.; Candan, F.; Ünsal, Ç.A.; Saru, H.D.G. Genetic Susceptibility to Multiple Sclerosis: The Role of FOXP3 Gene Polymorphism. *Noro Psikiyatri Arsivi* **2014**, *51*, 69–73. [CrossRef] [PubMed]

46. Flauzino, T.; Alfieri, D.F.; de Carvalho, J.; Pereira, W.L.; Oliveira, S.R.; Kallaur, A.P.; Lozovoy, M.A.B.; Kaimen-Maciel, D.R.; de Oliveira, K.B.; Simão, A.N.C.; et al. The rs3761548 FOXP3 variant is associated with multiple sclerosis and transforming growth factor β1 levels in female patients. *Inflamm. Res.* **2019**, *68*, 933–943. [CrossRef] [PubMed]
47. Zhang, Y.; Zhang, J.; Liu, H.; He, F.; Chen, A.; Yang, H.; Pi, B. Meta-analysis of FOXP3 gene rs3761548 and rs2232365 polymorphism and multiple sclerosis susceptibility. *Medicine* **2019**, *98*, e17224. [CrossRef]
48. He, Y.; Na, H.; Li, Y.; Qiu, Z.; Li, W. FoxP3 rs3761548 polymorphism predicts autoimmune disease susceptibility: A meta-analysis. *Hum. Immunol.* **2013**, *74*, 1665–1671. [CrossRef]

Review

Inherited Neuromuscular Disorders: Which Role for Serum Biomarkers?

Antonino Lupica *, Vincenzo Di Stefano, Andrea Gagliardo, Salvatore Iacono, Antonia Pignolo, Salvatore Ferlisi, Angelo Torrente, Sonia Pagano, Massimo Gangitano and Filippo Brighina

Department of Biomedicine, Neuroscience and advanced Diagnostic (BIND), University of Palermo, 90121 Palermo, Italy; vincenzo19689@gmail.com (V.D.S.); andrigl@gmail.com (A.G.); salvo.iak@gmail.com (S.I.); niettapignolo@gmail.com (A.P.); fransalvo1@gmail.com (S.F.); angelotorrente92@gmail.com (A.T.); sonia.pagano@policlinico.pa.it (S.P.); massimo.gangitano@unipa.it (M.G.); filippobrighina@gmail.com (F.B.)
* Correspondence: antlupica@gmail.com

Abstract: Inherited neuromuscular disorders (INMD) are a heterogeneous group of rare diseases that involve muscles, motor neurons, peripheral nerves or the neuromuscular junction. Several different lab abnormalities have been linked to INMD: sometimes they are typical of the disorder, but they usually appear to be less specific. Sometimes serum biomarkers can point out abnormalities in presymtomatic or otherwise asymptomatic patients (e.g., carriers). More often a biomarker of INMD is evaluated by multiple clinicians other than expert in NMD before the diagnosis, because of the multisystemic involvement in INMD. The authors performed a literature search on biomarkers in inherited neuromuscular disorders to provide a practical approach to the diagnosis and the correct management of INMD. A considerable number of biomarkers have been reported that support the diagnosis of INMD, but the role of an expert clinician is crucial. Hence, the complete knowledge of such abnormalities can accelerate the diagnostic workup supporting the referral to specialists in neuromuscular disorders.

Keywords: biomarkers; inherited neuromuscular disorders; rare diseases

Citation: Lupica, A.; Di Stefano, V.; Gagliardo, A.; Iacono, S.; Pignolo, A.; Ferlisi, S.; Torrente, A.; Pagano, S.; Gangitano, M.; Brighina, F. Inherited Neuromuscular Disorders: Which Role for Serum Biomarkers?. *Brain Sci.* **2021**, *11*, 398. https://doi.org/10.3390/brainsci11030398

Academic Editors: Boel De Paepe and Marcello Ciaccio

Received: 14 February 2021
Accepted: 18 March 2021
Published: 21 March 2021

1. Background and Aims

Inherited neuromuscular disorders (INMD) are rare diseases that involve muscles, motor neurons, peripheral nerves or the neuromuscular junction [1].

Even if the most important role in the diagnosis of INMD is played by clinical and subsequent examinations (e.g., neurophysiological, histo-morphological and genetic investigations), serum biomarkers may have a crucial role in the diagnosis and follow-up.

Indeed, it is a well-known fact that serum biomarkers have a crucial role in the diagnosis of many acquired neuromuscular conditions ranging from autoimmune, inflammatory and paraneoplastic diseases affecting the muscle and peripheral nerve [2–4], but their role appears to be less immediate and appreciated for INMD [1].

A particular mentioning for genetic investigations is worth in the diagnostic workup: indeed genetics is in rapid evolution and has radically modified the approach to rare diseases in recent years [1,5,6]. In particular, the recent availability of next generation sequencing has deeply revolutioned the scenario of INMD, compared to only a few years ago, when the diagnostic algorithms were based on the sequencing gene by gene (e.g., Sanger), based on clinical, neurophysiology, biochemical and morphological evaluations [1]. In this scenario, serum biomarkers retain an important role in both the diagnosis and follow-up of INMD [7–10]. Many lab abnormalities commonly evaluated in routine blood examinations have been reported "typical" of certain INMD [8,11,12], but their role is sometimes underestimated in current clinical practice [13]. Moreover, lab samples are usually cheap, easy-to-obtain and, in many circumstances, they are parts of other general clinical procedures. Conversely, some biomarkers are very unspecific (e.g., serum creatine kinase (CKO), but associated conditions may point out the underlining condition (e.g., concomitant factors; the entity and/or the persistence of the abnormalities) [8,14] or rule out other differential diagnoses [15]. Finally, in specific clinical conditions, some biomarkers commonly used in clinical practice are indicative only if the sample is collected in a specific condition [16].

Moreover, the correct use of biomarkers in INMD allows a specific time- and resource-sparing algorithm. In fact, INMD patients, if promptly evaluated through biomarkers, can reach the correct diagnosis earlier and receive consequently an adequate management and therapy [17,18]. Occasionally, lab essays are abnormal despite the patient does not display any clinical sign (e.g., carriers); in these cases, a serum abnormality encourages the beginning of a diagnostic workup allowing a prompt genetic counselling [10,19].

Multiple organs are usually involved in INMD, so that patients are often evaluated by multiple clinicians other than expert in NMD before the diagnosis [20,21].

In this paper, we present a review of serum biomarkers in INMD to provide a guide to address the diagnosis and the correct management. INMD have been classified depending on the main organ involved (i.e., spinal cord, peripheral nerve, muscle, etc.) and separately analyzed as neuropathies (axonal and demyelinating), neuromuscular junction's disorders, myopathies (muscular dystrophies, Ion channel myopathies, metabolic myopathies) and hereditary motor neurons diseases.

2. Muscular Dystrophies

Muscular dystrophies (MD) constitute a heterogeneous group of INMD with progressive muscle weakness. Duchenne muscular dystrophy (DMD) is the most common and most severe MD and it is characterized by a complete lack of dystrophin due to mutations in the DMD gene [19]. Serum creatine kinase (CK) concentration as well as transminases (Supplementary Materials: Table S1 and Diagram S1) are markedly increased in children with DMD prior to the appearance of any clinical sign of disease; increased values are even observed among newborns [22]. Of interest, the CK peak by age of two years is usually from 10 to 20 times the upper normal limit, even if higher values have been reported [23]; serum CK progressively fall about 25% per year, eventually reaching the normal range in most cases, as the skeleton muscle is replaced by fat and fibrosis. In patients with Becker muscular dystrophy (BMD), CK concentration is usually elevated above five or more times the normal limits [19]. Serum CK is usually modestly elevated in limb-girdle muscular dystrophy (LGMD). However, it can be very high in sarcoglycanopathy, dysferlinopathy, and caveolinopathy [24]. Of interest, in LGMD2L, a condition resulting from mutations in ANO5, the gene for anoctamin 5, CK are usually elevated to around 1500 to 4500 U/L (range 200–40,000 U/L) and tend to decrease over time [24]. Furthermore, in DMD patients carbonic anhydrase 3 (CAH3), microtubule-associated-protein-4 and collagen type I-alpha 1 chain concentrations decline constantly over time; of interest, myosin light chain 3 (MLC3), electron transfer flavoprotein A (ETFA), troponin T, malate dehydrogenase 2 (MDH2), lactate dehydrogenase B and nestin plateaus in early teens; ETFA correlates with the score of the 6-min-walking-test, whereas MDH2, MLC3, CAH3 and nestin correlate with respiratory capacity [25]. In addition, high serum concentrations of a set of muscle-enriched miRNAs, (miR-1, miR-133, miR-206, miR-208b, miR-208a and miR-499) were significantly elevated in DMD patients compared to healthy subjects [7,18].

Serum matrix metalloproteinase-9 (MMP-9) and its inhibitor (TIMP-1) have been proposed as markers of disease progression in DMD [26]. In DMD patients the decrease in creatinine and an increase in creatine serum concentrations is likely due to the insufficient creatine utilization by muscles. Similar profiles have been observed in other MD such as BMD, LGMD2A and LGMD2B. Creatine/creatinine ratio is particularly elevated in the older and more severely affected DMD patients, thus encouraging its use as a marker of disease progression. Reduced citrulline concentrations have been found in DMD patients [27]. There is a significant increase in fibronectin concentrations in DMD patients compared to age-matched controls, while a similar abnormality is not observable in BMD [28].

Patients with myotonic dystrophy type 1 and 2 (DM1 and 2) commonly present modestly elevated serum CK (less than 500 U/L) [29]. Low concentrations of anti-Müllerian hormone (AMH) demonstrates decreased ovarian reserve in women with DM1 [30]. Results of a study in a cohort of DM patients [14] showed a persistent elevation of high sensitivity cardiac troponin T (hs-cTnT) and CK; in contrast, high sensitivity cardiac troponin I (hs-cTnI) values were persistently normal throughout the study. In addition, a relevant quote of DM patients with cardiac conduction abnormalities and preserved systolic function presented abnormal NT-proBNP concentration [14]. DM1 patients had insulin-resistance and significantly higher triglycerides (TGs), glucose and Tumor Necrosis Factor A; conversely, they had significantly lower concentrations of total serum adiponectin with a selective, pronounced decrease of its high molecular weight oligomers [31]. Finally, sleep dysfunction, very common in DM1, may be mediated by a dysfunction of the hypothalamic hypocretin because of hypocretin1 reduction in CSF [32].

Facioscapulohumeral muscular dystrophy (FSHD), the third most common type of MD, is characterized by slowly progressive muscle weakness involving the facial, scapular, upper arm, lower leg and hip girdle muscles, usually with asymmetric involvement. Serum CK can be normal or modestly elevated in FSHD (usually less than five times the upper normal limit) [33]. Finally, an elevation of serum muscles (MM) and myocardial band (MB) isoforms of CK, as well as carbonic anhydrase III, and troponin I type 2 have a minor role to assess the severity and progression of FSHD [8].

In Emery-Dreifuss muscular dystrophy (EDMD), a modest elevation of serum CK concentration is typical, less than 1000 U/L; an increase up to 20 times the upper normal limit sometimes occurs, but it may be seen often higher in the early stages of the disease [34]. The concentrations of circulating tenascin-C (TN-C) are elevated in autosomic dominant-EDMD and X-linked-EDMD patients and in some X-linked-EDMD carriers, allowing an identification of EDMD patients at risk of dilated cardiomyopathy [35]. Transforming growth factor beta 2 (TGF b2) and interleukin 17 (IL-17) serum values are consistently elevated in EDMD type 2 and LGMD1B patients [36].

GNE myopathy, where "GNE" is an abbreviation for the mutated gene (*UDP-N-acetylglucosamine 2-epimerase/N-acetylmannosamine kinase*) codifying the key enzyme of sialic acid synthesis, is an adult-onset progressive myopathy, in which a mild serum CK elevation is reported [37]. The ratio of the Thomsen–Friedenreich (T)-antigen to its sialylated form, ST-antigen, detected by liquid chromatography—mass spectrometry and liquid chromatography—tandem mass spectrometry (LC-MS/MS), is a robust blood-based (serum or plasma) biomarker informative for diagnosis and possibly for response to therapy for this rare condition [38].

3. Metabolic Myopathies

Metabolic myopathies are INMD in which the pathogenetic mechanism compromises enzymes and proteins involved in intermediary metabolism of glucose and free fatty acids [16]. In this review, we focused on glycogenoses, disorders of muscular lipids metabolism and mitochondrial myopathies.

3.1. Glycogenoses

Muscle glycogenoses constitute a growing number of inborn errors of glycogen metabolism. Deficiencies of virtually all enzymes, which intervene in the synthesis or degradation of glycogen, may cause glycogen storage disease (GSD) because of aberrant storage or utilization of glycogen [39].

In this group McArdle and Pompe disease are the most common conditions [40,41]. CK is chronically elevated in nearly all cases of McArdle disease; patients can often show rabdomyolisis and hemoglobinuria after a short-intensity exercise, depicting a very typic feature of this disorder [16]. Apart from CK, serum lactate after forearm exercise is another simple tool that can be very useful to early recognize this disorder [42]. Patients with the most common McArdle disease mutation (R49X) have no protein expressed because of nonsense-mediated mRNA decay; consequently, they have a secondary deficiency of pyridoxine (vitamin B6), due to protein-to-protein interaction [16,43].

Other metabolic disorders such as Pompe disease (PD) present a less specific pattern due to different biochemical pathway involved [44]. HyperCKemia is usually found in PD, but the increment of CK is stably high and a muscular stretching impairment is often observed mimicking a LGMD [41]; moreover, dried blood spot offers an important tool to screen suspected patients for PD [45,46].

Further serum biomarkers have a less validated use in clinical practice: some authors reported that platelet derived grow factor BB (PDGF-BB) and transforming growth factor β1 (TGF-β1) concentrations are significantly lower in PD patients compared to the control group and serum; conversely, platelet derived grow factor AA (PDGF-AA) and connective tissue growth factor (CTGF) values are significantly higher compared to control samples. Interestingly, PDGF-BB level differs among PD and MD and between symptomatic and asymptomatic PD patients [47].

In addition, GSD can be associated with hemolysis/hemolytic anemia (e.g., GSD7, phosphoglycerate kinase 1 deficiency, GSD12). Finally, many glycogenolytic and glycolytic defects have been also linked to myogenic hyperuricemia, which can lead to gout [48].

3.2. Disorders of Muscular Lipids Metabolism

Disorders of muscular lipids metabolism may involve intramyocellular triglyceride degradation, carnitine uptake, long-chain fatty acids mitochondrial transport, or fatty acid β-oxidation [17]. In these conditions, blood testing for CK, lactate, and glucose is usually normal between the episodes of rhabdomyolysis [17]. However, during an acute bout of rhabdomyolysis, a significant increase of CK starts within a few hours, during which hyperkalemia and hypoketotic hypoglycemia can also occur [17]. Moreover, in more severe cases, an acute renal failure can occur with elevations of potassium, creatinine, and urea [17]. The acylcarnitine profile, performed by liquid chromatography tandem mass spectrometry, is the most sensitive and specific test for a fatty acid oxidation defects [16].

Carnitine deficiency (CD) is a potentially lethal but very treatable disorder due to a defect in the carnitine organication transporter (OCTN2), resulting in impaired fatty acid oxidation [49].

CD is clinically characterized by carnitine-responsive acute metabolic decompensation early in life, or in late onset forms, with skeletal and cardiac myopathy or sudden death from arrhythmia [50–52].

Hypoglycemia, with minimal or no ketones in urine, and hyperammonemia, with variably elevated liver function tests are typical features of this disorder [53].

The deficiency of carnitine–palmitoyl-transferase-II (CPT-II, an enzyme involved in the transport of fatty acids into the mitochondrial matrix) is clinically characterized by recurrent myoglobinuric attacks triggered by exercise, fasting, fever, and infection, which may be complicated by acute renal failure and respiratory failure [49].

Multiple acyl-CoA dehydrogenation deficiency (MADD) is a disorder of oxidative metabolism with a broad range of clinical severity [49]; symptoms and age of onset are highly variable and characterized by muscle involvement (myalgia and weakness recurrent episodes of lethargy, vomiting, hypoglycemia, metabolic acidosis, and hepatomegaly) [54].

The diagnosis is suggested by the acyl-carnitine profile and urinary organic acids, revealing low serum free carnitine but elevated acyl-carnitines [49].

Neutral lipid storage disease with myopathy (NLSD-M), increased CK concentrations, cardiomyopathy, diabetes mellitus, hepatic steatosis, and hypertriglyceridemia. Leuko-cytes show a characteristic accumulation in cytoplasm of triglycerides called "Jordan's anomaly" [49].

3.3. Mitochondrial Myopathies

Mitochondrial myopathies (MM) are disorders with mitochondrial function impairment, which usually present multisystem features and, sometimes specifical biomarkers have been recognized [55].

Elevated serum basal lactate have a sensitivity of approximately 65% and specificity of 90% [56], the abnormal lactate value is more evident after aerobic exercise test, an important tool in the clinical suspected of MM [55].

Serum CK is often normal in MM, but it might result elevated after rhabdomyolysis and in a minority of patients [55]. CK concentrations continually higher than three times the normal upper limit should arise a prompt consideration of a different diagnosis [16].

About 50% of patients with Mitochondrial encephalopathy with lactic acidosis and stroke-like episodes (MELAS) can manifest with type 2 diabetes mellitus or impaired oral glucose tolerance [16,57,58]. Plasma amino acid testing can reveal elevated alanine, especially after aerobic exercise. Urine organic acid analysis may reveal high 3-methylglutaconic acid or the tricarboxylic acid intermediates (fumarate, malate, citrate) [16]. Increased CSF proteins, lactic acidemia and hyperalaninemia are also common in mitochondrial neurogastrointestinal encephalopathy [59], an autosomal recessive condition due to mutations in the gene specifying thymidine phosphorylase with progressive gastrointestinal symptoms and muscle involvement [60]. The diagnosis is based on the detection of pathogenic mutation, reduction of thymidine phosphorylase enzyme activity or elevated plasmatic thymidine and deoxyuridine.

Cycle ergometry exercise testing in mitochondrial myopathies might demonstrate a low the maximum oxygen consumption a high respiratory exchange ratio (indicative of early lactate production), or both [16]. Moreover, the oxidative stress have a relevant pathogenic role in MM; hence, oxidative stress markers, such as advanced oxidation protein products have been successfully tested in MM [61].

4. Ion Channel Myopathies

Ion channel myopathies (ICM) include various rare INMD due to mutations in genes encoding ion channels Main disorders of this group are periodic paralyses (PP) and non-dystrophic myotonias (NDM) [62].

The diagnosis of such myopathies relies on the patient's personal and familial medical history and several laboratory findings, with genetic studies representing the diagnostic mainstay.

As far as hypo- and hyperkaliemic periodic paralyses (PP) are concerned, serum diagnostic supportive criteria are constituted by K^+ concentrations <3.5 and >4.5 mEq/L during weakness attacks, respectively [63–73]. Raised CK may be present too [74,75]. Moreover, thyroid function tests may be useful to distinguish HypoPP from thyrotoxic paralysis [76,77], because of HypoPP is reported as a rare complication of hypertyroidisim [78].

CK elevation is common in myotonia congenita [75,79–87] and may be present in paramyotonia congenita [88]. Inherited susceptibility to malignant hyperthermia syndrome (MHS) occurs in patients exposed to anesthetic agents causing increased body temperature, muscle rigidity or spasms, rhabdomyolysis, tachycardia, and other life-threatening symptoms [89]. Mutations in the ryanodine receptor RYR1, CACNA1S or STAC3 genes are encountered in MHS, resulting in excessive cytoplasmic calcium accumulation in response to both pharmacological and non-pharmacological triggers. As a result, patients experience recurrent episodes of hyperCKemia and rhabdomyolysis [11,89,90]. RYR1-related central core myopathy usually shows elevation of CK [91–94], and, in some cases, it can be also normal [95], AST, ALT and γGT may be elevated or normal [93,96].

5. Congenital Myasthenic Syndromes

Congenital myasthenic syndromes (CMS) refer to a heterogenic group of inherited neuromuscular transmission disorders (NTD) with exercise intolerance and muscular weakness [97]. There are no relevant biomarkers for CMS and the diagnosis is usually achieved through neurophysiology [97]. Indeed, CK concentrations are usually normal in CMS [97], but there are reported cases in which CK and aldolase concentrations can be mildly elevated [98,99]. In these cases, elevated CK is due to a phenomenon called endplate myopathy [99]. Circulating miRNAs [100] and plasmatic titin [101] have been proposed as possible biomarkers for autoimmune *Myasthenia Gravis*, but their role has not been explored yet in inherited CMS.

6. Motor Neurons Diseases

6.1. Spinal Muscular Atrophy

Spinal muscular atrophy (SMA) is characterized by motor neuron degeneration and progressive muscle atrophy and weakness caused by mutation of the survival motor neuron (*SMN*) genes. After neuronal degeneration or axonal injury specific cytoskeletal proteins of the neurons, including neurofilaments (NF), are released in the extracellular space, in Cerebrospinal fluid (CSF) and in the blood; about 80% of NF are phosphorylated (pNF) conferring more resistance in the serum. Several studies explored the role of NF as a biomarker of disease severity and progression. There are three different NF type distinguished by molecular weight: light (NF-L), medium and heavy (NF-H). High concentrations of NF-H have been detected in blood of amyotrophic lateral sclerosis, SMA, Alzheimer's disease, Parkinson disease and multiple sclerosis [102–107].

Plasma phosphorylated neurofilament heavy chain (pNF-H) in infants with SMA are about 10-fold higher than age-matched infants without SMA. Treatment with nusinersen (the first approved treatment for SMA favoring expression of normal SMN protein) induce a faster decline in pNF-H concentrations at two and ten months than sham controlled-treated infants, followed by a relative stabilization [108]. These data suggest NF as a promising biomarker for SMA severity and response to treatment.

Patients with SMA have been shown to have lower values of creatinine compared to age-matched controls [109]. In the skeletal muscles is stored most creatine of the body, and its metabolism is essential to maintain the muscle function; creatinine is a product of muscle creatine metabolism. Chronic medical conditions associated with cachexia (and muscle hypotrophy) could be associated with lower concentrations of creatinine. A recent study [109] showed significant differences of creatinine concentrations depending on the SMA subtype (with SMA 3 > SMA 2 > SMA 1 concentrations) and on clinical severity; furthermore, a positive association of serum creatinine concentrations and the number of *SMN2* gene copies has been documented; a positive association of creatinine concentrations was also related to denervation signs expressed by maximal compound muscle action potential amplitude evoked by distal stimulation of ulnar nerve, and to the motor unit number estimation. Considering that the muscle atrophy could be considered a confounding factor for low creatinine concentrations, all the reported associations with the severity of disease, still resulted statistically significant, after correction for lean mass [109].

6.2. Amyotrophic Lateral Sclerosis

Amyotrophic lateral sclerosis (ALS) causes upper and lower motor neuron degeneration, with progressive paresis of arms, legs and bulbar muscles finally resulting in death for respiratory failure.

Most ALS cases are sporadic and only about 10% are familiar as associated to various specific gene mutations (e.g., *SOD1, TDB-43, FUS, DPRs*, etc.).

Even if an aspecific marker serum CK is reported in an high percentage of ALS patients [110], markers of inflammation [111,112] as well as creatinine [113] have a prognostic role in ALS.

The most studied biomarker for SLA is the pNF-H which show increased concentrations both in CSF (sensitivity 71%, specificity 88%) [114] and in the blood (sensitivity 58%, specificity 89%) [115]. As showed in SMA and other neurodegenerative diseases, the axonal injury induces the release of these cytoskeletal proteins in the extracellular space, CSF and blood. However, this measure is not specific for familiar ALS respect to sporadic forms and it is not statistically different in Riluzole-treated vs. non-treated patients [115]. By the way, NF still represent a possible marker for upcoming therapies.

Testing the hypothesis of ALS as an autoimmune pathology, neuron glycolipids such as sulfoglucuronosyl paragloboside (SGPG), target of various neuropathies, have been explored in ALS and anti-SGPG were found in the sera of 13.3% ALS patients [116].

Other biological markers of ALS remain of uncertain significance in the pathobiology of disease, but include transthyretin, complement C3 protein and fetuin-A [117–119].

7. Axonal Neuropathies

This group includes disorders in which polyneuropathy (sensory-motor deficit associated with reduced jerks in the distribution of the main peripheral nerves) is the main feature and disorders with nerve involvement as part of multisystemic disease.

7.1. Charcot–Marie–Tooth 2 (CMT2)

Charcot–Marie–Tooth (CMT) is an autosomal dominant inherited motor and sensory neuropathy, characterized by peripheral nerve damage and classically described in CMT Type 1 and 2 depending on the neurophysiological features [120]. CMT2 is genetically heterogeneous and exists in several different subtypes (e.g., CMT2A, CMT2B, CMT2E, etc.) [20]. CMT2 is an axonal form of the disease in which the loss of axons reduces muscle innervations [120]. Neurofilament light chain (NfL) represent a possible biomarker of axonal damage in CMT2 correlating with disease severity [121].

7.2. Multisystemic Diseases

The term "amyloidosis" gathers several diseases associated with deposits of amyloid fibrils involving peripheral and autonomic nervous systems; among them about 10% is represented by inherited forms (hereditary amyloidosis, HA) [122]. HA includes Hereditary amyloidogenic transthyretin (hATTRv) [123], amyloidosis derived from a fibrinogen variant (AFib), that is the most frequent HA in Northern Europe [124], variants of apolipoproteins (AI, AII, C-II, CIII), gelsolin (AGel), and lysozyme (ALys) [125].

In a study on hATTR, serum retinol-binding protein 4 (RBP4) concentration was lower in patients with hATTR V122I amyloidosis; also a lower concentration of B-type natriuretic peptide (BNP) and transtiretin (TTR), as well as, a higher concentration of Troponin I (TrI) and hematocrit have been reported [126]. Increased serum concentrations of amino-terminal prohormone of brain natriuretic peptide (NT-proBNP) and Troponin T (TrT) are the mainstay in patients with polineurophaty compared to *TTRv* carriers and healthy subjects [127]. Of interest, serum neurofilament light chain (NfL) concentrations correlated with Troponin T in all patient groups, but not with NT-proBNP. Furthermore, higher cytokine concentrations (TNF-α, IL-1β, IL-8, IL-33, IFN-β, IL-10 and IL-12) have been reported in hATTR patients. In ATTRv [128]. Many study described elevated NfL also in early stage of hATTR, showing a correlation with the severity of polyneuropathy, thus proposing NfL as a diagnostic and prognostic biomarker [10,129,130]. A further study confirmed the importance of NfL as a biomarker for nerve damage showing a significant decrease after therapy with Patisiran [131].

Ataxia with oculomotor apraxia is a group of recessive ataxias: ataxia-telangectasia (AT), ataxia with oculomotor apraxia type 1 (AOA1) and type 2 (AOA2). All these conditions share axonal neuropathy and elevated serum a.lpha-fetoprotein (AFP) [132]. AOA1 is an autosomal recessive disease due to mutations in the *APTX* gene, which encodes for ataxin. During the course of AOA1, decreased serum albumin and increased serum total cholesterol can be seen [13,133]. In addition, hypoalbuminemia, hypercholesterolemia and increased serum creatine kinase (CK) were reported in association with increased AFP [134]. However, according to another study hypoalbuminemia and hypercholesterolemia with normal AFP are the hallmarks of AOA1 [135]. AOA2 is caused by mutations in the *SETX* gene, encoding for a DNA/RNA helicase protein [136]. Elevation of AFP is present in 99% of cases and is stable over the course of disease, but it usually do not correlate with disease severity [137]; elevated CK, mild hypoalbuminemia, or hypercholesterolemia can also be present in AOA2 patients [138]. Ataxia-telengectasia (AT) is caused by mutation in the *ATM* gene, that encodes a phosphatidylinositol-3 kinase protein (PI3K). Immunoglobulins (Ig) lare usually reduced in AT and elevation of AFP is progressive [138]. Conversely, in A-TLD, a rare form caused by mutations in the *hMRE11 gene*, patients have normal AFP serum concentrations [132,139].

Spinocerebellar ataxias (SCA) are a heterogenous group of diseases with CAG-trinucleotide expansions. Axonal polyneuropathy in SCA is one of the typical clinical features and depends on neuronal loss for inactivation of repair pathways [140]. Peripheral axonal neuropathy, tested by electrophysiology, is consistent among SCA1, 2, 3, 18, 25, 38, 43, 46 and specially SCA4 [5,141]. Neurofilament light polypeptide (NfL) is marker of neuronal damage. Serum concentrations of NfL are increased in patients with SCA, but the results have not already been validated [140,142].

Hereditary spastic paraplegias (HSP) are inherited neurodegenerative diseases, characterized by lower limbs weakness and spasticity [143]. Axonal neuropathy is frequent in SPG7, SPG11, and SPG15 and less commonly in SPG3A, SPG4, and SPG5 [143]. SPG5, a rare form of HSP, is caused by recessive mutation in the *CYP7B1* gene, encoding oxysterol-7α-hydroxylase [144]. CYP7B1 substrates, including 27-hydroxycholesterol (27-OHC), have been described cerebrospinal fluid (CSF) and in the sera of SPG5 patients; the latter were correlated with severity and duration of the disease [145].

Interestingly, serum 25-OHC and 27-OHC, have been validated as diagnostic biomarkers in SPG5 and it has been proved that atorvastatin can decrease serum 27-OHC by $\sim$30% [146].

Hepatic porphyrias (HP) are metabolic disorders of the heme metabolism; HP are caused by accumulation of heme intermediates due to enzymatic alteration in the heme biosynthesis pathway [147]. Acute intermittent porphyrias (AIP) is the most common subtype of HP and presents an autosomal dominant inherited pathology characterized by accumulation of porphyrin precursors, porphobilinogen (PBG) due to inherited deficiency of porphobilinogen deaminase [148]. Variegate porphyria (VP) is caused by deficiency of protoporphyrinogen oxidase (PPOX) that causes accumulation of δ -aminolevulinic acid (ALA) [147,148]. Porphyric attacks can be characterized by peripheral neuropathy, typically a motor neuropathy with symmetrical proximal muscle weakness [149]. Increased urinary porphobilinogen (PBG) and δ-aminolaevulinic acid (ALA) concentrations are usually detected during acute attacks, with normal interictal ranges; it has been reported that raised concentrations persist for many years in AIP patients [150].

GM2 gangliosidosis comprises Tay-Sachs disease, Sandhoff disease and GM2 gangliosidosis AB due to β-hexosaminidase A, β-hexosaminidase A+B and GM2 activator deficiency respectively. Several serums (i.e., lactic dehydrogenase malic dehydrogenase, fructose 1,6-diphosphate aldolase) and liquoral biomarkers (i.e., epithelial-derived neutrophil activating protein 78, monocyte chemotactic protein 1, macrophage inflammatory protein-1 alpha, macrophage inflammatory protein-1 beta, tumor necrosis factor receptor 2) were identified in gangliosidosis affected patients [151]. A recent study proposed serum aspartate transaminase (AST) as new potentially biomarker for gangliosidosis [151].

X-linked adrenoleukodystrophy (XALD) is the most common peroxisomal disorder due to mutations of the ABCD1 gene leading to very long-chain fatty acids (VLCFA) accumulation in blood and a variety of tissue [152]. In males with clinical suspicion, the diagnosis of XALD is allowed by elevated serum or plasma concentrations of hexacosanoic acid C26:0 and high ratio C24:0/C22:0 and C26:0/C22:0 [152]. An alternative and more accurate biomarker to detect the elevated serum VLCFA is the use of 1-hexacosanosyl-2-lyso-sn-3-glycero-phosphatidylcholine (26:0-lyso-PC) in DBS and it could be useful for newborn screening and diagnosis in women [153]. A recent study suggests new optimized cutoff values for both ratios C24:0/C22:0 and C26:0/C22:0, in combination with standard lipid profile considered that low-density lipid concentrations strongly correlate to all VLCFA [9].

Fabry disease (FD) is an X-linked inborn error of glycosphingolipid catabolism due to deficiency of α-galactosidase A (α-Gal A) leading accumulation of globotriaosylceramide in body fluids and lysosomes of vascular endothelium [154]. The diagnostic algorithm in FD is gender-specific indeed the measurement of blood α-Gal A activity is recommended in males, and optionally in females. Globotriaosysphingosine for identification of atypical FD variants and high- sensitive troponin T for identification of cardiac involvement are also important diagnostic biomarkers [6].

8. Demyelinating Neuropathies

Among this group of disorders, we report demyelinating CMT and inborn errors of metabolism.

8.1. Charcot–Marie–Tooth Spectrum Disorders

In the actual scenario there are no validated blood or liquoral biomarkers for CMT spectrum disorders, although a recent observational study suggested potentially new diagnostic and prognostic biomarkers in CMT1A patients [155]; indeed, the upregulation of the protein and lipid catabolism products plasma concentrations (i.e., dipeptide glutaminyl-serine, tryptophan, urobilinogen, polyamine acetylagmatine, sphingosine-1-phosphate, 6-hydroxysphingosine, lysophosphatidylcholine, 11-hydroxyeicosatetraenoate glyceryl ester) and downregulation of leucine plasma concentrations have been reported [155]. A more recent cross-sectional study showed the higher concentration of plasma neurofilament light chain in CMT patients correlating with disease severity [121]. In addition, it has been documented that some CMT1A patients can display serum PMP22 antibody and CSF protein content moderately elevated [156].

8.2. Inborn Errors of Metabolism

Inborn errors of metabolism (IEMs) are genetic disorders that cause disruption of a metabolic pathway leading to the accumulation of a toxic substrate [157]. IEMs may present as an apparently isolated peripheral neuropathy at any age or a polyneuropathy may be the only part of the clinical picture.

Refsum disease is an inborn error of lipid metabolism due to low phytanoyl-CoA hydroxylase activity leading to high phytanic acid plasma concentrations which represents the pathognomonic biomarker [158]. A relapsing-remitting polyneuropathy is the main feature with acute exacerbation in the setting of elevated phytanic acid blood concentrations caused by high phytanic acid intake, weight loss or pregnancy with elevated CSF protein concentration [159].

Tangier disease (TD) is an inherited lipid trafficking disorder characterized by severe deficiency or absence of high-density lipoprotein (HDL) in the circulation resulting in tissue accumulation of cholesteryl esters, particularly in the reticuloendothelial system with hyperplastic yellow-orange tonsil and hepatosplenomegaly [160]. As the high clinical suspicion, serum hypertriglyceridemia, very low HDL cholesterol and undetectable apolipoprotein A1 are pathognomonic for TD [161].

Metachromatic leukodystrophy (MLD) is caused by deficiency of lysosomal arylsulfatase A with cerebroside deficiency and sulfatides accumulation in various tissues. The diagnosis of MLD is confirmed by demonstrating deficient ASA activity in leukocytes and increased urinary sulfatide concentrations [162]. Some authors documented elevations of proinflammatory cytokines (e.g., IL8, IL-1Ra) and vascular endothelial growth factor in both CSF and plasma of MLD patients [163].

Krabbe disease (KD), also known as globoid cell leukodystrophy, is a rare lysosomal storage disorder caused by deficiency of galactocerebrosidase (GALC) leading to generation of psychosine, a cytotoxic agent that causes loss of oligodendrocytes and Schwann cells [164]. Newborn screenings are available for Krabbe disease with measurement of GALC enzymatic activity in DBS and measurement of psychosine in DBS as second-tier test may help to differentiate infantile from late-onset KD variants, as well as from GALC variant and pseudodeficiency carriers [165]. Psychosine concentrations have also been evaluated as biomarker to predict the severity of disease considering patients with psychosine greater concentrations at high risk to develop early-onset KD [166].

Zellweger spectrum disorders are caused by mutation in one of 13 different PEX genes resulting in different clinical phenotypes [167]. The higher plasma concentrations of very long-chain fatty acids (e.g., C26:0 lysophosphatidylcholine), polyunsaturated fatty acids, bile acid intermediates (i.e., di- and trihydroxycholestanoic acid), phytanic acid, pristanic acid, pipecolic acid and detectable urinary concentrations of bile acid intermediates or detectable pipecolic, glycolic, and oxalic acid may support the diagnosis [168].

Niemann–Pick disease is an autosomal recessive lipid storage disorder that comprises types A (NP-A) and B (NP-B) caused by deficiency of acid sphingomyelinase and type C (NP-C) caused by protein NPC1 defect leading to intracellular sphingomyelin and free cholesterol accumulation respectively. Reduced sphingomyelinase activity in circulating leukocytes, elevated both triglycerides and serum LDL-cholesterol and reduced HDL-cholesterol serum concentrations are the diagnostic hallmarks of the NP-B [169]. Cholesterol oxidation products such as 3β,5α,6β-triol and 7-ketocholesterol (7-KC) might represent a specific and sensitive blood-based biomarker for NP-C [170]. A further study identified two secreted proteins (i.e., galectin-3, cathepsin D) with increased serum concentrations in NPC1 patients correlating with disease severity [171]. Additionally, a recent study showed significantly decreased values of HSP70 in individuals with NP-C [172]. The combination of 7-KC, lysosphingomyelin and bile acid-408 improves the accuracy of NP-C diagnosis [173].

Abetalipoproteinemia typically presents in infancy with failure to thrive and malabsorption of fat and fat-soluble vitamins leading to multiorgan and neuromuscular involvement with cranial and peripheral nerve demyelination due to vitamin E deficiency. Common serums findings are extremely low LDL-cholesterol, triglyceride and apoB concentrations [174].

Beta-Mannosidosis is a lysosomal storage disease due to an isolated deficiency of the enzyme betas-mannosidase with various degrees of neurological deterioration. The diagnosis is supported by reduced enzyme activity in plasma and leukocyte [175]. Liquid chromatography-tandem mass spectrometry (UPLC-MS/MS) for urinary free oligosaccharides is a useful diagnostic tool [176].

Type 1 hyperoxaluria (PH1) is a rare autosomal recessive disease due to mutation in the *alanine-glyoxalate aminotransferase* gene leading to glyoxylate accumulation and its conversion to oxalate with calcium oxalate crystals deposition in various tissues [177]. The biochemical hallmarks of PH1 are hyperoxaluria with or without hyperglycolaturia; hence, 24 h urine collection measuring oxalate, creatinine and glycolate is recommended. Plasma oxalate concentrations are also useful in patients with renal failure, whereas higher ones are helpful to the diagnosis [12].

Tyrosinaemia type 1 (HT1) is a rare autosomal recessive disorder of amino acid metabolism resulting from fumarylacetoacetase 1 deficiency. The diagnosis is supported by increased plasma concentrations of tyrosine, methionine and phenylalanine, increased concentrations of hepatic transaminases and plasma alfa- fetoprotein [178]. As hypertyrosinemia is caused by other conditions, the diagnosis is allowed by the identification of the pathognomonic and more sensitive and specific biomarker succinylacetone on blood or urine with different cut-off reported [179,180].

Methylmalonic aciduria and homocystinuria cblC type are the most frequent inborn error of vitamin B12 metabolism. The biochemical hallmarks are elevated plasma and urine concentrations of methylmalonic acid, cystathionine and homocysteine with hypomethioninemia, whereas metabolic acidosis and hyperammonemia have been rarely reported [181].

Cerebrotendinous xanthomatosis (CTX) is an autosomal recessive inborn lipid storage disorder due to enzyme sterol 27-hydroxylase deficiency resulting in impaired bile primary acid synthesis, increased concentration of bile alcohols and increased accumulation of both cholestanol and cholesterol in plasma and in various tissues, particularly in the nervous system [182]. High plasmatic concentrations of cholestanol in the presence of normal or low plasma concentrations of cholesterol, decreased chenodeoxycholic acid and increased concentration of bile alcohols are the laboratoristic hallmark of CTX [183].

9. Conclusions

Neuromuscular disorders can be very complex with the involvement of various organs and systems. A number of biomarkers are to date available to support the diagnosis of INMD, but the role of expert clinician is crucial. Some markers are commonly altered in many conditions in similar way in different disorders, so that the physical examination should be the first step to raise the clinical suspicion. The complete neuromuscular examination, followed by subsequent evaluations is the easiest way to obtain diagnoses to date. On the other hand, some biomarkers are associated with specific disorders and their recognition can be crucial for the diagnosis. Indeed, it is not unusual to see clinicians who are not familiar with rare diseases facing with INMD, because of the very broad range of serum and urinary alterations encountered in clinical practice. Hence, the complete knowledge of such abnormalities can accelerate the diagnostic workup supporting the referral to specialists in neuromuscular disorders.

Supplementary Materials: The following are available online at https://www.mdpi.com/2076-3425/11/3/398/s1, Table S1: main biomarkers, Diagram S1: increase of serum CK.

Author Contributions: All authors have made substantial contribution in study design, data analysis, or interpretation, in drafting the manuscript, or critically contributing to or revising the manuscript, or enhancing its intellectual content. All authors have read and agreed to the published version of the manuscript. The manuscript has not been submitted elsewhere nor published elsewhere in whole or in part.

Funding: This research received no external funding.

Institutional Review Board Statement: Not applicable.

Informed Consent Statement: Not applicable.

Data Availability Statement: Data sharing not applicable. No new data were created or analyzed in this study. Data sharing is not applicable to this article.

Conflicts of Interest: The authors declare no conflict of interest.

References

1. Thompson, R.; Spendiff, S.; Roos, A.; Bourque, P.R.; Chardon, J.W.; Kirschner, J.; Horvath, R.; Lochmüller, H. Advances in the Diagnosis of Inherited Neuromuscular Diseases and Implications for Therapy Development. *Lancet Neurol.* **2020**, *19*, 522–532. [CrossRef]
2. Romi, F.; Helgeland, G.; Gilhus, N.E. Serum Levels of Matrix Metalloproteinases: Implications in Clinical Neurology. *Eur. Neurol.* **2012**, *67*, 121–128. [CrossRef] [PubMed]
3. Di Stefano, V.; Lupica, A.; Rispoli, M.G.; Di Muzio, A.; Brighina, F.; Rodolico, C. Rituximab in AChR Subtype of Myasthenia Gravis: Systematic Review. *J. Neurol. Neurosurg. Psychiatry* **2020**, *91*, 392–395. [CrossRef] [PubMed]
4. Di Stefano, V.; Barbone, F.; Ferrante, C.; Telese, R.; Vitale, M.; Onofrj, M.; Di Muzio, A. Inflammatory Polyradiculoneuropathies: Clinical and Immunological Aspects, Current Therapies, and Future Perspectives. *Eur. J. Inflamm.* **2020**, *18*. [CrossRef]
5. Bhandari, J.; Thada, P.K.; Samanta, D. *Spinocerebellar Ataxia*; StatPearls Publishing: Treasure Island, FL, USA, 2020.
6. Vardarli, I.; Herrmann, C.R.K.; Weidemann, F. Diagnosis and Screening of Patients with Fabry Disease. *Ther. Clin. Risk Manag.* **2020**, *16*, 551–558. [CrossRef]
7. Li, X.; Li, Y.; Zhao, L.; Zhang, D.; Yao, X.; Zhang, H.; Wang, Y.C.; Wang, X.Y.; Xia, H.; Yan, J.; et al. Circulating Muscle-specific miRNAs in Duchenne Muscular Dystrophy Patients. *Mol. Ther. Nucleic Acids* **2014**, *3*, e177. [CrossRef] [PubMed]
8. Petek, L.M.; Rickard, A.M.; Budech, C.; Poliachik, S.L.; Shaw, D.; Ferguson, M.R.; Tawil, R.; Friedman, S.D.; Miller, D.G. A Cross Sectional Study of Two Independent Cohorts Identifies Serum Biomarkers for Facioscapulohumeral Muscular Dystrophy (FSHD). *Neuromuscul. Disord.* **2016**, *26*, 405–413. [CrossRef] [PubMed]
9. Rattay, T.W.; Rautenberg, M.; Söhn, A.S.; Hengel, H.; Traschütz, A.; Röben, B.; Hayer, S.N.; Schüle, R.; Wiethoff, S.; Zeltner, L.; et al. Defining Diagnostic Cutoffs in Neurological Patients for Serum Very Long Chain Fatty Acids (VLCFA) in Genetically Confirmed X-Adrenoleukodystrophy. *Sci. Rep.* **2020**, *10*. [CrossRef] [PubMed]

10. Maia, L.F.; Maceski, A.; Conceição, I.; Obici, L.; Magalhães, R.; Cortese, A.; Leppert, D.; Merlini, G.; Kuhle, J.; Saraiva, M.J. Plasma Neurofilament Light Chain: An Early Biomarker for Hereditary ATTR Amyloid Polyneuropathy. *Amyloid* **2020**, *27*, 97–102. [CrossRef]

11. Kraeva, N.; Sapa, A.; Dowling, J.J.; Riazi, S. Malignant Hyperthermia Susceptibility in Patients with Exertional Rhabdomyolysis: A Retrospective Cohort Study and Updated Systematic Review. *Can. J. Anaesth.* **2017**, *64*, 736–743. [CrossRef]

12. Bouzidi, H.; Majdoub, A.; Daudon, M.; Najjar, M.F. Hyperoxalurie Primitive: Une Revue de la Littérature. *Nephrol. Ther.* **2016**, *12*, 431–436. [CrossRef] [PubMed]

13. Renaud, M.; Moreira, M.C.; Ben Monga, B.; Rodriguez, D.; Debs, R.; Charles, P.; Chaouch, M.; Ferrat, F.; Laurencin, C. Clinical, Biomarker, and Molecular Delineations and Genotype-Phenotype Correlations of Ataxia with Oculomotor Apraxia Type 1. *JAMA Neurol.* **2018**, *75*, 495–502. [CrossRef] [PubMed]

14. Valaperta, R.; De Siena, C.; Cardani, R.; Lombardia, F.; Cenko, E.; Rampoldi, B.; Fossati, B.; Rigolini, R.; Gaia, P. Cardiac Involvement in Myotonic Dystrophy: The Role of Troponins and N-terminal pro B-type Natriuretic Peptide. *Atherosclerosis* **2017**, *267*, 110–115. [CrossRef] [PubMed]

15. Uchio, N.; Taira, K.; Ikenaga, C.; Kadoya, M.; Unuma, A.; Yoshida, K.; Nakatani-Enomoto, S.; Hatanaka, Y.; Sakurai, Y.; Shiio, Y.; et al. Inflammatory Myopathy with Myasthenia Gravis: Thymoma Association and Polymyositis Pathology. *Neurol. Neuroimmunol. NeuroInflammation* **2019**, *6*. [CrossRef] [PubMed]

16. Tarnopolsky, M.A. Metabolic Myopathies. *Contin. Lifelong Learn. Neurol.* **2016**, *22*, 1829–1851. [CrossRef]

17. Laforêt, P.; Vianey-Saban, C. Disorders of Muscle Lipid Metabolism: Diagnostic and Therapeutic Challenges. *Neuromuscul. Disord.* **2010**, *20*, 693–700. [CrossRef]

18. Coenen-Stass, A.M.L.; Wood, M.J.A.; Roberts, T.C. Biomarker Potential of Extracellular miRNAs in Duchenne Muscular Dystrophy. *Trends Mol. Med.* **2017**, *23*, 989–1001. [CrossRef]

19. Darras, B.T.; Urion, D.K.; Ghosh, P.S. *Dystrophinopathies*; Adam, M.P., Ardinger, H.H., Pagon, R.A., Wallace, S.E., Bean, L.J.H., Mirzaa, G., Amemiya, A., Eds.; University of Washington, Seattle: Seattle, WA, USA, 1993.

20. Kleopa, K.A.; Scherer, S.S. Inherited Neuropathies. *Neurol. Clin.* **2002**, *20*, 679–709. [CrossRef]

21. Cortese, A.; Vegezzi, E.; Lozza, A.; Alfonsi, E.; Montini, A.; Moglia, A.; Merlini, G.; Obici, L. Diagnostic Challenges in Hereditary Transthyretin Amyloidosis with Polyneuropathy: Avoiding Misdiagnosis of a Treatable Hereditary Neuropathy. *J. Neurol. Neurosurg. Psychiatry* **2017**, *88*, 457–458. [CrossRef]

22. Darras, B.T.; Urion, D.K.; Ghosh, P.S. *GeneReviews®Dystrophinopathies–Last Update: April 26, 2018*; University of Washington, Seattle: Seattle, WA, USA, 2000.

23. Mendell, J.R.; Shilling, C.; Leslie, N.D.; Flanigan, K.M.; Al-Dahhak, R.; Gastier-Foster, J.; Kneile, K.; Dunn, D.M.; Duval, B.; Aoyagi, A.; et al. Evidence-based Path to Newborn Screening for Duchenne Muscular Dystrophy. *Ann. Neurol.* **2012**, *71*, 304–313. [CrossRef]

24. Wicklund, M.P. The Limb-Girdle Muscular Dystrophies. *Continuum* **2019**, *25*, 1599–1618. [CrossRef]

25. Strandberg, K.; Ayoglu, B.; Roos, A.; Reza, M.; Niks, E.; Signorelli, M.; Fasterius, E.; Pontén, F.; Lochmüller, H.; Domingos, J.; et al. Blood-derived Biomarkers Correlate with Clinical Progression in Duchenne Muscular Dystrophy. *J. Neuromuscul. Dis.* **2020**, *7*, 231–246. [CrossRef]

26. Nadarajah, V.D.; van Putten, M.; Chaouch, A.; Straub, V.; Lochmüller, H.; Ginjaar, H.B.; Aartsma-Rus, A.M.; van Ommen, G.J.; den Dunnen, J.T.; Hoen, P.A. Serum Matrix Metalloproteinase-9 (MMP-9) as a Biomarker for Monitoring Disease Progression in Duchenne Muscular Dystrophy (DMD). *Neuromuscul. Disord.* **2011**, *21*, 569–578. [CrossRef]

27. Spitali, P.; Hettne, K.; Tsonaka, R.; Sabir, E.; Seyer, A.; Hemerik, J.B.A.; Goeman, J.J.; Picillo, E.; Ergoli, M.; Politano, L.; et al. Cross-sectional Serum Metabolomic Study of Multiple Forms of Muscular Dystrophy. *J. Cell. Mol. Med.* **2018**, *22*, 2442–2448. [CrossRef]

28. Martin, F.C.; Hiller, M.; Oonk, S.; Dalebout, H.; Palmblad, M.; Chaouch, A.; Guglieri, M.; Straub, V.; Lochmüller, H.; Niks, E.H.; et al. Fibronectin is a Serum Biomarker for Duchenne Muscular Dystrophy. *Proteomics. Clin. Appl.* **2014**, *8*, 269–278. [CrossRef] [PubMed]

29. Johnson, N.E. Myotonic Muscular Dystrophies. *Continuum* **2019**, *25*, 1682–1695. [CrossRef] [PubMed]

30. Parmova, O.; Vlckova, E.; Hulova, M.; Mensova, L.; Crha, I.; Stradalova, P.; Kralickova, E.; Jurikova, L.; Podborska, M.; Mazanec, R.; et al. Anti-Müllerian Hormone as an Ovarian Reserve Marker in Women with the Most Frequent Muscular Dystrophies. *Medicine* **2020**, *99*, e20523. [CrossRef]

31. Daniele, A.; De Rosa, A.; De Cristofaro, M.; Monaco, M.; Masullo, M.; Porcile, C.; Capasso, M.; Tedeschi, G.; Oriani, G.; Di Costanzo, A. Decreased Concentration of Adiponectin Together with a Selective Reduction of Its High Molecular Weight Oligomers is Involved in Metabolic Complications of Myotonic Dystrophy Type 1. *Eur. J. Endocrinol.* **2011**, *165*, 969–975. [CrossRef] [PubMed]

32. Martinez-Rodríguez, J.E.; Lin, L.; Iranzo, A.; Genis, D.; Martí, M.J.; Santamaria, J.; Mignot, E. Decreased Hypocretin-1 (orexin-A) Levels in the Cerebrospinal Fluid of Patients with Myotonic Dystrophy and Excessive Daytime Sleepiness. *Sleep* **2003**, *26*, 287–290. [CrossRef]

33. Preston, M.K.; Tawil, R.; Wang, L.H. *Facioscapulohumeral Muscular Dystrophy*; University of Washington, Seattle: Seattle, WA, USA, 1993.
34. Bialer, M.G.; Bruns, D.E.; Kelly, T.E. Muscle Enzymes and Isoenzymes in Emery-Dreifuss Muscular Dystrophy. *Clin. Chem.* **1990**, *36*, 427–430. [CrossRef]
35. Niebroj-Dobosz, I.; Madej-Pilarczyk, A.; Marchel, M.; Sokołowska, B.; Hausmanowa-Petrusewicz, I. Circulating Tenascin-C Levels in Patients with Dilated Cardiomyopathy in the Course of Emery-Dreifuss Muscular Dystrophy. *Clin. Chim. Acta* **2011**, *412*, 1533–1538. [CrossRef] [PubMed]
36. Bernasconi, P.; Carboni, N.; Ricci, G.; Siciliano, G.; Politano, L.; Maggi, L.; Mongini, T.; Vercelli, L.; Rodolico, C.; Biagini, E.; et al. Elevated TGF β2 Serum Levels in Emery-Dreifuss Muscular Dystrophy: Implications for Myocyte and Tenocyte Differentiation and Fibrogenic Processes. *Nucleus* **2018**, *9*, 292–304. [CrossRef] [PubMed]
37. Carrillo, N.; Malicdan, M.C.; Huizing, M. GNE Myopathy: Etiology, Diagnosis, and Therapeutic Challenges. *Neurotherapeutics* **2018**, *15*, 900–914. [CrossRef] [PubMed]
38. Leoyklang, P.; Malicdan, M.C.; Yardeni, T.; Celeste, F.; Ciccone, C.; Li, X.; Jiang, R.; Gahl, W.A.; Carrillo-Carrasco, N.; He, M.; et al. Sialylation of Thomsen-Friedenreich Antigen is a Noninvasive Blood-based Biomarker for GNE Myopathy. *Biomark. Med.* **2014**, *8*, 641–652. [CrossRef]
39. Laforêt, P.; Malfatti, E.; Vissing, J. Update on New Muscle Glycogenosis. *Curr. Opin. Neurol.* **2017**, *30*, 449–456. [CrossRef]
40. Bruno, C.; Cassandrini, D.; Martinuzzi, A.; Toscano, A.; Moggio, M.; Morandi, L.; Servidei, S.; Mongini, T.; Angelini, C.; Musumeci, O.; et al. McArdle disease: The Mutation Spectrum of PYGM in a Large Italian Cohort. *Hum. Mutat.* **2006**, *27*, 718. [CrossRef]
41. Lukacs, Z.; Nieves Cobos, P.; Wenninger, S.; Willis, T.A.; Guglieri, M.; Roberts, M.; Quinlivan, R.; Hilton-Jones, D.; Evangelista, T.; Zierz, S. Prevalence of Pompe Disease in 3,076 Patients with Hyperckemia and Limb-girdle Muscular Weakness. *Neurology* **2016**, *87*, 295–298. [CrossRef]
42. Kazemi-Esfarjani, P.; Skomorowska, E.; Jensen, T.D.; Haller, R.G.; Vissing, J. A Nonischemic Forearm Exercise Test for McArdle Disease. *Ann. Neurol.* **2002**, *52*, 153–159. [CrossRef]
43. Beynon, R.J.; Leyland, D.M.; Evershed, R.P.; Edwards, R.H.T.; Coburn, S.P. Measurement of the Turnover of Glycogen Phosphorylase by GC/MS Using Stable Isotope Derivatives of Pyridoxine (Vitamin B6). *Biochem. J.* **1996**, *317*, 613–619. [CrossRef]
44. Preisler, N.; Laforet, P.; Madsen, K.L.; Hansen, R.S.; Lukacs, Z.; Ørngreen, M.C.; Lacour, A.; Vissing, J. Fat and Carbohydrate Metabolism during Exercise in Late-onset Pompe Disease. *Mol. Genet. Metab.* **2012**, *107*, 462–468. [CrossRef] [PubMed]
45. Musumeci., O.; La Marca, G.; Spada, M.; Mondello, S.; Danesino, C.; Comi, G.P.; Pegoraro, E.; Antonini, G.; Marrosu, G.; Liguori, R.; et al. LOPED Study: Looking for an Early Diagnosis in a Late-onset Pompe Disease High-risk Population. *J. Neurol. Neurosurg. Psychiatry* **2016**, *87*, 5–11. [CrossRef]
46. van der Ploeg, A.T.; Kruijshaar, M.E.; Toscano, A.; Laforêt, P.; Angelini, C.; Lachmann, R.H.; Pascual Pascual, S.I.; Roberts, M.; Rösler, K.; Stulnig, T.; et al. European Consensus for Starting and Stopping Enzyme Replacement Therapy in Adult Patients with Pompe Disease: A 10-year Experience. *Eur. J. Neurol.* **2017**, *24*, 768-e31. [CrossRef]
47. Fernández-Simón, E.; Carrasco-Rozas, A.; Gallardo, E.; Figueroa-Bonaparte, S.; Belmonte Pedrosa, I.; Montiel, E.; Suárez-Calvet, X.; Alonso-Pérez, J.; Segovia, S.; Nuñez-Peralta, C.; et al. Spanish Pompe Study Group; Díaz-Manera, J. PDGF-BB Serum Levels are Decreased in Adult Onset Pompe Patients. *Sci. Rep.* **2019**, *9*. [CrossRef]
48. Mineo, I.; Tarui, S. Myogenic Hyperuricemia: What Can We Learn from Metabolic Myopathies? *Muscle Nerve* **1995**, *18*, S75–S81. [CrossRef]
49. Angelini, C.; Nascimbeni, A.C.; Cenacchi, G.; Tasca, E. Lipolysis and Lipophagy in Lipid Storage Myopathies. *Biochim. Biophys. Acta Mol. Basis Dis.* **2016**, *1862*, 1367–1373. [CrossRef]
50. Rose, E.C.; di San Filippo, C.A.; Ndukwe Erlingsson, U.C.; Ardon, O.; Pasquali, M.; Longo, N. Genotype-phenotype Correlation in Primary Carnitine Deficiency. *Hum. Mutat.* **2012**, *33*, 118–123. [CrossRef]
51. Wang, Y.; Ye, J.; Ganapathy, V.; Longo, N. Mutations in the Organic Cation/carnitine Transporter OCTN2 in Primary Carnitine Deficiency. *Proc. Natl. Acad. Sci. USA* **1999**, *96*, 2356–2360. [CrossRef]
52. Wang, Y.; Kelly, M.A.; Cowan, T.M.; Longo, N. A Missense Mutation in the OCTN2 Gene Associated with Residual Carnitine Transport Activity. *Hum. Mutat.* **2000**, *15*, 238–245. [CrossRef]
53. Longo, N.; Frigeni, M.; Pasquali, M. Carnitine Transport and Fatty Acid Oxidation. *Biochim. Biophys. Acta Mol. Cell Res.* **2016**, *1863*, 2422–2435. [CrossRef] [PubMed]
54. Angelini, C.; Tavian, D.; Missaglia, S. Heterogeneous Phenotypes in Lipid Storage Myopathy due to ETFDH Gene Mutations. *JIMD Rep.* **2018**, *38*, 33–40. [PubMed]
55. Lindholm, H.; Löfberg, M.; Somer, H.; Näveri, H.; Sovijärvi, A. Abnormal Blood Lactate Accumulation after Exercise in Patients with Multiple Mitochondrial DNA Deletions and Minor Muscular Symptoms. *Clin. Physiol. Funct. Imaging* **2004**, *24*, 109–115. [CrossRef]
56. Tarnopolsky, M.; Stevens, L.; MacDonald, J.R.; Rodriguez, C.; Mahoney, D.; Rush, J.; Maguire, J. Diagnostic Utility of a Modified Forearm Ischemic Exercise Test and Technical Issues Relevant to Exercise Testing. *Muscle Nerve* **2003**, *27*, 359–366. [CrossRef] [PubMed]

57. Yeung, R.O.; Al Jundi, M.; Gubbi, S.; Bompu, M.E.; Sirrs, S.; Tarnopolsky, M.; Hannah-Shmouni, F. Management of Mitochondrial Diabetes in the Era of Novel Therapies. *J. Diabetes Complicat.* **2021**, *35*. [CrossRef] [PubMed]
58. Chae, H.W.; Na, J.H.; Kim, H.S.; Lee, Y.M. Mitochondrial Diabetes and Mitochondrial DNA Mutation Load in MELAS Syndrome. *Eur. J. Endocrinol.* **2020**, *163*, 505–512. [CrossRef] [PubMed]
59. Nishino, I.; Spinazzola, A.; Hirano, M. Thymidine Phosphorylase Gene Mutations in MNGIE, a Human Mitochondrial Disorder. *Science* **1999**, *283*, 689–692. [CrossRef] [PubMed]
60. Spagnoli, C.; Pisani, F.; Di Mario, F.; Leandro, G.; Gaiani, F.; De' Angelis, G.; Fusco, C. Peripheral Neuropathy and Gastroenterologic Disorders: An Overview on an Underrecognized Association. *Acta Biomedica.* **2018**, *89*, 22–32. [CrossRef]
61. Mancuso, M.; Orsucci, D.; Logerfo, A.; Rocchi, A.; Petrozzi, L.; Nesti, C.; Galetta, F.; Santoro, G.; Murri, L.; Siciliano, G. Oxidative Stress Biomarkers in Mitochondrial Myopathies, Basally and after Cysteine Donor Supplementation. *J. Neurol.* **2010**, *257*, 774–781. [CrossRef]
62. Horga, A.; Raja Rayan, D.L.; Matthews, E.; Sud, R.; Fialho, D.; Durran, S.C.; Burge, J.A.; Portaro, S.; Davis, M.B.; Haworth, A.; et al. Prevalence Study of Genetically Defined Skeletal Muscle Channelopathies in England. *Neurology* **2013**, *80*, 1472–1475. [CrossRef]
63. Statland, J.M.; Fontaine, B.; Hanna, M.G.; Johnson, N.E.; Kissel, J.T.; Sansone, V.A.; Shieh, P.B.; Tawil, R.N.; Trivedi, J.; Cannon, S.C.; et al. Review of the Diagnosis and Treatment of Periodic Paralysis. *Muscle Nerve* **2018**, *57*, 522–530. [CrossRef]
64. Cannon, S.C. Channelopathies of Skeletal Muscle Excitability. *Compr. Physiol.* **2015**, *5*, 761–790. [CrossRef]
65. Alhasan, K.A.; Abdallah, M.S.; Kari, J.A.; Bashiri, F.A. Hypokalemic Periodic Paralysis due to CACNA1S Gene Mutation. *Neurosciences* **2019**, *24*, 225–230. [CrossRef]
66. Bayless-Edwards, L.; Winston, V.; Lehmann-Horn, F.; Arinze, P.; Groome, J.R.; Jurkat-Rott, K. NaV1.4 DI-S4 Periodic Paralysis Mutation R222W Enhances Inactivation and Promotes Leak Current to Attenuate Action Potentials and Depolarize Muscle Fibers. *Sci. Rep.* **2018**, *8*. [CrossRef]
67. Gun Bilgic, D.; Aydin Gumus, A.; Gerik Celebi, H.B.; Bilgic, A.; Unaltuna Erginel, N.; Cam, F.S. A New Clinical Entity in T704M Mutation in Periodic Paralysis. *J. Clin. Neurosci.* **2020**, *78*, 203–206. [CrossRef]
68. Yang, B.; Yang, Y.; Tu, W.; Shen, Y.; Dong, Q. A Rare Case of Unilateral Adrenal Hyperplasia Accompanied by Hypokalaemic Periodic Paralysis Caused by a Novel Dominant Mutation in CACNA1S: Features and Prognosis after Adrenalectomy. *BMC Urol.* **2014**, *14*. [CrossRef]
69. Statland, J.M.; Barohn, R.J. Muscle Channelopathies: The Nondystrophic Myotonias and Periodic Paralyses. *Contin. Lifelong Learn. Neurol.* **2013**, *19*, 1598–1614. [CrossRef]
70. Kokunai, Y.; Dalle, C.; Vicart, S.; Sternberg, D.; Pouliot, V.; Bendahhou, S.; Fournier, E.; Chahine, M.; Fontaine, B.; Nicole, S. A204E Mutation in Nav1.4 DIS3 Exerts Gain- and Loss-of-function Effects That Lead to Periodic Paralysis Combining Hyper- with Hypo-kalaemic Signs. *Sci. Rep.* **2018**, *8*. [CrossRef]
71. Luo, S.; Sampedro Castañeda, M.; Matthews, E.; Sud, R.; Hanna, M.G.; Sun, J.; Song, J.; Lu, J.; Qiao, K.; Zhao, C.; et al. Hypokalaemic Periodic Paralysis and Myotonia in a Patient with Homozygous Mutation p.R1451L in NaV1.4. *Sci. Rep.* **2018**, *8*. [CrossRef] [PubMed]
72. Weber, F.; Lehmann-Horn, F. *Hypokalemic Periodic Paralysis*; Adam, M.P., Ardinger, H.H.P., Pagon, R.A., Wallace, S.E., Bean, L.J.H., Mirzaa, G., Amemiya, A., Eds.; University of Washington, Seattle: Seattle, WA, USA, 1993.
73. Weber, F.; Jurkat-Rott, K.; Lehmann-Horn, F. *Hyperkalemic Periodic Paralysis*; Adam, M.P., Ardinger, H.H.P., Pagon, R.A., Wallace, S.E., Bean, L.J.H., Mirzaa, G., Amemiya, A., Eds.; University of Washington, Seattle: Seattle, WA, USA, 1993.
74. Fan, C.; Mao, N.; Lehmann-Horn, F.; Bürmann, J.; Jurkat-Rott, K. Effects of S906T Polymorphism on the Severity of a Novel Borderline Mutation I692M in Nav1.4 Cause Periodic Paralysis. *Clin. Genet.* **2017**, *91*, 859–867. [CrossRef]
75. Hirano, M.; Kokunai, Y.; Nagai, A.; Nakamura, Y.; Saigoh, K.; Kusunoki, S.; Takahashi, M.P. A Novel Mutation in the Calcium Channel Gene in a Family with Hypokalemic Periodic Paralysis. *J. Neurol. Sci.* **2011**, *309*, 9–11. [CrossRef] [PubMed]
76. Spillane, J.; Fialho, D.; Hanna, M.G. Diagnosis of Skeletal Muscle Channelopathies. *Expert Opin. Med. Diagn.* **2013**, *7*, 517–529. [CrossRef] [PubMed]
77. Fialho, D.; Griggs, R.C.; Matthews, E. Periodic Paralysis. *Handb. Clin. Neurol.* **2018**, *148*, 505–520. [PubMed]
78. Mcfadzean, A.J.S.; Yeung, R. Periodic Paralysis Complicating Thyrotoxicosis in Chinese. *Br. Med. J.* **1967**, *1*, 451–455. [CrossRef] [PubMed]
79. Miao, J.; Wei, X.J.; Liu, X.M.; Kang, Z.X.; Gao, Y.L.; Yu, X.F. A Case Report: Autosomal Recessive Myotonia Congenita Caused by a Novel Splice Mutation (c.1401 + 1G > A) in CLCN1 Gene of a Chinese Han Patient. *BMC Neurol.* **2018**, *18*. [CrossRef] [PubMed]
80. Passeri, E.; Sansone, V.A.; Verdelli, C.; Mendola, M.; Corbetta, S. Asymptomatic Myotonia Congenita Unmasked by Severe Hypothyroidism. *Neuromuscul. Disord.* **2014**, *24*, 365–367. [CrossRef] [PubMed]
81. Nagamitsu, S.; Matsuura, T.; Khajavi, M.; Armstrong, R.; Gooch, C.; Harati, Y.; Ashizawa, T. A 'Dystrophic' Variant of Autosomal Recessive Myotonia Congenita Caused by Novel Mutations in the CLCN1 Gene. *Neurology* **2000**, *55*, 1697–1703. [CrossRef] [PubMed]
82. Ulzi, G.; Lecchi, M.; Sansone, V.; Redaelli, E.; Corti, E.; Saccomanno, D.; Pagliarani, S.; Corti, S.; Magri, F.; Raimondi, M.; et al. Myotonia Congenita: Novel Mutations in CLCN1 Gene and Functional Characterizations in Italian Patients. *J. Neurol. Sci.* **2012**, *318*, 65–71. [CrossRef] [PubMed]

83. Portaro, S.; Altamura, C.; Licata, N.; Camerino, G.M.; Imbrici, P.; Musumeci, O.; Rodolico, C.; Conte Camerino, D.; Toscano, A.; Desaphy, J.F. Clinical, Molecular, and Functional Characterization of CLCN1 Mutations in Three Families with Recessive Myotonia Congenita. *NeuroMolecular Med.* **2015**, *17*, 285–296. [CrossRef]

84. Lyons, M.J.; Duron, R.; Molinero, I.; Sangiuolo, F.; Holden, K.R. Novel CLCN1 Mutation in Carbamazepine-Responsive Myotonia Congenita. *Pediatr. Neurol.* **2010**, *42*, 365–368. [CrossRef]

85. Gurgel-Giannetti, J.; Senkevics, A.S.; Zilbersztajn-Gotlieb, D.; Yamamoto, L.U.; Muniz, V.P.; Pavanello, R.C.; Oliveira, A.B.; Zatz, M.; Vainzof, M. Thomsen or Becker Myotonia? A Novel Autosomal Recessive Nonsense Mutation in the CLCN1 Gene Associated with a Mild Phenotype. *Muscle Nerve* **2012**, *45*, 279–283. [CrossRef]

86. Sahin, I.; Erdem, H.B.; Tan, H.; Tatar, A. Becker's Myotonia: Novel Mutations and Clinical Variability in Patients Born to Consanguineous Parents. *Acta Neurol. Belg.* **2018**, *118*, 567–572. [CrossRef]

87. Portaro, S.; Cacciola, A.; Naro, A.; Milardi, D.; Morabito, R.; Corallo, F.; Marino, S.; Bramanti, A.; Mazzon, E.; Calabrò, R.S. A Case Report of Recessive Myotonia Congenita and Early Onset Cognitive Impairment. *Medicine* **2018**, *97*. [CrossRef] [PubMed]

88. Kim, D.-S.; Kim, E.J.; Jung, D.S.; Park, K.H.; Kim, I.J.; Kwak, K.Y.; Kim, C.M.; Ko, H.Y. A Korean Family with Arg1448Cys Mutation of SCN4A Channel Causing Paramyotonia Congenita: Electrophysiologic, Histopathologic, and Molecular Genetic Studies. *J. Korean Med. Sci.* **2002**, *17*, 856–860. [CrossRef]

89. Kruijt, N.; van den Bersselaar, L.R.; Wijma, J.; Verbeeck, W.; Coenen, M.J.H.; Neville, J.; Snoeck, M.; Kamsteeg, E.J.; Jungbluth, H.; Kramers, C.; et al. HyperCKemia and Rhabdomyolysis in the Neuroleptic Malignant and Serotonin Syndromes: A Literature Review. *Neuromuscul. Disord.* **2020**, *30*, 949–958. [CrossRef]

90. Voermans, N.C.; Snoeck, M.; Jungbluth, H. RYR1-related Rhabdomyolysis: A Common but Probably Underdiagnosed Manifestation of Skeletal Muscle Ryanodine Receptor Dysfunction. *Rev. Neurol.* **2016**, *172*, 546–558. [CrossRef]

91. Chan, B.; Chen, S.P.; Wong, W.C.; Mak, C.M.; Wong, S.; Chan, K.Y.; Chan, A.Y. RYR1-related central core myopathy in a Chinese adolescent boy. *Hong Kong Med. J.* **2011**, *17*, 67–70.

92. Lawal, T.A.; Todd, J.J.; Meilleur, K.G. Ryanodine Receptor 1-Related Myopathies: Diagnostic and Therapeutic Approaches. *Neurother. J. Am. Soc. Exp. Neurother.* **2018**, *15*, 885–899. [CrossRef] [PubMed]

93. Taylor, A.; Lachlan, K.; Manners, R.M.; Lotery, A.J. A Study of a Family with the Skeletal Muscle RYR1 Mutation (c.7354C>T) Associated with Central Core Myopathy and Malignant Hyperthermia Susceptibility. *J. Clin. Neurosci. Off. J. Neurosurg. Soc. Australas.* **2012**, *19*, 65–70. [CrossRef]

94. Løseth, S.; Voermans, N.C.; Torbergsen, T.; Lillis, S.; Jonsrud, C.; Lindal, S.; Kamsteeg, E.J.; Lammens, M.; Broman, M.; Dekomien, G.; et al. A Novel Late-onset Axial Myopathy Associated with Mutations in the Skeletal Muscle Ryanodine Receptor (RYR1) Gene. *J. Neurol.* **2013**, *260*, 1504–1510. [CrossRef] [PubMed]

95. Zhou, H.; Lillis, S.; Loy, R.E.; Ghassemi, F.; Rose, M.R.; Norwood, F.; Mills, K.; Al-Sarraj, S.; Lane, R.J.; Feng, L.; et al. Multi-minicore Disease and Atypical Periodic Paralysis Associated with Novel Mutations in the Skeletal Muscle Ryanodine Receptor (RYR1) Gene. *Neuromuscul. Disord.* **2010**, *20*, 166–173. [CrossRef] [PubMed]

96. Laforgia, N.; Capozza, M.; De Cosmo, L.; Di Mauro, A.; Baldassarre, M.E.; Mercadante, F.; Torella, A.L.; Nigro, V.; Resta, N. A Rare Case of Severe Congenital RYR1-Associated Myopathy. *Case Rep. Genet.* **2018**, *2018*, 6184185. [CrossRef] [PubMed]

97. Ciafaloni, E. Myasthenia Gravis and Congenital Myasthenic Syndromes. *Contin. Lifelong Learn. Neurol.* **2019**, *25*. [CrossRef]

98. Della Marina, A.; Wibbeler, E.; Abicht, A.; Kölbel, H.; Lochmüller, H.; Roos, A.; Schara, U. Long Term Follow-Up on Pediatric Cases With Congenital Myasthenic Syndromes—A Retrospective Single Centre Cohort Study. *Front. Hum. Neurosci.* **2020**, *14*. [CrossRef]

99. Farmakidis, C.; Pasnoor, M.; Barohn, R.J.; Dimachkie, M.M. Congenital Myasthenic Syndromes: A Clinical and Treatment Approach. *Curr. Treat. Options Neurol.* **2018**, *20*. [CrossRef]

100. Sabre, L.; Maddison, P.; Wong, S.H.; Sadalage, G.; Ambrose, P.A.; Plant, G.T.; Punga, A.R. miR-30e-5p as Predictor of Generalization in Ocular Myasthenia Gravis. *Ann. Clin. Transl. Neurol.* **2019**, *6*. [CrossRef] [PubMed]

101. Matsuo, M.; Awano, H.; Maruyama, N.; Nishio, H. Titin Fragment in Urine: A Noninvasive Biomarker of Muscle Degradation. *Adv. Clin. Chem.* **2019**, *90*, 1–23. [PubMed]

102. Weydt, P.; Oeckl, P.; Huss, A.; Müller, K.; Volk, A.E.; Kuhle, J.; Knehr, A.; Andersen, P.M.; Prudlo, J.; Steinacker, P.; et al. Neurofilament Levels as Biomarkers in Asymptomatic and Symptomatic Familial Amyotrophic Lateral Sclerosis. *Ann. Neurol.* **2016**, *79*, 152–158. [CrossRef]

103. Teunissen, C.E.; Khalil, M. Neurofilaments as Biomarkers in Multiple Sclerosis. *Mult. Scler. J.* **2012**, *18*, 552–556. [CrossRef] [PubMed]

104. Gresle, M.; Liu, Y.; Dagley, L.F.; Haartsen, J.; Pearson, F.; Purcell, A.W.; Laverick, L.; Petzold, A.; Lucas, R.M.; Van der Walt, A.; et al. Serum Phosphorylated Neurofilament-heavy Chain Levels in Multiple Sclerosis Patients. *J. Neurol. Neurosurg. Psychiatry* **2014**, *85*, 1209–1213. [CrossRef]

105. Kuhle, J.; Leppert, D.; Petzold, A.; Regeniter, A.; Schindler, C.; Mehling, M.; Anthony, D.C.; Kappos, L.; Lindberg, R.L. Neurofilament Heavy Chain in CSF Correlates with Relapses and Disability in Multiple Sclerosis. *Neurology* **2011**, *76*, 1206–1213. [CrossRef] [PubMed]

106. Constantinescu, R.; Zetterberg, H.; Holmberg, B.; Rosengren, L. Levels of Brain Related Proteins in Cerebrospinal fluid: An Aid in the Differential Diagnosis of Parkinsonian Disorders. *Park. Relat. Disord.* **2009**, *15*, 205–212. [CrossRef]
107. Xu, Z.; Henderson, R.D.; David, M.; McCombe, P.A. Neurofilaments as Biomarkers for Amyotrophic Lateral Sclerosis: A Systematic Review and Meta-analysis. *PLoS ONE* **2016**, *11*. [CrossRef]
108. Darras, B.T.; Crawford, T.O.; Finkel, R.S.; Mercuri, E.; De Vivo, D.C.; Oskoui, M.; Tizzano, E.F.; Ryan, M.M.; Muntoni, F.; Zhao, G.; et al. Neurofilament as a Potential Biomarker for Spinal Muscular Atrophy. *Ann. Clin. Transl. Neurol.* **2019**. [CrossRef] [PubMed]
109. Alves, C.R.R.; Zhang, R.; Johnstone, A.J.; Garner, R.; New, P.H.; Siranosian, J.J.; Swoboda, K.J. Serum Creatinine is a Biomarker of Progressive Denervation in Spinal Muscular Atrophy. *Neurology* **2020**. [CrossRef] [PubMed]
110. Lima, A.F.; Evangelista, T.; de Carvalho, M. Increased creatine kinase and spontaneous activity on electromyography, in amyotrophic lateral sclerosis. *Electromyogr. Clin. Neurophysiol.* **2003**, *43*, 189–192.
111. Tortelli, R.; Zecca, C.; Piccininni, M.; Benmahamed, S.; Dell'Abate, M.T.; Barulli, M.R.; Capozzo, R.; Battista, P.; Logroscino, G. Plasma Inflammatory Cytokines Are Elevated in ALS. *Front. Neurol.* **2020**, *11*. [CrossRef]
112. Lunetta, C.; Lizio, A.; Maestri, E.; Sansone, V.A.; Mora, G.; Miller, R.G.; Appel, S.; Chiò, A. Serum C-reactive Protein as a Prognostic Biomarker in Amyotrophic Lateral Sclerosis. *JAMA Neurol.* **2017**, *74*, 660–667. [CrossRef]
113. Van Eijk, R.P.A.; Eijkemans, M.J.C.; Ferguson, T.A.; Nikolakopoulos, S.; Veldink, J.H.; Van Den Berg, L.H. Monitoring Disease Progression with Plasma Creatinine in Amyotrophic Lateral Sclerosis Clinical Trials. *J. Neurol. Neurosurg. Psychiatry* **2018**, *89*, 156–161. [CrossRef] [PubMed]
114. Brettschneider, J.; Petzold, A.; Süßmuth, S.D.; Ludolph, A.C.; Tumani, H. Axonal Damage Markers in Cerebrospinal Fluid are Increased in ALS. *Neurology* **2006**. [CrossRef] [PubMed]
115. Boylan, K.; Yang, C.; Crook, J.; Overstreet, K.; Heckman, M.; Wang, Y.; Borchelt, D.; Shaw, G. Immunoreactivity of the Phosphorylated Axonal Neurofilament H Subunit (pNF-H) in Blood of ALS Model Rodents and ALS Patients: Evaluation of Blood pNF-H as a Potential ALS Biomarker. *J. Neurochem.* **2009**. [CrossRef]
116. Li, D.; Usuki, S.; Quarles, B.; Rivner, M.H.; Ariga, T.; Yu, R.K. Anti-Sulfoglucuronosyl Paragloboside Antibody: A Potential Serologic Marker of Amyotrophic Lateral Sclerosis. *ASN Neuro* **2016**. [CrossRef]
117. Goldknopf, I.L.; Sheta, E.A.; Folsom, B.; Wilson, C.; Duty, J.; Yen, A.A.; Appel, S.H. Complement C3c and Related Protein Biomarkers in Amyotrophic Lateral Sclerosis and Parkinson's Disease. *Biochem. Biophys. Res. Commun.* **2006**. [CrossRef] [PubMed]
118. Brettschneider, J.; Lehmensiek, V.; Mogel, H.; Pfeifle, M.; Dorst, J.; Hendrich, C.; Ludolph, A.C.; Tumani, H. Proteome Analysis Reveals Candidate Markers of Disease Progression in Amyotrophic Lateral Sclerosis (ALS). *Neurosci. Lett.* **2010**. [CrossRef] [PubMed]
119. Brettschneider, J.; Mogel, H.; Lehmensiek, V.; Ahlert, T.; Süssmuth, S.; Ludolph, A.C.; Tumani, H. Proteome Analysis of Cerebrospinal Fluid in Amyotrophic Lateral Sclerosis (ALS). *Neurochem. Res.* **2008**. [CrossRef]
120. Barreto, L.C.; Oliveira, F.S.; Nunes, P.S.; de França Costa, I.M.; Garcez, C.A.; Goes, G.M.; Neves, E.L.; de Souza Siqueira Quintans, J.; de Souza Araújo, A.A. Epidemiologic Study of Charcot-Marie-Tooth Disease: A Systematic Review. *Neuroepidemiology* **2016**, *46*, 157–165. [CrossRef]
121. Sandelius, Å.; Zetterberg, H.; Blennow, K.; Adiutori, R.; Malaspina, A.; Laura, M.; Reilly, M.M.; Rossor, A.M. Plasma Neurofilament Light Chain Concentration in the Inherited Peripheral Neuropathies. *Neurology* **2018**, *90*, e518–e524. [CrossRef] [PubMed]
122. Picken, M.M. The Pathology of Amyloidosis in Classification: A Review. *Acta Haematol.* **2020**, *143*, 322–334. [CrossRef]
123. Adams, D.; Ando, Y.; Beirão, J.M.; Coelho, T.; Gertz, M.A.; Gillmore, J.D.; Hawkins, P.N.; Lousada, I.; Suhr, O.B.; Merlini, G. Expert Consensus Recommendations to Improve Diagnosis of ATTR Amyloidosis with Polyneuropathy. *J. Neurol.* **2020**, 1–14. [CrossRef] [PubMed]
124. Stangou, A.J.; Banner, N.R.; Hendry, B.M.; Rela, M.; Portmann, B.; Wendon, J.; Monaghan, M.; Maccarthy, P.; Buxton-Thomas, M.; Mathias, C.J.; et al. Hereditary Fibrinogen a α-chain Amyloidosis: Phenotypic Characterization of a Systemic Disease and the Role of Liver Transplantation. *Blood* **2010**, *115*, 2998–3007. [CrossRef]
125. Benson, M.D. *The Hereditary Amyloidoses*; Humana Press: Totowa, NJ, USA, 2015; pp. 65–80.
126. Arvanitis, M.; Koch, C.M.; Chan, G.G.; Torres-Arancivia, C.; LaValley, M.P.; Jacobson, D.R.; Berk, J.L.; Connors, L.H.; Ruberg, F.L. Identification of Transthyretin Cardiac Amyloidosis Using Serum Retinol-binding Protein 4 and a Clinical Prediction Model. *JAMA Cardiol.* **2017**, *2*, 305–313. [CrossRef]
127. Sapio, M.R.; Goswami, S.C.; Gross, J.R.; Mannes, A.J.; Iadarola, M.J. Transcriptomic Analyses of Genes and Tissues in Inherited Sensory Neuropathies. *Exp. Neurol.* **2016**, *283*, 375–395. [CrossRef]
128. Azevedo, E.P.; Guimaraes-Costa, A.B.; Bandeira-Melo, C.; Chimelli, L.; Waddington-Cruz, M.; Saraiva, E.M.; Palhano, F.L.; Foguel, D. Inflammatory Profiling of Patients with Familial Amyloid Polyneuropathy. *BMC Neurol.* **2019**, *19*, 146. [CrossRef]
129. Louwsma, J.; Brunger, A.F.; Bijzet, J.; Kroesen, B.J.; Roeloffzen, W.W.H.; Bischof, A.; Kuhle, J.; Drost, G.; Lange, F.; Kuks, J.B.M.; et al. Neurofilament Light Chain, a Biomarker for Polyneuropathy in Systemic Amyloidosis. *Amyloid* **2020**, 1–6. [CrossRef]
130. Kapoor, M.; Foiani, M.; Heslegrave, A.; Zetterberg, H.; Lunn, M.P.; Malaspina, A.; Gillmore, J.D.; Rossor, A.M.; Reilly, M.M. Plasma Neurofilament Light Chain Concentration is Increased and Correlates with the Severity of Neuropathy in Hereditary Transthyretin Amyloidosis. *J. Peripher. Nerv. Syst.* **2019**, *24*, 314–319. [CrossRef] [PubMed]

131. Ticau, S.; Sridharan, G.V.; Tsour, S.; Cantley, W.L.; Chan, A.; Gilbert, J.A.; Erbe, D.; Aldinc, E.; Reilly, M.M.; Adams, D.; et al. Neurofilament Light Chain as a Biomarker of Hereditary Transthyretin-Mediated Amyloidosis. *Neurology* **2021**, *96*, e412–e422. [CrossRef] [PubMed]

132. Nanetti, L.; Cavalieri, S.; Pensato, V.; Erbetta, A.; Pareyson, D.; Panzeri, M.; Zorzi, G.; Antozzi, C.; Moroni, I.; Gellera, C.; et al. SETX Mutations are a Frequent Genetic Cause of Juvenile and Adult Onset Cerebellar Ataxia with Neuropathy and Elevated Serum Alpha-fetoprotein. *Orphanet, J. Rare Dis.* **2013**, *8*, 123. [CrossRef] [PubMed]

133. Le Ber, I.; Moreira, M.C.; Rivaud-Péchoux, S.; Chamayou, C.; Ochsner, F.; Kuntzer, T.; Tardieu, M.; Saïd, G.; Habert, M.O.; Demarquay, G.; et al. Cerebellar Ataxia with Oculomotor Apraxia Type 1: Clinical and Genetic Studies. *Brain* **2003**, *126*, 2761–2772. [CrossRef] [PubMed]

134. Castellotti, B.; Mariotti, C.; Rimoldi, M.; Fancellu, R.; Plumari, M.; Caimi, S.; Uziel, G.; Nardocci, N.; Zorzi, G.; Pareyson, D.; et al. Ataxia with Oculomotor Apraxia Type1 (AOA1): Novel and Recurrent Aprataxin Mutations, Coenzyme Q10 Analyses, and Clinical Findings in Italian Patients. *Neurogenetics* **2011**, *12*, 193–201. [CrossRef]

135. Coutinho, P.; Barbot, C. Ataxia with Oculomotor Apraxia-type 1. *Curr. Clin. Neurol.* **2012**, *36*, 224–225. [CrossRef]

136. Vermeer, S.; van de Warrenburg, B.P.; Willemsen, M.A.; Cluitmans, M.; Scheffer, H.; Kremer, B.P.; Knoers, N.V. Autosomal Recessive Cerebellar Ataxias: The Current State of Affairs. *J. Med. Genet.* **2011**, *48*, 651–659. [CrossRef] [PubMed]

137. Anheim, M.; Monga, B.; Fleury, M.; Charles, P.; Barbot, C.; Salih, M.; Delaunoy, J.P.; Fritsch, M.; Arning, L.; Synofzik, M.; et al. Ataxia with Oculomotor Apraxia Type 2: Clinical, Biological and Genotype/phenotype Correlation Study of a Cohort of 90 Patients. *Brain* **2009**, *132*, 2688–2698. [CrossRef]

138. Paucar, M.; Taylor, A.M.R.; Hadjivassiliou, M.; Fogel, B.L.; Svenningsson, P. Progressive Ataxia with Elevated Alpha-fetoprotein: Diagnostic Issues and Review of the Literature. *Tremor. Other Hyperkinetic Mov.* **2019**, *9*, 1–4. [CrossRef]

139. Taylor, A.M.R.; Groom, A.; Byrd, P.J. Ataxia-telangiectasia-like Disorder (ATLD)–Its Clinical Presentation and Molecular Basis. *DNA Repair* **2004**, *3*, 1219–1225. [CrossRef]

140. Linnemann, C.; Tezenas du Montcel, S.; Rakowicz, M.; Schmitz-Hübsch, T.; Szymanski, S.; Berciano, J.; van de Warrenburg, B.P.; Pedersen, K.; Depondt, C.; Rola, R.; et al. Peripheral Neuropathy in Spinocerebellar Ataxia Type 1, 2, 3, and 6. *Cerebellum* **2016**, *15*, 165–173. [CrossRef] [PubMed]

141. Sullivan, R.; Yau, W.Y.; O'Connor, E.; Houlden, H. Spinocerebellar ataxia: An update. *J. Neurol.* **2019**, *266*, 533–544. [CrossRef] [PubMed]

142. Wilke, C.; Bender, F.; Hayer, S.N.; Brockmann, K.; Schöls, L.; Kuhle, J.; Synofzik, M. Serum Neurofilament Light is Increased in Multiple System Atrophy of Cerebellar Type and in Repeat-expansion Spinocerebellar Ataxias: A pilot study. *J. Neurol.* **2018**, *265*, 1618–1624. [CrossRef]

143. Harding, A.E. Classification of the Hereditary Ataxias and Paraplegias. *Lancet* **1983**, *321*, 1151–1155. [CrossRef]

144. Tsaousidou, M.K.; Ouahchi, K.; Warner, T.T.; Yang, Y.; Simpson, M.A.; Laing, N.G.; Wilkinson, P.A.; Madrid, R.E.; Patel, H.; Hentati, F.; et al. Sequence Alterations within CYP7B1 Implicate Defective Cholesterol Homeostasis in Motor-Neuron Degeneration. *Am. J. Hum. Genet.* **2008**, *82*, 510–515. [CrossRef]

145. Schöls, L.; Rattay, T.W.; Martus, P.; Meisner, C.; Baets, J.; Fischer, I.; Jägle, C.; Fraidakis, M.J.; Martinuzzi, A.; Saute, J.A.; et al. Hereditary Spastic Paraplegia Type 5: Natural History, Biomarkers and a Randomized Controlled Trial. *Brain* **2017**, *140*, 3112–3127. [CrossRef]

146. Marelli, C.; Lamari, F.; Rainteau, D.; Lafourcade, A.; Banneau, G.; Humbert, L.; Monin, M.L.; Petit, E.; Debs, R.; Castelnovo, G.; et al. Plasma Oxysterols: Biomarkers for Diagnosis and Treatment in Spastic Paraplegia Type 5. *Brain* **2018**, *141*, 72–84. [CrossRef]

147. Kazamel, M.; Desnick, R.J.; Quigley, J.G. Porphyric Neuropathy: Pathophysiology, Diagnosis, and Updated Management. *Curr. Neurol. Neurosci. Rep.* **2020**, *20*. [CrossRef]

148. Arora, S.; Young, S.; Kodali, S.; Singal, A.K. Hepatic Porphyria: A Narrative Review. *Indian J. Gastroenterol.* **2016**, *35*, 405–418. [CrossRef]

149. Albers, J.W.; Fink, J.K. Porphyric Neuropathy. *Muscle Nerve* **2004**, *30*, 410–422. [CrossRef]

150. Marsden, J.T.; Rees, D.C. Urinary Excretion of Porphyrins, Porphobilinogen and–δaminolaevulinic Acid Following an Attack of Acute Intermittent Porphyria. *J. Clin. Pathol.* **2014**, *67*, 60–65. [CrossRef]

151. Kılıç, M.; Kasapkara, Ç.S.; Kılavuz, S.; Mungan, N.Ö.; Biberoğlu, G. A Possible Biomarker of Neurocytolysis in Infantile Gangliosidoses: Aspartate Transaminase. *Metab. Brain Dis.* **2019**, *34*, 495–503. [CrossRef]

152. Mahmood, A.; Raymond, G.V.; Dubey, P.; Peters, C.; Moser, H.W. Survival Analysis of Haematopoietic Cell Transplantation for Childhood Cerebral X-linked Adrenoleukodystrophy: A Comparison Study. *Lancet Neurol.* **2007**, *6*, 687–692. [CrossRef]

153. Huffnagel, I.C.; van de Beek, M.C.; Showers, A.L.; Orsini, J.J.; Klouwer, F.C.C.; Dijkstra, I.M.E.; Schielen, P.C.; van Lenthe, H.; Wanders, R.J.A.; Vaz, F.M.; et al. Comparison of C26:0-carnitine and C26:0-lysophosphatidylcholine as Diagnostic Markers in Dried Blood Spots from Newborns and Patients with Adrenoleukodystrophy. *Mol. Genet. Metab.* **2017**, *122*, 209–215. [CrossRef] [PubMed]

154. Ferraz, M.J.; Kallemeijn, W.W.; Mirzaian, M.; Herrera Moro, D.; Marques, A.; Wisse, P.; Boot, R.G.; Willems, L.I.; Overkleeft, H.S.; Aerts, J.M. Gaucher Disease and Fabry Disease: New Markers and Insights in Pathophysiology for Two Distinct Glycosphingolipidoses. *Biochim. Biophys. Acta Mol. Cell Biol. Lipids* **2014**, *1841*, 811–825. [CrossRef]

155. Soldevilla, B.; Cuevas-Martín, C.; Ibáñez, C.; Santacatterina, F.; Alberti, M.A.; Simó, C.; Casasnovas, C.; Márquez-Infante, C.; Sevilla, T.; Pascual, S.I.; et al. Plasma Metabolome and Skin Proteins in Charcot-Marie-Tooth 1A Patients. *PLoS ONE* **2017**, *12*. [CrossRef]

156. Gabriel, C.M.; Gregson, N.A.; Wood, N.W.; Hughes, R.A.C. Immunological Study of Hereditary Motor and Sensory Neuropathy Type 1 a (HMSN 1 a). *J. Neurol. Neurosurg. Psychiatry* **2002**, *72*, 230–235. [CrossRef] [PubMed]

157. Ferreira, C.R.; van Karnebeek, C.D.M. Inborn Errors of Metabolism. *Handb. Clin. Neurol.* **2019**, *162*, 449–481. [PubMed]

158. Wierzbicki, A.S.; Lloyd, M.D.; Schofield, C.J.; Feher, M.D.; Gibberd, F.B. Refsum's Disease: A Peroxisomal Disorder Affecting Phytanic Acid α-oxidation. *J. Neurochem.* **2002**, *80*, 727–735. [CrossRef]

159. Wills, A.J.; Manning, N.J.; Reilly, M.M. Refsum's Disease. *QJM Mon. J. Assoc. Physicians* **2001**, *94*, 403–406. [CrossRef]

160. Burnett, J.R.; Hooper, A.J.; McCormick, S.P.; Hegele, R.A. *Tangier Disease*; University of Washington, Seattle: Seattle, WA, USA, 1993.

161. Sedel, F.; Barnerias, C.; Dubourg, O.; Desguerres, I.; Lyon-Caen, O.; Saudubray, J.M. Peripheral Neuropathy and Inborn Errors of Metabolism in Adults. *J. Inherit. Metab. Dis.* **2007**, *30*, 642–653. [CrossRef] [PubMed]

162. Beerepoot, S.; Nierkens, S.; Boelens, J.J.; Lindemans, C.; Bugiani, M.; Wolf, N.I. Peripheral Neuropathy in Metachromatic Leukodystrophy: Current Status and Future Perspective. *Orphanet, J. Rare Dis.* **2019**, *14*. [CrossRef]

163. Thibert, K.A.; Raymond, G.V.; Tolar, J.; Miller, W.P.; Orchard, P.J.; Lund, T.C. Cerebral Spinal Fluid Levels of Cytokines are Elevated in Patients with Metachromatic Leukodystrophy. *Sci. Rep.* **2016**, *6*. [CrossRef]

164. Li, Y.; Xu, Y.; Benitez, B.A.; Nagree, M.S.; Dearborn, J.T.; Jiang, X.; Guzman, M.A.; Woloszynek, J.C.; Giaramita, A.; Yip, B.K.; et al. Genetic Ablation of Acid Ceramidase in Krabbe Disease Confirms the Psychosine Hypothesis and Identifies a New Therapeutic Target. *Proc. Natl. Acad. Sci. USA* **2019**, *116*, 20097–20103. [CrossRef] [PubMed]

165. Guenzel, A.J.; Turgeon, C.T.; Nickander, K.K.; White, A.L.; Peck, D.S.; Pino, G.B.; Studinski, A.L.; Prasad, V.K.; Kurtzberg, J.; Escolar, M. The Critical Role of Psychosine in Screening, Diagnosis, and Monitoring of Krabbe Disease. *Genet. Med.* **2020**, *22*, 1108–1118. [CrossRef]

166. Herbst, Z.; Turgeon, C.T.; Biski, C.; Khaledi, H.; Shoemaker, N.B.; DeArmond, P.D.; Smith, S.; Orsini, J.; Matern, D.; Gelb, M.H. Achieving Congruence among Reference Laboratories for Absolute Abundance Measurement of Analytes for Rare Diseases: Psychosine for Diagnosis and Prognosis of Krabbe Disease. *Int. J. Neonatal Screen.* **2020**, *6*, 29. [CrossRef]

167. Berendse, K.; Engelen, M.; Ferdinandusse, S.; Majoie, C.B.; Waterham, H.R.; Vaz, F.M.; Koelman, J.H.; Barth, P.; Wanders, R.J.; Poll-The, B.T. Zellweger Spectrum Disorders: Clinical Manifestations in Patients Surviving into Adulthood. *J. Inherit. Metab. Dis.* **2016**, *39*, 93–106. [CrossRef] [PubMed]

168. Klouwer, F.C.; Huffnagel, I.C.; Ferdinandusse, S.; Waterham, H.R.; Wanders, R.J.; Engelen, M.; Poll-The, B.T. Clinical and Biochemical Pitfalls in the Diagnosis of Peroxisomal Disorders. *Neuropediatrics* **2016**, *47*, 205–220. [CrossRef]

169. Schuchman, E.H.; Desnick, R.J. Types A and B Niemann-Pick Disease. *Mol. Genet. Metab.* **2017**, *120*, 27–33. [CrossRef]

170. Porter, F.D.; Scherrer, D.E.; Lanier, M.H.; Langmade, S.J.; Molugu, V.; Gale, S.E.; Olzeski, D.; Sidhu, R.; Dietzen, D.J.; Fu, R.; et al. Cholesterol Oxidation Products are Sensitive and Specific Blood-based Biomarkers for Niemann-Pick C1 Disease. *Sci. Transl. Med.* **2010**, *2*. [CrossRef]

171. Cluzeau, C.V.M.; Watkins-Chow, D.E.; Fu, R.; Borate, B.; Yanjanin, N.; Dail, M.K.; Davidson, C.D.; Walkley, S.U.; Ory, D.S.; Wassif, C.A.; et al. Microarray Expression Analysis and Identification of Serum Biomarkers for Niemann-Pick Disease, Type c1. *Hum. Mol. Genet.* **2012**, *21*, 3632–3646. [CrossRef] [PubMed]

172. Mengel, E.; Bembi, B.; Del Toro, M.; Deodato, F.; Gautschi, M.; Grunewald, S.; Grønborg, S.; Héron, B.; Maier, E.M.; Roubertie, A.; et al. Clinical Disease Progression and Biomarkers in Niemann–Pick Disease Type C: A Prospective Cohort Study. *Orphanet, J. Rare Dis.* **2020**, *15*. [CrossRef] [PubMed]

173. Wu, C.; Iwamoto, T.; Hossain, M.; Akiyama, K.; Igarashi, J.; Miyajima, T.; Eto, Y. A Combination of 7-ketocholesterol, Lysosphingomyelin and Bile Acid-408 to Diagnose Niemann-Pick Disease Type C Using LC-MS/MS. *PLoS ONE* **2020**, *15*. [CrossRef] [PubMed]

174. Burnett, J.R.; Hooper, A.J.; Hegele, R.A. *Abetalipoproteinemia*; University of Washington, Seattle: Seattle, WA, USA, 1993.

175. Levade, T.; Graber, D.; Flurin, V.; Delisle, M.B.; Pieraggi, M.T.; Testut, M.F.; Carrière, J.P.; Salvayre, R. Human β-mannosidase Deficiency Associated with Peripheral Neuropathy. *Ann. Neurol.* **1994**, *35*, 116–119. [CrossRef]

176. Huang, R.; Cathey, S.; Pollard, L.; Wood, T. UPLC-MS/MS Analysis of Urinary Free Oligosaccharides for Lysosomal Storage Diseases: Diagnosis and Potential Treatment Monitoring. *Clin. Chem.* **2018**, *64*, 1772–1779. [CrossRef]

177. Berini, S.E.; Tracy, J.A.; Engelstad, J.K.; Lorenz, E.C.; Milliner, D.S.; Dyck, P.J. Progressive Polyradiculoneuropathy due to Intraneural Oxalate Deposition in Type 1 Primary Hyperoxaluria. *Muscle Nerve* **2015**, *51*, 449–454. [CrossRef]

178. Morrow, G.; Tanguay, R.M. Biochemical and Clinical Aspects of Hereditary Tyrosinemia Type 1. *Adv. Exp. Med. Biol.* **2017**, *959*, 9–21. [PubMed]

179. Chinsky, J.M.; Singh, R.; Ficicioglu, C.; van Karnebeek, C.D.M.; Grompe, M.; Mitchell, G.; Waisbren, S.E.; Gucsavas-Calikoglu, M.; Wasserstein, M.P. Diagnosis and Treatment of Tyrosinemia Type I: A US and Canadian Consensus Group Review and Recommendations. *Genet. Med. Off. J. Am. Coll. Med. Genet.* **2017**, *19*. [CrossRef] [PubMed]

180. Stinton, C.; Geppert, J.; Freeman, K.; Clarke, A.; Johnson, S.; Fraser, H.; Sutcliffe, P.; Taylor-Phillips, S. Newborn Screening for Tyrosinemia Type 1 Using Succinylacetone–a Systematic Review of Test Accuracy. *Orphanet, J. Rare Dis.* **2017**, *12*, 48. [CrossRef] [PubMed]
181. Carrillo-Carrasco, N.; Chandler, R.J.; Venditti, C.P. Combined Methylmalonic Acidemia and Homocystinuria, cblC Type. I. Clinical Presentations, Diagnosis and Management. *J. Inherit. Metab. Dis.* **2012**, *35*, 91–102. [CrossRef] [PubMed]
182. Federico, A.; Dotti, M.T.; Gallus, G.N. Cerebrotendinous Xanthomatosis. In *GeneReviews®*; Adam, M.P., Ardinger, H.H., Pagon, R.A., Wallace, S.E., Bean, L.J.H., Mirzaa, G., Amemiya, A., Eds.; University of Washington, Seattle: Seattle, MA, USA, 2008.
183. Cali, J.J.; Hsieh, C.L.; Francke, U.; Russell, D.W. Mutations in the Bile Acid Biosynthetic Enzyme Sterol 27-hydroxylase Underlie Cerebrotendinous Xanthomatosis. *J. Biol. Chem.* **1991**, *266*. [CrossRef]

Review

The Role of Vitamin D as a Biomarker in Alzheimer's Disease

Giulia Bivona [1], Bruna Lo Sasso [1,2], Caterina Maria Gambino [1], Rosaria Vincenza Giglio [1], Concetta Scazzone [1], Luisa Agnello [1] and Marcello Ciaccio [1,2,*]

[1] Institute of Clinical Biochemistry, Clinical Molecular Medicine and Laboratory Medicine, Department of Biomedicine, Neurosciences and Advanced Diagnostics, University of Palermo, 90127 Palermo, Italy; giulia.bivona@unipa.it (G.B.); bruna.losasso@unipa.it (B.L.S.); cmgambino@libero.it (C.M.G.); rosaria.vincenza.giglio@alice.it (R.V.G.); concetta.scazzone@unipa.it (C.S.); luisa.agnello@unipa.it (L.A.)
[2] Department of Laboratory Medicine, AOUP "P. Giaccone", 90127 Palermo, Italy
* Correspondence: marcello.ciaccio@unipa.it

Abstract: Vitamin D and cognition is a popular association, which led to a remarkable body of literature data in the past 50 years. The brain can synthesize, catabolize, and receive Vitamin D, which has been proved to regulate many cellular processes in neurons and microglia. Vitamin D helps synaptic plasticity and neurotransmission in dopaminergic neural circuits and exerts anti-inflammatory and neuroprotective activities within the brain by reducing the synthesis of pro-inflammatory cytokines and the oxidative stress load. Further, Vitamin D action in the brain has been related to the clearance of amyloid plaques, which represent a feature of Alzheimer Disease (AD), by the immune cell. Based on these considerations, many studies have investigated the role of circulating Vitamin D levels in patients affected by a cognitive decline to assess Vitamin D's eventual role as a biomarker or a risk factor in AD. An association between low Vitamin D levels and the onset and progression of AD has been reported, and some interventional studies to evaluate the role of Vitamin D in preventing AD onset have been performed. However, many pitfalls affected the studies available, including substantial discrepancies in the methods used and the lack of standardized data. Despite many studies, it remains unclear whether Vitamin D can have a role in cognitive decline and AD. This narrative review aims to answer two key questions: whether Vitamin D can be used as a reliable tool for diagnosing, predicting prognosis and response to treatment in AD patients, and whether it is a modifiable risk factor for preventing AD onset.

Keywords: Alzheimer's Disease; Vitamin D; 25(OH)D levels; biomarker; Vitamin D deficiency

check for
updates

Citation: Bivona, G.; Lo Sasso, B.; Gambino, C.M.; Giglio, R.V.; Scazzone, C.; Agnello, L.; Ciaccio, M. The Role of Vitamin D as a Biomarker in Alzheimer's Disease. *Brain Sci.* **2021**, *11*, 334. https://doi.org/10.3390/brainsci11030334

Academic Editor: Chiara Villa

Received: 15 February 2021
Accepted: 2 March 2021
Published: 6 March 2021

Publisher's Note: MDPI stays neutral with regard to jurisdictional claims in published maps and institutional affiliations.

1. Introduction

If one searches for the keywords "Vitamin D" and "Cognition" in Pubmed.com, one finds over 1000 articles that have been published with no break in continuity for the past 50 years. The idea of a possible link between Vitamin D metabolism and brain function has been successfully proposed and then proved by a remarkable body of data. When assessing the Vitamin D circulating levels in Mild Cognitive Impairment and Alzheimer Disease (AD) patients, an association has yet been found. Nevertheless, the attempt to use Vitamin D as a biomarker of cognitive decline systematically failed and, furthermore, Vitamin D supplementation in these patients yielded controversial results. Many reasons can explain this debacle. First, the studies assessing Vitamin D levels and its serum biomarker 25(OH)D in AD patients have some limitations (different assay methods; heterogeneity of Vitamin D cut-offs; discrepancies among the measures used to define the cognitive function), which sharply limit the robustness of findings achieved. Second, discrepancies in the cut-offs and methods used to measure 25(OH)D across the studies, due to the lack of 25(OH)D measurement standardization, made the results difficult to interpret. Third, a specific biomarker's clinical usefulness is defined as its capability to influence clinicians to diagnose the disease, predict prognosis, and guide treatment, which is nothing Vitamin D can do.

Indeed, well-established diagnostic biomarkers for AD are currently available. Hence, there is no need for a marker for diagnosis, and, on the other hand, effective treatment for AD lacks so far. Therefore, it is unclear whether Vitamin D circulating levels can impact AD patients' outcome until a question is addressed: is AD onset preventable by reaching the optimal Vitamin D levels? Based on the available literature data, this review aims to explain why this question's answer could be no.

Vitamin D is a steroid hormone that can be synthesized endogenously. Primarily known to regulate calcium-phosphorus metabolism, it exerts several biological activities, counting brain function and immune response regulation [1–3].

In humans, Vitamin D is produced in a multi-step process that involves the ultraviolet B (UVB) rays irradiation of a cutaneous compound, the 7-dehydro-cholesterol (7-DHC).

Once UVB rays act on 7-DHC, the cholecalciferol is produced, needing two sequential hydroxylation steps to form the active Vitamin D. First hydroxylation occurs in the liver, by a 25 hydroxylase generating 25(OH)D, while the second mainly depends on a renal 1,25 hydroxylase, producing 1,25(OH)$_2$D. 1,25 hydroxylase is present within various organs and cells; thus, Vitamin D's active form can be produced in several tissues, including the lung, brain, prostate, placenta, and immune system cells. CYP2R1, CYP3A4, and CYP27A1 enzymes have 25-hydroxylase activities, while CYP27B1 is responsible for 1,25 hydroxylation. Kidney CYP27B1 gives rise to a hormone involved in calcium-phosphorus metabolism. Non-renal active Vitamin D is implicated in regulating some cellular processes, including cell differentiation and proliferation. While CYP27B1 is regulated by the parathyroid hormone (PTH), the fibroblast growth factor (FGF23) and 1,25(OH)$_2$D, extra-renal CYP27B1 is regulated by interferon γ (IFN-γ) and tumour necrosis factor (TNF) [4,5].

Vitamin D binding protein (VDBP) conveys both 25(OH)D and 1,25(OH)$_2$D from the liver and kidney to other tissues, where active Vitamin D binds the nuclear Vitamin D Receptor (VDR) [3,6–9], leading to the genomic and non-genomic actions (for more details on Vitamin D genomic and non-genomic actions see reference 1).

CYP24A1 enzyme, displaying 24 hydroxylase activity, carries out Vitamin D catabolism.

Vitamin D status is typically evaluated by measuring serum 25(OH)D [9]. A consensus on which 25(OH)D levels define Vitamin D sufficiency, deficiency, and insufficiency is lacking, also due to the standardization dearth in the past decades [10]. Most of the studies performed on Vitamin D's role in various diseases report unstandardized data, and AD is no exception. Thus, Vitamin D's reliability as a serum biomarker in AD has been considered a debatable issue, leading to controversial opinions across the scientific community [10].

2. Vitamin D and Alzheimer Disease

A growing interest in Vitamin D role in both brain development and function in adulthood led several authors to investigate the 25(OH)D circulating levels in AD patients [11–26]. The brain displays the capability to produce and receive Vitamin D's active form, which is deemed to support neurotransmission, synaptic plasticity, and neuroprotection [1,2,10]. From a pathophysiologic point of view, the relation between Vitamin D and AD onset and progression has been explained by impressive in vitro and in vivo studies. Given that amyloid plaques, along with neurofibrillary tangles, represent features of AD, it has been shown that 1,25(OH)$_2$D can help the amyloid plaques phagocytosis and clearance by the innate immune cells [1,2,27–31]. For instance, MCI and AD patient-derived macrophages show enhanced capability to eliminate amyloid plaques after 1,25(OH)$_2$D treatment [30], and a Vitamin D-enriched diet can decrease the number of plaques in AβPP-PS1 transgenic mice, an AD animal model [31]. Also, amyloid protein precursor (APP) metabolism involves some transcription factors, counting SMAD and transforming growth factor-beta (TGF-β), that, in turn, interact with VDR/ligand complex in the nucleus [29,32,33]. Finally, it should be considered that Vitamin D has a role in reducing cerebral microenvironment inflammation and oxidative stress, which are regarded as possible mechanisms underlying neurodegeneration and AD pathogenesis [1,10,29]. Table 1 summarises the characteristic of the studies considered.

Table 1. Characteristics of studies included in the analysis of vitamin D deficiency and the risk developing Alzheimer Disease.

Author & Publication Year	Ref.	Study Type	No. Patients (Total)	Follow-Up Duration	Vitamin D Deficiency Cut-Off	Vitamin D Assessment Method	Use of Procedure NIST	Conclusion
Afzal, 2014, Denmark	[19]	Prospective	10186	30 years	25 nmol/L	ECLIA	Not reported	Lower vitamin D concentrations increase the risk of developing AD
Aguilar-Navarro, 2019, Mexico	[21]	Cross-sectional	208	Not reported	20 ng/mL	CMIA	Not reported	Vitamin D deficiency is associated with AD
Buell, 2010, France	[16]	Cross-sectional	318	Not reported	10 ng/mL	RIA	Not reported	Vitamin D deficiency is associated with AD
Duchaine, 2020, Canada	[11]	Prospective	661	5.4 years	50 nmol/L	CLIA	Not reported	No association between 25(OH)D and AD
Feart, 2017, France	[17]	Prospective	916	12 years	25 nmol/L	CMIA	Not reported	Association between lower vitamin D concentrations and increased risk of AD
Karakis, 2016,	[25]	Prospective	1663	9 years	12 ng/mL	RIA	Not reported	No associations between vitamin D levels and incident of AD
Lee, 2020, Korea	[13]	Prospective	2990	Not reported	10 nmol/L	CMIA	Not reported	No direct correlation between VitD deficiency and cognitive impairment
Licher, 2017, Netherlands	[15]	Prospective	6220	13.3 years	25 nmol/L	ECLIA	Not reported	Lower vitamin D concentrations increase the risk of developing AD
Littlejohns, 2014, US	[14]	Prospective	1658	5.6 years	50 nmol/L	LC-MS	SRM certified by NIST	Vitamin D deficiency increases the risk of developing AD
Manzo, 2016, Italy	[12]	Cross-sectional	132	Not reported	10 ng/mL	Not reported	Not reported	No association between vitamin D deficiency and cognitive impairment
Olsson, 2017, Sweden	[24]	Prospective	1182	18 years	50 nmol/L	HPLC-MS	Not reported	No association between baseline vitamin D status and long-term risk of dementia
Shih, 2020, China	[22]	Cross-sectional	146	Not reported	20 ng/mL	RIA	Not reported	Reduced serum 25(OH)D levels are associated with lower MMSE scores in patients with mild AD

CLIA: Chemiluminescence-immunoassay; CMIA: ChemiluminescentMicroparticle immunoassay; ECLIA: electrochemiluminescent immunoassay; HPLC: High-performance liquid chromatography-mass spectrometry; LC-MS: Liquid chromatography tandem mass spectrometry; MMSE: Mini-Mental State Examination; NIST: National Institute for Standard and Technology; RIA: Radioimmunoassay; SRM:standard reference materials.

2.1. Observational Studies on 25(OH)D Serum Levels in AD Patients

Based on these considerations, Littlejohns et al. [14] enrolled, in 2014, 1658 subjects, of which 171 developed dementia (102 AD out of 171 all-cause dementia) over a 5.6-year follow-up period. Findings revealed that subjects with 25(OH)D serum levels <25 nm/L had a two-fold risk of AD onset compared to those with >50 nm/L. Authors defined Vitamin D deficiency <50 nm/L, distinguishing between deficiency and severe deficiency (25 to 50 nmol/L and <25 nm/L, respectively). The strength of the study was the use of procedures and materials certified by NIST. Within the Rotterdam Study [15], Licher et al. evaluated the role of Vitamin D levels as a risk factor for developing AD. Authors found that subjects with vitamin D <25 nmol/L (defined as the deficiency) had an increased risk of developing dementia, compared to those with ≥50 nmol/L (sufficiency), but this finding did not achieve statistical significance. However, the longitudinal analyses (follow-up period 13.3 years) revealed that the lower the baseline 25(OH)D levels, the higher the risk of developing AD. The Licher's study has various plus points, consisting of robust methods: for instance, the first 5 year follow-up period was excluded from the analysis to avoid reverse causation; a sensitivity analysis excluding patients with stroke was performed; each analysis was adjusted for several confounders. Nonetheless, an electrochemiluminescence binding assay was used to measure Vitamin D, while liquid chromatography-tandem mass spectrometry (LC/MS-MS) is recommended as the gold standard assay method; also, the adoption of NIST-certified procedures and materials has been not reported.

Opposite results were obtained by Ulstein et al. [23], who reported no association between vitamin D levels and AD development. To note that the Ulstein study sample size was small (73 AD patients and 63 controls). Karakis et al. [25] analyzed 1663 non-demented subjects for a 9-years follow-up period, documenting that no association exists between 25(OH)D levels and incident AD. In this study, Vitamin D deficiency, insufficiency, and sufficiency were defined as <12 ng/mL, 12 to <20 ng/mL, and 20 to <50 ng/mL, respectively.

As it can be noted, a high heterogeneity among the cut-offs used to define Vitamin D status exists, as it has been confirmed by Balion et al. [26], who documented an association between 25(OH)D concentrations and the risk of developing AD in a meta-analysis of 35,000 subjects. However, the authors highlighted remarkable discrepancies among the studies reviewed, undermining the findings obtained.

The interpretation of the studies mentioned above should take into account some considerations. First, many drawbacks weaken the results of the studies performed, including the differences among the assay methods used to measure Vitamin D, the heterogeneity among the cut-offs used to define Vitamin D deficiency and insufficiency, the lack of internationally recognized procedures and materials, and the discrepancies among the measures used to define the cognitive function.

2.2. Interventional Studies

To establish a role for Vitamin D in AD, a key question is whether AD onset is preventable by increasing Vitamin D serum levels, since diagnostic biomarkers for AD are available, and predicting prognosis and treatment response is difficult due to the lack of effective therapies [34]. Randomized controlled trials (RCTs) are suitable tools to address this question, but, unfortunately, they are few and achieved debatable conclusions [35–45].

Generally, it could be stated that Vitamin D supplementation failed to prevent AD onset [35–38,42,43,45,46]. It is worth mentioning Rossom et al. on 4143 older women free from dementia, receiving 400 IU or placebo, reporting a similar cognitive decline incidence between the treatment and placebo groups. Authors proved that exogenous Vitamin D has no impact on dementia development risk [37]. Although Jia et al. gained opposite findings, it should be noted that the sample size of the Jia study was smaller (210 patients) and the follow-up period short (12 months vs. 7.8 years in Rossom's study) [39]. Some authors reported that Vitamin D could improve cognitive function combined with other compounds, like memantine [40] and medium-chain triglycerides plus L-leucine-rich amino acids [41], but also these studies had limited populations and follow-up duration. Oppositely, the Cochrane Database of Systematic Reviews has recently published the exciting findings of Rutjes et al., who performed a meta-analysis to assess the impact of vitamins supplementation on cognition in healthy individuals. Authors found no evidence of a significant influence of vitamin supplementation in the risk of cognitive decline, and, importantly, revealed that many studies reporting an effect of Vitamin D in cognitive performance had a low grade of certainty, that is a marked difference between the estimated effect and the true one [45]. In 2020, Bischoff-Ferrari et al. carried out an RCT in 1900 subjects within the DO-HEALTH RTC, evaluating the impact of Vitamin D supplement on the Montreal Cognitive Assessment (MoCA) in a 3-year follow-up. Authors conclude that Vitamin D has no impact on cognitive function improvement [42]. Although other authors gained different results in the same year [44], here again, the study sample and follow-up sharply differ between the two studies, having Bischoff-Ferrari's RCT a larger population and a longer follow-up.

When evaluating interventional studies, the impact of AD lengthy latency period should be taken into account, which further hinders univocal interpretation of the potential role for vitamin D in this disease. Indeed, during the course of AD, modifications of the mechanisms underlying the progression occur, which increases intricacy in understanding the pathophysiology and, in turn, of the candidate risk factors of the disease.

Taken together, RCTs suggest that Vitamin D supplementation does not influence cognition, regardless of the dose of the administration [46].

3. Conclusions

There is no uncertainty that Vitamin D takes part in normal brain function, and low Vitamin D levels can occur among demented patients. However, this finding's clinical and laboratory significance remains unclear, also due to several drawbacks of the available studies, weakening their results and hampering concluding. Taken the evidence of past and recent literature with the appropriate cautions, Vitamin D cannot be considered a reliable biomarker of AD, since measuring the biomarker does not improve diagnosis and prognosis in these patients. Also, no clear evidence on the role of low Vitamin D levels as a risk factor for the disease exists since interventional studies on this topic are few and findings are inconsistent. Preventing the onset of AD by modifying Vitamin D levels seems too good to be true.

Author Contributions: G.B. conceived and wrote the manuscript; B.L.S. and C.S. contributed to the literature research; C.M.G., R.V.G. and L.A. contributed to writing the manuscript; M.C. revised the manuscript and supervised the entire process. All authors have read and agreed to the published version of the manuscript.

Funding: This research received no external funding.

Conflicts of Interest: The authors declare no conflict of interest.

References

1. Bivona, G.; Gambino, C.M.; Iacolino, G.; Ciaccio, M. Vitamin D and the nervous system. *Neurol. Res.* **2019**, *41*, 827–835. [CrossRef]
2. Bivona, G.; Agnello, L.; Bellia, C.; Iacolino, G.; Scazzone, C.; Lo Sasso, B.; Ciaccio, M. Non-Skeletal Activities of Vitamin D: From Physiology to Brain Pathology. *Medicina* **2019**, *55*, 341. [CrossRef]
3. Bikle, D.; Christakos, S. New aspects of vitamin D metabolism and action—Addressing the skin as source and target. *Nat. Rev. Endocrinol.* **2020**, *16*, 234–252. [CrossRef] [PubMed]
4. Bikle, D.D.; Patzek, S.; Wang, Y. Physiologic and pathophysiologic roles of extra renal CYP27b1: Case report and review. *Bone Rep.* **2018**, *8*, 255–267. [CrossRef]
5. Bivona, G.; Agnello, L.; Ciaccio, M. The immunological implication of the new vitamin D metabolism. *Cent. Eur. J. Immunol.* **2018**, *43*, 331–334. [CrossRef] [PubMed]
6. Bikle, D.D.; Schwartz, J. Vitamin D binding protein, total and free vitamin D levels in different physiological and pathophysiological conditions. *Front. Endocrinol.* **2019**, *10*, 317. [CrossRef] [PubMed]
7. Prabhu, A.V.; Luu, W.; Li, D.; Sharpe, L.J.; Brown, A.J. DHCR7: A vital enzyme switch between cholesterol and vitamin D production. *Prog. Lipid Res.* **2016**, *64*, 138–151. [CrossRef] [PubMed]

8. Meyer, M.B.; Benkusky, N.A.; Kaufmann, M.; Lee, S.M.; Onal, M.; Jones, G.; Pike, J.W. A kidney-specific genetic control module in mice governs endocrine regulation of the cytochrome P450 gene Cyp27b1 essential for vitamin D3 activation. *J. Biol. Chem.* **2017**, *292*, 17541–17558. [CrossRef] [PubMed]

9. Holick, M.F.; Binkley, N.C.; Bischoff-Ferrari, H.A.; Gordon, C.M.; Hanley, D.A.; Heaney, R.P.; Murad, M.H.; Weaver, C.M. Evaluation, treatment, and prevention of vitamin D deficiency: An Endocrine Society clinical practice guideline. *J. Clin. Endocrinol. Metab.* **2011**, *96*, 1911–1930. [CrossRef] [PubMed]

10. Bivona, G.; Lo Sasso, B.; Iacolino, G.; Gambino, C.M.; Scazzone, C.; Agnello, L.; Ciaccio, M. Standardized measurement of circulating vitamin D [25(OH)D] and its putative role as a serum biomarker in Alzheimer's disease and Parkinson's disease. *Clin. Chim. Acta* **2019**, *497*, 82–87. [CrossRef]

11. Duchaine, C.S.; Talbot, D.; Nafti, M.; Giguère, Y.; Dodin, S.; Tourigny, A.; Carmichael, P.H.; Laurin, D. Vitamin D status, cognitive decline and incident dementia: The Canadian Study of Health and Aging. *Can. J. Public Health* **2020**, *111*, 312–321. [CrossRef]

12. Manzo, C.; Castagna, A.; Palummeri, E.; Traini, E.; Cotroneo, A.M.; Fabbo, A.; Natale, M.; Gareri, P.; Putignano, S. Relationship between 25-hydroxy vitamin D and cognitive status in older adults: The COGNIDAGE study. *Recenti Prog. Med.* **2016**, *107*, 75–83. [CrossRef]

13. Lee, D.H.; Chon, J.; Kim, Y.; Seo, Y.K.; Park, E.J.; Won, C.W.; Soh, Y. Association between vitamin D deficiency and cognitive function in the elderly Korean population: A Korean frailty and aging cohort study. *Medicine* **2020**, *99*, e19293. [CrossRef]

14. Littlejohns, T.J.; Henley, W.E.; Lang, I.A.; Annweiler, C.; Beauchet, O.; Chaves, P.H.; Fried, L.; Kestenbaum, B.R.; Kuller, L.H.; Langa, K.M.; et al. Vitamin D and the risk of dementia and Alzheimer disease. *Neurology* **2014**, *83*, 920–928. [CrossRef] [PubMed]

15. Licher, S.; de Bruijn, R.F.A.G.; Wolters, F.J.; Zillikens, M.C.; Ikram, M.A.; Ikram, M.K. Vitamin D and the risk of dementia: The Rotterdam study. *J. Alzheimers Dis.* **2017**, *60*, 989–997. [CrossRef] [PubMed]

16. Buell, J.S.; Dawson-Hughes, B.; Scott, T.M.; Weiner, D.E.; Dallal, G.E.; Qui, W.Q.; Bergethon, P.; Rosenberg, I.H.; Folstein, M.F.; Patz, S.; et al. 25-Hydroxyvitamin D, dementia, and cerebrovascular pathology in elders receiving home services. *Neurology* **2010**, *74*, 18–26. [CrossRef]

17. Feart, C.; Helmer, C.; Merle, B.; Herrmann, F.R.; Annweiler, C.; Dartigues, J.F.; Delcourt, C.; Samieri, C. Associations of lower vitamin D concentrations with cognitive decline and long-term risk of dementia and Alzheimer's disease in older adults. *Alzheimers Dement.* **2017**, *13*, 1207–1216. [CrossRef]

18. Ouma, S.; Suenaga, M.; BölükbaşıHatip, F.F.; Hatip-Al-Khatib, I.; Tsuboi, Y.; Matsunaga, Y. Serum vitamin D in patients with mild cognitive impairment and Alzheimer's disease. *Brain Behav.* **2018**, *8*, e00936. [CrossRef]

19. Afzal, S.; Bojesen, S.E.; Nordestgaard, B.G. Reduced 25-hydroxyvitamin D and risk of Alzheimer's disease and vascular dementia. *Alzheimers Dement.* **2014**, *10*, 296–302. [CrossRef] [PubMed]

20. Ertilav, E.; Barcin, N.E.; Ozdem, S. Comparison of Serum Free and Bioavailable 25-Hydroxyvitamin D Levels in Alzheimer's Disease and Healthy Control Patients. *Lab. Med.* **2020**, lmaa066. [CrossRef]

21. Aguilar-Navarro, S.G.; Mimenza-Alvarado, A.J.; Jiménez-Castillo, G.A.; Bracho-Vela, L.A.; Yeverino-Castro, S.G.; Ávila-Funes, J.A. Association of Vitamin D with Mild Cognitive Impairment and Alzheimer's Dementia in Older Mexican Adults. *Rev. Investig. Clin.* **2019**, *71*, 381–386. [CrossRef]

22. Shih, E.J.; Lee, W.J.; Hsu, J.L.; Wang, S.J.; Fuh, J.L. Effect of vitamin D on cognitive function and white matter hyperintensity in patients with mild Alzheimer's disease. *Geriatr. Gerontol. Int.* **2020**, *20*, 52–58. [CrossRef]

23. Ulstein, I.; Bøhmer, T. Normal Vitamin Levels and Nutritional Indices in Alzheimer's Disease Patients with Mild Cognitive Impairment or Dementia with Normal Body Mass Indexes. *J. Alzheimers Dis.* **2017**, *55*, 717–725. [CrossRef] [PubMed]

24. Olsson, E.; Byberg, L.; Karlström, B.; Cederholm, T.; Melhus, H.; Sjögren, P.; Kilander, L. Vitamin D is not associated with incident dementia or cognitive impairment: An 18-y follow-up study in community-living old men. *Am. J. Clin. Nutr.* **2017**, *105*, 936–943. [CrossRef]

25. Karakis, I.; Pase, M.P.; Beiser, A.; Booth, S.L.; Jacques, P.F.; Rogers, G.; DeCarli, C.; Vasan, R.S.; Wang, T.J.; Himali, J.J.; et al. Association of serum vitamin D with the risk of incident dementia and subclinical indices of brain aging: The Framingham Heart Study. *J. Alzheimers Dis.* **2016**, *51*, 451–461. [CrossRef] [PubMed]

26. Balion, C.; Griffith, L.E.; Strifler, L.; Henderson, M.; Patterson, C.; Heckman, G.; Llewellyn, D.J.; Raina, P. Vitamin D, cognition, and dementia: A systematic review and meta-analysis. *Neurology* **2012**, *79*, 1397–1405. [CrossRef]

27. Scazzone, C.; Agnello, L.; Ragonese, P.; Lo Sasso, B.; Bellia, C.; Bivona, G.; Schillaci, R.; Salemi, G.; Ciaccio, M. Association of CYP2R1 rs10766197 with MS risk and disease progression. *J. Neurosci. Res.* **2018**, *96*, 297–304. [CrossRef]

28. Mizwicki, M.T.; Menegaz, D.; Zhang, J.; Barrientos-Durán, A.; Tse, S.; Cashman, J.R.; Griffin, P.R.; Fiala, M. Genomic and non genomic signaling induced by 1α,25(OH)2-vitamin D3 promotes the recovery of amyloid-β phagocytosis by Alzheimer's disease macrophages. *J. Alzheimers Dis.* **2012**, *29*, 51–62. [CrossRef]

29. Banerjee, A.; Khemka, V.K.; Ganguly, A.; Roy, D.; Ganguly, U.; Chakrabarti, S. Vitamin D and Alzheimer's Disease: Neurocognition to Therapeutics. *Int. J. Alzheimers. Dis.* **2015**, *2015*, 192747. [CrossRef]

30. Masoumi, A.; Goldenson, B.; Ghirmai, S.; Avagyan, H.; Zaghi, J.; Abel, K.; Zheng, X.; Espinosa-Jeffrey, A.; Mahanian, M.; Liu, P.T.; et al. 1α,25-dihydrox-yvitamin D3 interacts with curcuminoids to stimulate amyloid-βclearance by macrophages of alzheimer's disease patients. *J. Alzheimers Dis.* **2009**, *17*, 703–717. [CrossRef]

31. Yu, J.; Gattoni-Celli, M.; Zhu, H.; Bhat, N.R.; Sambamurti, K.; Gattoni-Celli, S.; Kindy, M.S. Vitamin D3-enriched diet correlates with a decrease of amyloid plaques in the brain of AβPP transgenic mice. *J. Alzheimers Dis.* **2011**, *25*, 295–307. [CrossRef]

32. Carlberg, C. The concept of multiple vitamin D signaling pathways. *J. Investig. Dermatol. Symp. Proc.* **1996**, *1*, 10–14.
33. Yanagisawa, J.; Yanagi, Y.; Masuhiro, Y.; Suzawa, M.; Watanabe, M.; Kashiwagi, K.; Toriyabe, T.; Kawabata, M.; Miyazono, K.; Kato, S. Convergence of transforming growth factor-βand vitamin D signaling pathways on SMAD transcriptional coactivators. *Science* **1999**, *283*, 1317–1321. [CrossRef] [PubMed]
34. Agnello, L.; Piccoli, T.; Vidali, M.; Cuffaro, L.; Lo Sasso, B.; Iacolino, G.; Giglio, R.V.; Lupo, F.; Alongi, P.; Bivona, G.; et al. Diagnostic accuracy of cerebrospinal fluid biomarkers measured by chemiluminescent enzyme immunoassay for Alzheimer disease diagnosis. *Scand. J. Clin. Lab. Investig.* **2020**, *80*, 313–317. [CrossRef]
35. Stein, M.S.; Scherer, S.C.; Ladd, K.S.; Harrison, L.C. A randomized controlled trial of high-dose vitamin D2 followed by intranasal insulin in Alzheimer's disease. *J. Alzheimers Dis.* **2011**, *26*, 477–484. [CrossRef]
36. Przybelski, R.; Agrawal, S.; Krueger, D.; Engelke, J.A.; Walbrun, F.; Binkley, N. Rapid correction of low vitamin D status in nursing home residents. *Osteoporos Int.* **2008**, *19*, 1621–1628. [CrossRef]
37. Rossom, R.C.; Espeland, M.A.; Manson, J.E.; Dysken, M.W.; Johnson, K.C.; Lane, D.S.; LeBlanc, E.S.; Lederle, F.A.; Masaki, K.H.; Margolis, K.L. Calcium and vitamin D supplementation and cognitive impairment in the women's health initiative. *J. Am. Geriatr. Soc.* **2012**, *60*, 2197–2205. [CrossRef] [PubMed]
38. Moran, C.; Scotto di Palumbo, A.; Bramham, J.; Moran, A.; Rooney, B.; De Vito, G.; Egan, B. Effects of a six-month multi-ingredient nutrition supplement intervention of omega-3 polyunsaturated fatty acids, vitamin D, resveratrol, and whey protein on cognitive function in older adults: A randomized, double-blind, controlled trial. *J. Prev. Alzheimers Dis.* **2018**, *5*, 175–183. [CrossRef] [PubMed]
39. Jia, J.; Hu, J.; Huo, X.; Miao, R.; Zhang, Y.; Ma, F. Effects of vitamin D supplementation on cognitive function and blood Aβ-related biomarkers in older adults with Alzheimer's disease: A randomized, double-blind, placebo-controlled trial. *Neurol. Neurosurg. Psychiatry* **2019**, *90*, 1347–1352. [CrossRef] [PubMed]
40. Annweiler, C.; Herrmann, F.R.; Fantino, B.; Brugg, B.; Beauchet, O. Effectiveness of the combination of memantine plus vitamin D on cognition in patients with Alzheimer disease: A pre-post pilot study. *Cogn. Behav. Neurol.* **2012**, *25*, 121–127. [CrossRef]
41. Abe, S.; Ezaki, O.; Suzuki, M. Medium-Chain Triglycerides in Combination with Leucine and Vitamin D Benefit Cognition in Frail Elderly Adults: A Randomized Controlled Trial. *J. Nutr. Sci. Vitaminol.* **2017**, *63*, 133–140. [CrossRef] [PubMed]
42. Bischoff-Ferrari, H.A.; Vellas, B.; Rizzoli, R.; Kressig, R.W.; da Silva, J.A.P.; Blauth, M.; Felson, D.T.; McCloskey, E.V.; Watzl, B.; Hofbauer, L.C.; et al. Effect of Vitamin D Supplementation, Omega-3 Fatty Acid Supplementation, or a Strength-Training Exercise Program on Clinical Outcomes in Older Adults: The DO-HEALTH Randomized Clinical Trial. *JAMA* **2020**, *324*, 1855–1868. [CrossRef]
43. Jorde, R.; Kubiak, J.; Svartberg, J.; Fuskevåg, O.M.; Figenschau, Y.; Martinaityte, I.; Grimnes, G. Vitamin D supplementation has no effect on cognitive performance after four months in mid-aged and older subjects. *J. Neurol. Sci.* **2019**, *396*, 165–171. [CrossRef] [PubMed]
44. Yang, T.; Wang, H.; Xiong, Y.; Chen, C.; Duan, K.; Jia, J.; Ma, F. Vitamin D Supplementation Improves Cognitive Function Through Reducing Oxidative Stress Regulated by Telomere Length in Older Adults with Mild Cognitive Impairment: A 12-Month Randomized Controlled Trial. *J. Alzheimers Dis.* **2020**, *78*, 1509–1518. [CrossRef] [PubMed]
45. Rutjes, A.W.; Denton, D.A.; Di Nisio, M.; Chong, L.Y.; Abraham, R.P.; Al-Assaf, A.S.; Anderson, J.L.; Malik, M.A.; Vernooij, R.W.; Martínez, G.; et al. Vitamin and mineral supplementation for maintaining cognitive function in cognitively healthy people in mid and late life. *Cochrane Database Syst. Rev.* **2018**, *12*, CD011906. [CrossRef] [PubMed]
46. Bode, L.E.; McClester Brown, M.; Hawes, E.M. Vitamin D Supplementation for Extraskeletal Indications in Older Persons. *J. Am. Med. Dir. Assoc.* **2020**, *21*, 164–171. [CrossRef] [PubMed]

Article

Prognostic Role of CSF β-amyloid 1–42/1–40 Ratio in Patients Affected by Amyotrophic Lateral Sclerosis

Tiziana Colletti [1], Luisa Agnello [2], Rossella Spataro [3], Lavinia Guccione [1], Antonietta Notaro [1], Bruna Lo Sasso [2], Valeria Blandino [4], Fabiola Graziano [4], Caterina Maria Gambino [2], Rosaria Vincenza Giglio [2], Giulia Bivona [2], Vincenzo La Bella [1], Marcello Ciaccio [2,†] and Tommaso Piccoli [4,*,†]

[1] ALS Clinical Research Center and Laboratory of Neurochemistry, Department of Biomedicine, Neuroscience and Advanced Diagnostics, University of Palermo, 90129 Palermo, Italy; tizianacolletti@gmail.com (T.C.); lavyguccione@hotmail.it (L.G.); antonietta.notaro@libero.it (A.N.); vincenzo.labella@unipa.it (V.L.B.)

[2] Institute of Clinical Biochemistry, Clinical Molecular Medicine and Laboratory Medicine, Department of Biomedicine, Neurosciences and Advanced Diagnostics, University of Palermo, 90127 Palermo, Italy; luisa.agnello@unipa.it (L.A.); bruna.losasso@unipa.it (B.L.S.); cmgambino@libero.it (C.M.G.); rosaria.vincenza.giglio@alice.it (R.V.G.); giulia.bivona@unipa.it (G.B.); marcello.ciaccio@unipa.it (M.C.)

[3] IRCSS Bonino Pulejo, Presidio Pisani, 90129, Palermo, Italy; rossellaspataro@libero.it

[4] Unit of Neurology, Department of Biomedicine, Neurosciences and Advanced Diagnostics, University of Palermo, 90129, Palermo, Italy; valeriabl@libero.it (V.B.); fabiolagr93@gmail.com (F.G.)

*   Correspondence: tommaso.piccoli@unipa.it; Tel.: +39-091-6555-159

†   These authors shared the senior authorship and contributed equally to this paper.

Citation: Colletti, T.; Agnello, L.; Spataro, R.; Guccione, L.; Notaro, A.; Lo Sasso, B.; Blandino, V.; Graziano, F.; Gambino, C.M.; Giglio, R.V.; et al. Prognostic Role of CSF β-amyloid 1–42/1–40 Ratio in Patients Affected by Amyotrophic Lateral Sclerosis. *Brain Sci.* **2021**, *11*, 302. https://doi.org/10.3390/brainsci11030302

Academic Editor: Melissa Bowerman

Received: 25 January 2021
Accepted: 24 February 2021
Published: 27 February 2021

Publisher's Note: MDPI stays neutral with regard to jurisdictional claims in published maps and institutional affiliations.

Abstract: The involvement of β-amyloid (Aβ) in the pathogenesis of amyotrophic lateral sclerosis (ALS) has been widely discussed and its role in the disease is still a matter of debate. Aβ accumulates in the cortex and the anterior horn neurons of ALS patients and seems to affect their survival. To clarify the role of cerebrospinal fluid (CSF) Aβ 1–42 and Aβ 42/40 ratios as a potential prognostic biomarker for ALS, we performed a retrospective observational study on a cohort of ALS patients who underwent a lumbar puncture at the time of the diagnosis. CSF Aβ 1–40 and Aβ 1–42 ratios were detected by chemiluminescence immunoassay and their values were correlated with clinical features. We found a significant correlation of the Aβ 42/40 ratio with age at onset and Mini Mental State Examination (MMSE) scores. No significant correlation of Aβ 1–42 or Aβ 42/40 ratios to the rate of progression of the disease were found. Furthermore, when we stratified patients according to Aβ 1–42 concentration and the Aβ 42/40 ratio, we found that patients with a lower Aβ 42/40 ratio showed a shorter survival. Our results support the hypothesis that Aβ 1–42 could be involved in some pathogenic mechanism of ALS and we suggest the Aβ 42/40 ratio as a potential prognostic biomarker.

Keywords: ALS; biomarker; beta amyloid

1. Introduction

Amyotrophic lateral sclerosis (ALS) is the most common degenerative motor neuron disease, which results in progressive muscle weakness and causes death in a few years. The pathogenesis of ALS is not fully understood and several pathological processes have been proposed such as abnormal protein aggregation, mitochondrial dysfunction and oxidative stress [1]. To date, the diagnosis of ALS is on clinical features and electrophysiological parameters, indicating the degeneration of both upper and lower motor neurons [2]. Heterogeneity in terms of clinical presentation often makes an early and accurate diagnosis a real challenge for clinicians. For this reason, there has been a growing interest in identifying candidate biomarkers for ALS, which can help make an early diagnosis and predict disease progression. Among these, the role of a neurofilament (NF) phosphorylated heavy chain (pNF-H) and light chain (NF-L) as potential biomarkers for ALS is defining. NFs have a non-specific and not fully clarified role in the pathogenesis of ALS. The abnormal accumulation of NF aggregates was observed in perycaria and proximal axons of motoneurons both in ALS murine models and patients that seemed to be related to an impairment of intracellular transport [3]. Recently, a few authors have shown that the aggregation of NFs is related to their hyperphosphorylation state [4]. pNF-H and NF-L are increased in the cerebrospinal fluid (CSF) of ALS patients in comparison with control groups [5,6] and the higher levels are associated with a more rapidly evolving disease and shorter survival [7]. The role of other candidate biomarkers (such as Tau proteins) is still under investigation [8–10].

ALS shares common pathways with other neurodegenerative disorders. For example, C9 or f72 repeat expansions and TAR DNA-binding protein (TARDBP) mutations have been described in ALS and frontotemporal lobar degeneration (FTLD), modifying the idea of ALS as a disease confined to the motor system to the extreme phenotypic expression of a clinical/pathological continuum with FTLD [11–13]. Furthermore, the presence of β-amyloid (Aβ) deposits at the cortical level, hippocampus and spinal cord motor neurons have been described in ALS patients [14–16], suggesting the possibility of some overlapping features between ALS and Alzheimer's disease (AD).

AD is the most common cause of dementia, characterized by Aβ and Tau deposition, respectively, in senile plaques and neurofibrillary tangles as a result of a complex mechanism known as the amyloid cascade [17]. The amyloid precursor protein (APP) is processed by α-secretase into a soluble form α of the APP (sAPPα) and carbossi-terminal fragment α (CTFα) and by β-secretase sAPPβ and CTFβ. Subsequently, CTFβ is cleaved into Aβ 1–40 or Aβ 1–42 by γ-secretase and the imbalance of this process leads to an over-expression of Aβ 1–42 that precipitates, forming the senile plaques. The consequence is the hyperphosphorylation of the Tau protein and the formation of neurofibrillary tangles [18]. CSF Aβ 1–42 levels combined with total Tau (tTau) and phosphorylated Tau (pTau) are currently used as diagnostic biomarkers for AD with a high sensitivity and specificity [19–22], ameliorating the diagnostic accuracy in the very early stages of the disease.

Due to the pathogenetic similarities among neurodegenerative diseases, possible common pathways between AD and ALS have been investigated. Preclinical studies demonstrated the interaction between superoxide dismutase (SOD) and Aβ and evidence of the amyloid cascade has been reported [23] with an increase of sAPP in the CSF from ALS patients [24] and the post-mortem evidence of the over-expression of APP and Aβ in the hippocampi of ALS patients [17]. On the other hand, it is known that APP regulates glial cell-derived neurotrophic factor (GDNF) expression, having a role on neuromuscular junction formation and probably also in neuromuscular degenerative diseases [23].

Whether or not Aβ has a role in the pathogenesis of ALS is far from being clear but it has been recently proposed that the CSF Aβ 1–42 protein concentration is higher in ALS patients and that it is related to disease severity at the time of diagnosis [25].

Our aim is to evaluate the role of the CSF Aβ 1–42 and Aβ 1–40 concentration and the Aβ 42/40 ratio as a potential predictor factor for progression and overall survival in ALS.

2. Patients and Methods

2.1. Patients

Ninety-three (93) ALS patients (M/F: 1.11) were enrolled from the ALS Clinical and Research Center, Department of Biomedicine, Neuroscience and advanced Diagnostics (Bi.N.D.), University of Palermo, Italy, from January 2001 to October 2020. All ALS patients were diagnosed according to El-Escorial revised criteria [2] combined with the neurophysiological ones [26]. The revised ALS Functional Rating Scale (ALSFRS-R) [27] was used to score the severity of the symptoms of ALS patients; a higher score indicated normality and a lower score defined a locked-in condition. ΔFS ((ALSFRS-R at onset–ALSFRS-R at time of diagnosis)/diagnostic delay) was used to define the disease progression [28]. According to the ΔFS, patients could be classified in three groups: slow progression (ΔFS < 0.5), intermediate progression (ΔFS $\geq$ 0.5 < 1) and rapid progression (ΔFS $\geq$ 1) [28]. We considered co-morbidities for each patient.

All patients underwent a cognitive/behavioral assessment and the administration of neuropsychological tests such as the Frontal Systems Behavioral Scale (FrSBe), Mini Mental State Examination (MMSE) and Edinburgh Cognitive and Behavioral ALS Screen (ECAS) (S-TAB.1). Fewer than 30% showed some degree of behavioral/cognitive impairment according to the Italian Validation of ECAS but none of them were demented. All patients were tested for the most common ALS-related genes and no known mutations associated with ALS were detected.

ALS patients underwent a lumbar puncture (LP) and a CSF analysis as routine procedures of the diagnostic work-up. For the biomarker analysis, ALS patients were subdivided into three subgroups according to the rate of progression based on ΔFS (i.e., slow: ALS-s; intermediate: ALS-i; rapid: ALS-r). All demographic and clinical features of the selected ALS patients are shown in Table 1. None of the patients enrolled assumed any specific drug for ALS treatment at the time of the LP and all of them started riluzole immediately after the diagnosis was made. None of them participated in clinical trials.

Table 1. Demographic and clinical characteristics of the total cohort and amyotrophic lateral sclerosis (ALS) patients stratified based on their rate of progression: slow (ALS-s), intermediate (ALS-i) and rapid (ALS-r). Data are expressed as a median with an interquartile range (IQR).

Variables	ALS tot (n = 93)	ALS-s (n = 19)	ALS-i (n = 31)	ALS-r (n = 35)	p
Age at onset (years)	67 (63–72)	63 (61–67)	67 (64–72)	70 (64–74)	< 0.001 *
M/F	1.11	2.16	1	1.18	0.43 ** χ^2 = 1.67 with 2 DF
Education (years)	5 (5–13)	13 (5–13)	8 (5–8)	5 (5–9)	0.283 *
Type of onset familiar, % sporadic, %	3.2% 96.6%	5.2% 94.8%	3.3% 96.7%	2.8% 97.2%	0.90 ** χ^2 = 0.22 with 2 DF
Site of onset Spinal, % Bulbar, %	70.3% 29.7%	89.5% 10.5%	63.3% 36.7%	62.3% 37.2%	0.09 ** χ^2 = 4.80 with 2 DF
Diagnostic delay (months)	12 (9–20)	25 (18–37)	12 (10–24)	7 (4–9.5)	< 0.001 *
Rate of progression (ΔFS) [A]	0.8 (0.5–1.3)	–	–	–	–
FVC [a] (%)	81 (55–93)	84 (59–98)	83 (60–92)	67 (46–93)	0.334 *
BMI [b] (kg/m^2)	24.8 (21.5–27.1)	25 (21–28)	24 (22–27)	25.7 (46–83)	0.355 *
Survival (months)	30 (20–46)	57 (37–67)	35 (27–47)	17 (13–26)	< 0.001 *

[A] ΔFS at diagnosis = (ALSFRS-R at onset–ALSFRS-R at diagnosis)/diagnostic delay (months). [a] Forced vital capacity. [b] Body mass index. * Kruskal–Wallis one way analysis of variance on ranks. ** chi-squared test. Bold font indicates a statistical significance (p < 0.05).

All patients gave informed written consent. The study was approved by the local Ethics Committee. All of the clinical and biological assessments were carried out in accordance with the World Medical Association Declaration of Helsinki.

2.2. CSF Collection and Analytical Techniques

All CSF samples were collected in the morning hours and then sent to the Central Hospital Laboratory for a routine analysis. For biomarker detection, the CSF samples were centrifuged in case of blood contamination, aliquoted in polypropylene tubes and stored at –80 °C within one hour until further analysis according to international guidelines [29]. The CSF routine chemical parameters are shown in Table 2.

Table 2. Cerebrospinal parameters in ALS patients and in patients with slow (ALS-s), intermediate (ALS-i) and rapid (ALS-r) progression. Data are expressed as a median with an interquartile range (IQR).

Parameters	ALS tot (*n* = 93)	ALS-s (*n* = 19)	ALS-i (*n* = 31)	ALS-r (*n* = 35)	*p* *
Proteins (mg/dL)	39 (28–51)	37 (19–52)	39 (32–62)	37 (26–48)	0.524
Glucose (mg/dL)	60 (55–66)	58 (55–63)	56 (51–66)	62 (57–72)	0.103
Cells (lymphocytes)	0.8 (0.6–1.8)	0.8 (0.6–2.3)	1 (0.6–2.9)	0.8 (0.4–1.6)	0.371
Oligoclonal bands (y/n)**	17/76	4/15	4/27	7/28	

* Kruskal–Wallis one way analysis of variance on ranks. ** y = yes, n = no

The CSF Aβ 1–42 and Aβ 1–40 were measured by a chemiluminescent immunoassay CLEIA (Lumipulse G b-amyloid 1–40, Lumipulse G b-amyloid 1–42, Fujirebio Inc. Europe, Gent, Belgium) on a fully automatic platform (Lumipulse G1200 analyzer, Fujirebio Inc. Europe, Gent, Belgium). We used as reference cut-off for the Aβ 1–42 value and the Aβ 42/40 ratio < 650 pg/ML and < 0.055, respectively, as suggested by the manufacturer.

2.3. Statistical Analyses

All statistical analyses were performed using SIGMAPLOT 12.0 software package (Systat Software Inc., San Jose, CA, USA).

A Shapiro–Wilk test was performed to test the normality of the data. We expressed demographic, clinical and biochemical variables as a median with interquartile ranges (IQR). We performed Kruskal–Wallis one way analysis of variance on ranks to compare non-parametric data, a one way ANOVA to compare parametric data and a chi-squared test to assess differences between the groups. We analyzed non-parametric data with Spearman's rank correlation coefficient and parametric data with Pearson's correlation coefficient, considering *p* values < 0.05 as significant.

A survival analysis was performed with the Kaplan–Meier method and survival curves were compared with the log-rank test. Univariate and multivariate Cox regression analyses were performed to predict risk factors for overall survival.

3. Results

A retrospective observational study was performed on 93 ALS patients to analyze the role of Aβ 1–42, Aβ 1–40 and the Aβ 42/40 ratio as candidate biomarkers for ALS. As a few studies have shown that the CSF Aβ levels are correlated with the rate of progression [21], we stratified ALS patients into three subgroups: ALS-s (*n* = 19; M/F: 2.16), ALS-i (*n* = 31; M/F: 1) and ALS-r (*n* = 35; M/F: 1.18).

In our study, the total cohort of ALS patients had a median age at onset of 67 years. 96.6% of ALS patients were sporadic with a spinal onset in 70.3% of the whole cohort. At the time of diagnosis, ALS patients showed median values of a forced vital capacity (FVC)% of 80.5 (IQR = 54.75–93.25), of a body mass index (BMI) of 24.8 kg/m2 (IQR = 21.5–27.12) and of a ΔFS of 0.81 (IQR = 0.5–1.33). The Kruskal–Wallis one way ANOVA with the rate of progression (ΔFS) as a factor showed statistically significant differences in the age of onset (lower in the ALS-s group, *p* < 0.001), diagnostic delay (longer in the ALS-s group, *p* < 0.001) and survival (longer in the ALS-s group, *p* < 0.001); no statistically significant differences were found for the M/F ratio, education, FVC% and BMI (Table 1). The CSF biochemical profile was similar in the three subgroups (Table 2). The neuropsychological assessments with FrSBe, MMSE and ECAS showed no cognitive/behavioral impairments (Table S1).

Analyzing data with the Shapiro–Wilk test, we found that the CSF Aβ 42/40 values were normally distributed while the Aβ 1–42 and Aβ 1–40 ones were not. As shown in Figure 1, the median values of the CSF Aβ 1–42 concentration and the mean values of the Aβ 42/40 ratio resulted above the reference cut-off (< 650 pg/mL and < 0.05, respectively).

We found no significant differences among the three ALS subgroups (Aβ 1–42: $p = 0.685$; Aβ 1–40: $p = 0.340$; Aβ 42/40 ratio: $p = 0.426$).

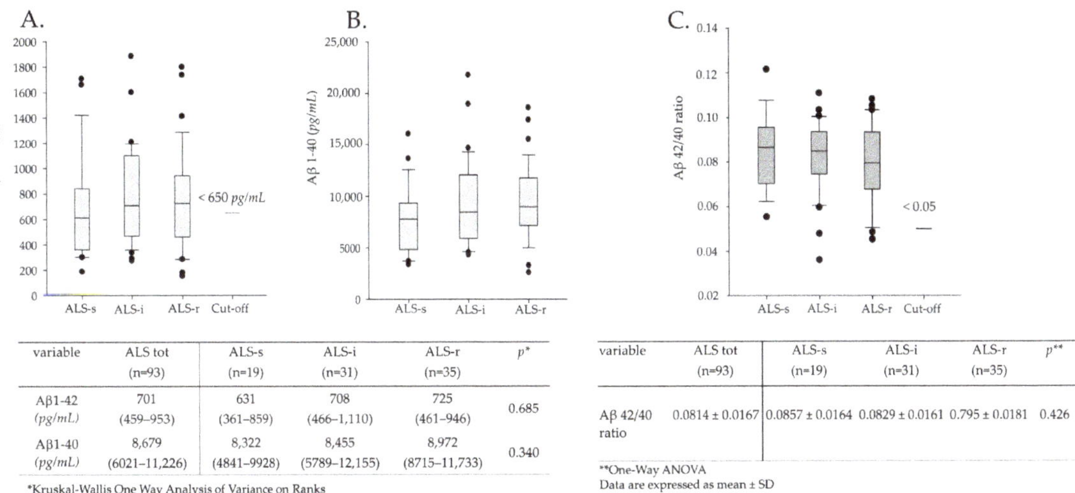

variable	ALS tot (n=93)	ALS-s (n=19)	ALS-i (n=31)	ALS-r (n=35)	p^*
Aβ1-42 (pg/mL)	701 (459–953)	631 (361–859)	708 (466–1,110)	725 (461–946)	0.685
Aβ1-40 (pg/mL)	8,679 (6021–11,226)	8,322 (4841–9928)	8,455 (5789–12,155)	8,972 (8715–11,733)	0.340

*Kruskal-Wallis One Way Analysis of Variance on Ranks
Data are expressed as median with interquartile range (IQR)

variable	ALS tot (n=93)	ALS-s (n=19)	ALS-i (n=31)	ALS-r (n=35)	p^{**}
Aβ 42/40 ratio	0.0814 ± 0.0167	0.0857 ± 0.0164	0.0829 ± 0.0161	0.795 ± 0.0181	0.426

**One-Way ANOVA
Data are expressed as mean ± SD

Figure 1. Cerebrospinal fluid (CSF) Aβ levels in ALS patients and in patients with slow (ALS-s), intermediate (ALS-i) and rapid (ALS-r) progression. (**A**) Aβ 1–42; (**B**) Aβ 1–40, (**C**) Aβ 42/40 ratio. Solid dots in A–C represent known outliers.

Spearman's correlation analyses for the CSF Aβ 1–42 and Aβ 1–40 levels showed no significant correlations (Table 3) while the Pearson's correlation analysis showed a significant correlation of Aβ 42/40 ratio values with the age at onset ($r^2 = -0.274$, $p = 0.008$) and MMSE scores ($r^2 = 0.396$, $p = 0.019$) (Table 4).

Table 3. Spearman's correlation of the CSF Aβ 1–42 and Aβ 1–40 with demographic, clinical and neuropsychological features of ALS patients.

	Aβ 1–42	**Aβ 1–40**
Age at onset (years)	$r = -0.041, p = 0.695$	$r = 0.312, p = 0.208$
Diagnostic delay (months)	$r = -0.140, p = 0.189$	$r = -0.163, p = 0.126$
Rate of progression (ΔFS)	$r = 0.008\ p = 0.936$	$r = 0.103, p = 0.347$
FVC (%)	$r = 0.141, p = 0.237$	$r = 0.116, p = 0.330$
FrSBe	$r = 0.125, p = 0.370$	$r = 0.185, p = 0.196$
MMSE	$r = -0.240, p = 0.146$	$r = -0.429, p = 0.007$
ECAS	$r = 0.005, p = 0.979$	$r = -0.049, p = 0.304$
Survival (months)	$r = 0.106, p = 0.933$	$r = -0.119, p = 0.350$

Frontal Systems Behavioral Scale (FrSBe). Mini Mental State Examination (MMSE). Edinburgh Cognitive and Behavioral ALS Screen (ECAS). Bold font indicates a statistical significance ($p < 0.05$).

Table 4. Pearson's correlation of the CSF Aβ 42/40 ratio with demographic, clinical and neuropsychological features of ALS patients.

Parameters	r^2	p
Age at onset (years)	−0.274	0.008
Diagnostic delay (months)	0.038	0.719
ΔFS	−0.086	0.432
FVC(%)	0.198	0.095
FrSBe	−0.076	0.695
MMSE	0.396	0.019
ECAS	0.054	0.792
Survival (months)	0.164	0.196

Bold font indicates a statistical significance ($p < 0.05$).

To verify if the CSF Aβ proteins could affect the survival of ALS patients, we stratified ALS patients according to the median values of Aβ 1–42, Aβ 1–40 and the Aβ 42/40 ratio, obtaining three subgroups for each analyzed protein: L-1-Q (i.e., patients with values lower than the first quartile), IQR (i.e., patients with values between the first and the third quartiles) and U-3-Q (i.e., patients with values upper than the third quartile). Only for the Aβ 42/40 ratio did the Kaplan–Meier analysis with a Holm–Sidak post-hoc test show that L-1-Q patients had a significantly shorter survival (27 (IQR: 17–41) months) in comparison with U-3-Q (39 (IQR: 26–60) months) (log-rank = 6.617; p = 0.037) (Figure 2).

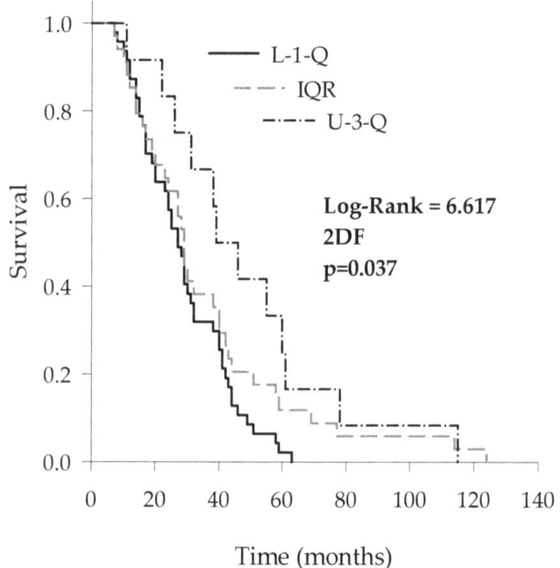

Figure 2. Kaplan–Meier survival curves of ALS patients stratified according to the median CSF values of the Aβ 42/40 ratio: lower than the first quartile (L-1-Q), interquartile range (IQR) and upper than the third quartile (U-3-Q).

Interestingly, patients in the L-1-Q showed a higher median age in comparison with other subgroups (L-1-Q: 71 (66.5–75.25); IQR: 66.5 (63–71.75); U-3-Q: 65.5 (61–70.5); p = 0.019).

Subsequently, we performed univariate and multivariate Cox regression analyses to test the predictor role of different demographic and clinical features of ALS patients and the CSF levels of Aβ 1–42, Aβ 1–40 and the Aβ 42/40 ratio. As shown in Table 5, at the univariate regression analysis, the age at onset (p = 0.001), diagnostic delay (p = 0.001), ΔFS at diagnosis (p < 0.001) and Aβ 42/40 ratio (p = 0.026) were significantly associated with overall survival. We then considered variables that were positively related to survival at the univariate analysis for the multivariate Cox regression analysis. As shown in Table 6, the diagnostic delay (p =0.025), ΔFS at diagnosis (p = 0.032) and Aβ 42/40 ratio (p = 0.015) were independent predictors of overall survival. Furthermore, the multivariate Cox regression analysis was performed to investigate the role of co-morbidities in overall survival but no significant data were obtained (Table S2).

Table 5. Univariate Cox regression analysis for the overall survival for ALS patients.

Parameters	b	± SE	p	HR	95% CI
Gender (M vs. F)	0.073	0.251	0.772	1.075	0.658–1.757
Age at onset	0.062	0.019	**0.001**	1.064	1.024–1.105
Site of onset (spinal vs. bulbar)	−0.443	0.271	0.102	0.642	0.378–1.092
Diagnostic delay	−0.038	0.012	**0.001**	0.963	0.000–0.986
FVC%	0.006	0.006	0.308	0.994	0.982–1.006
ΔFS at diagnosis	0.443	0.105	**< 0.001**	1.557	1.266–1.914
Aβ 1–42	0.000	0.000	0.862	1	0.999–1.001
Aβ 1–40	0.000	0.000	0.275	1	1–1
Aβ 42/40 ratio	−18.137	8.164	**0.026**	1.33×10^{-8}	0.000–0.118

b = regression coefficient; SE = standard error; HR = hazard ratio; CI = confidence interval. Bold font indicates a statistical significance ($p < 0.05$).

Table 6. Multivariate Cox regression analysis for overall survival for ALS patients. Significative variables believed to be significant at the univariate analysis were considered for multivariate analysis.

Parameters	b	± SE	p	HR	95% CI
Age at onset	0.038	0.022	0.08	1.038	0.994–1.085
Diagnostic delay	−0.032	0.014	**0.025**	0.968	0.000–0.996
ΔFS at diagnosis	0.301	0.140	**0.032**	1.351	1.026–1.779
Aβ 42/40 ratio	−20.662	8.504	**0.015**	1.6×10^{-9}	0.000–0.018

b = regression coefficient; SE = standard error; HR = hazard ratio; CI = confidence interval. Bold font indicates a statistical significance ($p < 0.05$).

4. Discussion

Our study was aimed at exploring the potential role of Aβ as a prognostic biomarker in ALS. For this purpose, we designed a retrospective observational study that included 93 patients. The CSF Aβ 1–42 and Aβ 1–40 levels and the Aβ 42/40 ratio were determined and correlated with demographic, clinical and neuropsychological features of ALS patients.

In recent years, CSF Aβ levels have been investigated to define their role as potential diagnostic and prognostic biomarkers for ALS and many studies in this field have been reported. However, the two largest studies about this topic found contrasting results. On one hand, higher CSF Aβ 1–42 levels were associated with a poorer prognosis [25] while, on the other hand, an interesting correlation of a higher concentration in patients with better performance was found, reporting increased CSF levels compared with a control group [26].

Even though the CSF Aβ and especially the Aβ 42/40 ratio represent a specific diagnostic biomarker for AD, the idea of shared mechanisms among different neurodegenerative disorders has led many authors to investigate the role of Aβ as a potential modulator of their rate of progression and overall survival. The CSF Aβ 1–42 levels were correlated to conversion from mild cognitive impairment to dementia and the progression of cognitive deficits in AD [27] as well as with the progression of cognitive impairments in Parkinson's disease (PD) [28]. Indeed, lower CSF Aβ 1–42 levels are related to a progressive deposition of Aβ in senile plaques at the cortical level [29]. An intracellular deposition of Aβ 1–42 was also detected in the anterior horn of motor neurons of patients affected by motor neuron disease (MND) [13] while extracellular aggregates of Aβ 1–42 were detected in the hippocampus of ALS and ALS-FTD patients [17]. Studies on murine models of ALS (i.e., SOD1 G93A mice) correlated the overexpression of Aβ with an earlier onset of motor symptoms [30]. Furthermore, few cases of co-morbidity between ALS and AD in a patient showing an overlapping clinical picture have been reported [15,25].

In our study, ALS patients were subdivided into three subgroups (i.e., ALS-s, ALS-i and ALS-r) to analyze the contribution of the CSF Aβ levels on the rate of progression. No statistically significant differences were detected among three subgroups for the CSF Aβ 1–42, Aβ 1–40 and the Aβ 1–42/40 ratio. Indeed, no statistically significant correlation between the Aβ and clinical features including the rate of progression was found. We found a significant correlation between the Aβ 42/40 ratio with the age at onset and MMSE scores.

When analyzing the contribution of the CSF Aβ 1–42, Aβ 1–40 and the Aβ 1–42/40 ratio on overall survival of ALS patients we found that patients with lower Aβ 42/40 ratio values showed a shorter survival in comparison with those with higher values. This finding was confirmed by univariate and multivariate Cox regression analyses, which showed that the Aβ 42/40 ratio could act as an independent predictor for overall survival for ALS patients. A decrease in the CSF Aβ 42/40 ratio values could be indicative of a decrease of the CSF Aβ 1–42 levels because it might deposit in different districts of the central nervous system (CNS) as previously described [29]. In ALS, the presence of intracellular or extracellular aggregates of Aβ 1–42 is probably related to an accumulation of APP following neuronal injury. This accumulation could be due to an impairment of axonplasmatic transport or enhanced biosynthesis of APP, representing an early neuroprotective phase to contrast extracellular and intracellular stresses. As the neuronal injuries continue, a shift toward a neurotoxic phase can occur. APP could be subjected to cleavage in Aβ by alternative mechanisms: caspase 3 giving rise to intracellular aggregates, an accumulation of which gives an increase in oxidative stress, while β-secretase contributes to extracellular deposition [31]. All of these mechanisms may contribute to a decrease of the CSF Aβ 1–42. Studies on murine models of ALS correlated the production of Aβ by β-secretase and consequently deposition as a key event that could improve motor functions and survival [32]. For this purpose, those authors treated asymptomatic and symptomatic SOD-1 G93A mice with a monoclonal antibody able to interfere with β-secretase activity, avoiding the formation of intracellular or extracellular Aβ aggregates: treated asymptomatic ALS mice showed a delay of the onset of symptoms, motor failure and death; however, the same effects were not obtained in treated symptomatic ALS mice.

Another interesting result that we obtained was related to the evidence that ALS patients with lower Aβ 42/40 ratio values presented a higher age at onset than those with higher values. These data were enforced with the finding that there was a statistically significant correlation of Aβ 42/40 ratio values with the age at onset. Considering that the age at onset was considered a strong prognostic factor for ALS [33] and that in cognitively normal subjects the concentration of the CSF AD biomarkers, including the Aβ 42/40 ratio, is associated with age [34], the decrease of the Aβ 42/40 ratio in the CSF of ALS patients might indicate that the triggering of the Aβ cascade could represent an early event that leads to an asymptomatic form of dementia that fails to fully become symptomatic as death occurs. Thus, the coexistence of an elevated age at onset and a low CSF Aβ 42/40 could represent a more severe prognostic condition that could influence survival time. The evidence that ALS patients with a low CSF Aβ 42/40 ratio had a high median age at onset make us speculate that the intracellular or extracellular deposition of Aβ would accelerate the course of the disease, worsening their survival. Indeed, we cannot exclude that a few patients in our population could have a preclinical condition of AD or the presence of age-related amyloid deposition.

Our study had a few limitations. First, the sample size. We studied 96 patients but, for our analyses, we stratified these into three groups resulting in quite small numbers. Related to that is the difference of the age of onset among groups that could partially reduce the significance of our results. Another limitation was the lack of a follow-up for the cognitive formal evaluation. This could have been useful for a correlation with the clinical progression but we found no correlation at the baseline and none of our patients developed clinically significant dementia. Finally, the lack of a control population represented a further limitation. Our goal was to assess the role of the CSF Aβ 1–42, Aβ 1–40 and the Aβ 42/40 ratio in the clinical progression of patients affected by ALS and a comparison with a control group could have enriched our work but we did not consider this to be mandatory.

5. Conclusion

In our study, we aimed to evaluate the potential role of Aβ in predicting the prognosis in ALS patients. We found that Aβ 42/40 is an independent predictor for survival and could be proposed as a potential prognostic biomarker as suggested by previous reports. Further studies are needed to confirm our findings in a larger population but we consider that we have added a piece to the understanding and management of the disease.

Supplementary Materials: The following are available online at https://www.mdpi.com/2076-342 5/11/3/302/s1, Table S1: Neuropsychologic assessment of ALS patients by FrSBe, MMSE and ECAS at the time of diagnosis (baseline). Table S2: Multivariate Cox regression analysis for overall survival for ALS patients considering their co-morbidities.

Author Contributions: T.C. collected samples, performed statistical analyses, reviewed the data and drafted the manuscript; L.A. made the assays, reviewed the data and the manuscript; R.S. performed clinical evaluations, reviewed the data and the manuscript; L.G. performed neuropsychologic evaluations, reviewed the data and the manuscript; A.N. collected samples, reviewed the data; V.B. performed neuropsychologic evaluations, reviewed the data and the manuscript; F.G. performed clinical evaluations, reviewed the data and the manuscript; B.L.S. made the assays, reviewed the data and the manuscript; C.M.G. made the assays, reviewed the data and the manuscript; R.V.G. made the assays, reviewed the data and the manuscript; G.B. made the assays, reviewed the data and the manuscript; V.L.B. designed the study, performed clinical evaluations and reviewed the data and the manuscript; M.C. designed the study, wrote the original draft, reviewed the data and the manuscript; T.P. designed the study, performed statistical analyses, wrote the original draft, reviewed the data and the manuscript. All authors have read and agreed to the published version of the manuscript.

Funding: This research received no external funding.

Institutional Review Board Statement: The study was conducted according to the guidelines of the Declaration of Helsinki and approved by the Ethics Committee of University Hospital "Paolo Giaccone" (Institutional Ethic Committee Palermo 1 n° 07/2017).

Informed Consent Statement: Informed consent was obtained from all subjects involved in the study. The informed consent contained a statement that the biological material may also be used for research purposes.

Data Availability Statement: The data presented in this study are available on request from the corresponding author. The data are not publicly available due to current privacy laws.

Acknowledgments: This study performed at University Hospital of Palermo is the result of transdisciplinary collaboration of ALS Research and Clinical Center of Palermo, Institute of Clinical Biochemistry, Clinical Molecular Medicine and Laboratory Medicine and IRCSS Bonino Pulejo. The Authors feel grateful for such and lively collaboration.

Conflicts of Interest: The authors declare no conflict of interest.

References

1. Van Es, M.A.; Hardiman, O. Amyotrophic lateral sclerosis. *Lancet* **2017**, *390*, 2084–2098. [CrossRef]
2. Brooks, B.R.; Miller, R.G. El Escorial revisited: Revised criteria for the diagnosis of amyotrophic lateral sclerosis. *Amyotroph. Lateral. Scler. Other Motor. Neuron. Disord.* **2000**, *1*, 293–299. [CrossRef] [PubMed]
3. Julien, J.P. A role for neurofilaments in the pathogenesis of amyotrophic lateral sclerosis. *Biochem. Cell Biol.* **1995**, *73*, 593–597. [CrossRef]
4. Didonna, A.; Opal, P. The role of neurofilament aggregation in neurodegeneration: Lessons from rare inherited neurological disorders. *Mol. Neurodegener.* **2019**, *14*, 19. [CrossRef]
5. Steinacker, P.; Feneberg, E. Neurofilaments in the diagnosis of motoneuron diseases: A prospective study on 455 patients. *J. Neurol. Neurosurg. Psychiatry* **2016**, *87*, 12–20. [CrossRef]
6. Rossi, D.; Volanti, P. CSF neurofilament proteins as diagnostic and prognostic biomarkers for amyotrophic lateral sclerosis. *J. Neurol.* **2018**, *265*, 510–521. [CrossRef]
7. Paladino, P.; Valentino, F. Cerebrospinal fluid tau protein is not a biological marker in amyotrophic lateral sclerosis. *Eur. J. Neur.* **2009**, *16*, 257–261. [CrossRef]
8. Grossman, M.; Elman, L.; McCluskey, L. Phosphorylated tau as a candidate biomarker for amyotrophic lateral sclerosis. *JAMA Neurol.* **2014**, *71*, 442–448. [CrossRef] [PubMed]

9. Wilke, C.; Deuschle, C. Total tau is increased, but phosphorylated tau not decreased, in cerebrospinal fluid in amyotrophic lateral sclerosis. *Neurobiol. Aging* **2015**, *36*, 1072–1074. [CrossRef] [PubMed]

10. Burrell, J.R.; Halliday, G.M. The frontotemporal dementia-motor neuron disease continuum. *Lancet* **2016**, *388*, 919–931. [CrossRef]

11. DeJesus-Hernandez, M.; Mackenzie, I.R. Expanded GGGGCC hexanucleotide repeat in noncoding region of C9ORF72 causes chromosome 9p-linked FTD and ALS. *Neuron* **2011**, *72*, 245–256. [CrossRef]

12. Renton, A.E.; Majounie, E. A hexanucleotide repeat expansion in C9ORF72 is the cause of chromosome 9p21-linked ALS-FTD. *Neuron* **2011**, *72*, 257–268. [CrossRef] [PubMed]

13. Calingasan, N.Y.; Chen, J. B-Amyloid 42 Accumulation in the Lumbar Spinal Cord Motor Neurons of Amyotrophic Lateral Sclerosis Patients. *Neurobiol. Dis.* **2005**, *19*, 340–347. [CrossRef]

14. Farid, K.; Carter, S.F.; Rodriguez-Vieitez, E. Case Report of Complex Amyotrophic Lateral Sclerosis with Cognitive Impairment and Cortical Amyloid Deposition. *JAD* **2015**, *47*, 661–667. [CrossRef] [PubMed]

15. Sasaki, S.; Iwata, M. Immunoreactivity of beta-amyloid precursor protein in amyotrophic lateral sclerosis. *Acta Neuropathol.* **1999**, *97*, 463–468. [CrossRef]

16. Gómez-Pinedo, U.; Villar-Quiles, R.N. Immununochemical Markers of the Amyloid Cascade in the Hippocampus in Motor Neuron Diseases. *Front Neurol.* **2016**, *7*, 195. [CrossRef]

17. Hardy, J. Alzheimer's disease: The amyloid cascade hypothesis—An update and reappraisal. *JAD* **2009**, *9*, 151–153. [CrossRef]

18. Jack, C.R.; Bennett, D.A. NIA-AA Research Framework: Toward a biological definition of Alzheimer's disease. *Alzheimers Dement.* **2018**, *14*, 535–562. [CrossRef]

19. Albert, M.S.; DeKosky, S.T. The diagnosis of mild cognitive impairment due to Alzheimer's disease: Recom- mendations from the National Institute on Aging-Alzheimer's Association workgroups on diagnostic guidelines for Alzheimer's disease. *Alzheimers Dement.* **2011**, *7*, 270–279. [CrossRef]

20. McKhann, G.M.; Knopman, D.S. The diagnosis of dementia due to Alzheimer's disease: Recommendations from the National Institute on Aging-Alzheimer's Association work- groups on diagnostic guidelines for Alzheimer's disease. *Alzheimers Dement.* **2011**, *7*, 263–269. [CrossRef] [PubMed]

21. Eun, J.Y.; Hyo-Jin, P. Intracellular amyloid beta interacts with SOD1 and impairs the enzymatic activity of SOD1: Implications for the pathogenesis of amyotrophic lateral sclerosis. *Exp. Mol. Med.* **2009**, *41*, 611–617.

22. Stanga, S.; Brambilla, L. A role for GDNF and soluble APP as biomarkers of amyotrophic lateral sclerosis pathophysiology. *Front. Neurol.* **2018**, *9*, 1–9. [CrossRef]

23. Steinacker, P.; Hendrich, C. Concentrations of beta-amyloid precursor protein processing products in cerebrospinal fluid of patients with amyotrophic lateral sclerosis and frontotemporal lobar degeneration. *J. Neural. Transm.* **2009**, *116*, 1169–1178. [CrossRef] [PubMed]

24. Rusina, R.; Ridzon, P. Relationship between ALS and the degree of cognitive impairment, markers of neurodegeneration and predictors for poor outcome. A prospective study. *Eur. J. Neurol.* **2010**, *17*, 23–30. [CrossRef]

25. Lanznaster, D.; Hergesheimer, R.C. Aβ1-42 and Tau as Potential Biomarkers for Diagnosis and Prognosis of Amyotrophic Lateral Sclerosis. *Int. J. Mol. Sci.* **2020**, *21*, 2911. [CrossRef]

26. Hadjichrysanthou, C.; Evans, S. The dynamics of biomarkers across the clinical spectrum of Alzheimer's disease. *Alzheimers Res. Ther.* **2020**, *12*, 74. [CrossRef]

27. Hall, S.; Surova, Y. CSF biomarkers and clinical progression of Parkinson disease. *Neurology* **2015**, *84*, 57–63. [CrossRef]

28. Hansson, O.; Seibyl, J. CSF biomarkers of Alzheimer's disease concord with amyloid-β PET and predict clinical progression: A study of fully automated immunoassays in BioFINDER and ADNI cohorts. *Alzheimers Dement* **1470**, *14*, 1470–1481. [CrossRef]

29. Li, Q.X.; Mok, S.S. Overexpression of Abeta is associated with acceleration of onset of motor impairment and superoxide dismutase 1 aggregation in an amyotrophic lateral sclerosis mouse model. *Aging Cell.* **2006**, *5*, 153–165. [CrossRef] [PubMed]

30. Sjögren, M.; Davidsson, P. Decreased CSF-beta-amyloid 42 in Alzheimer's disease and amyotrophic lateral sclerosis may reflect mismetabolism of beta-amyloid induced by disparate mechanisms. *Dement. Geriatr. Cogn. Disord.* **2002**, *13*, 112–118. [CrossRef] [PubMed]

31. Rabinovich-Toidman, P.; Becker, M. Inhibition of amyloid precursor protein beta-secretase cleavage site affects survival and motor functions of amyotrophic lateral sclerosis transgenic mice. *Neurodegener. Dis.* **2012**, *10*, 30–33. [CrossRef] [PubMed]

32. Chiò, A.; Logroscino, G. Prognostic factors in ALS: A critical review. *Amyotroph. Lateral. Scler.* **2009**, *10*, 310–323. [CrossRef] [PubMed]

33. Paternicò, D.; Galluzzi, S. Cerebrospinal fluid markers for Alzheimer's disease in a cognitively healthy cohort of young and old adults. *Alzheimer's Dement.* **2012**, *8*, 520–527. [CrossRef] [PubMed]

34. Rodrigue, K.M.; Kennedy, K.M. Beta-amyloid deposition and the aging brain. *Neuropsychol. Rev.* **2009**, *19*, 436–450. [CrossRef] [PubMed]

Review

COVID-19 and Alzheimer's Disease

Marcello Ciaccio [1,2,*], Bruna Lo Sasso [1,2], Concetta Scazzone [1], Caterina Maria Gambino [1], Anna Maria Ciaccio [3], Giulia Bivona [1], Tommaso Piccoli [4], Rosaria Vincenza Giglio [1,†] and Luisa Agnello [1,†]

1   Institute of Clinical Biochemistry, Clinical Molecular Medicine and Laboratory Medicine, Department of Biomedicine, Neurosciences and Advanced Diagnostics, University of Palermo, 90127 Palermo, Italy; bruna.losasso@unipa.it (B.L.S.); concetta.scazzone@unipa.it (C.S.); cmgambino@libero.it (C.M.G.); giulia.bivona@unipa.it (G.B.); rosaria.vincenza.giglio@alice.it (R.V.G.); luisa.agnello@unipa.it (L.A.)
2   Department of Laboratory Medicine, University Hospital "P. Giaccone", 90127 Palermo, Italy
3   Unit of Clinical Biochemistry, University of Palermo, 90127 Palermo, Italy; amciaccio21@gmail.com
4   Unit of Neurology, Department of Biomedicine, Neurosciences and Advanced Diagnostics, University of Palermo, 90127 Palermo, Italy; tommaso.piccoli@unipa.it
*   Correspondence: marcello.ciaccio@unipa.it; Tel.: +39-0916-553-296
†   Both authors contributed equally to this work.

Abstract: The Severe Acute Respiratory Syndrome Coronavirus 2 (SARS-CoV-2) is a neurotropic virus with a high neuroinvasive potential. Indeed, more than one-third of patients develop neurological symptoms, including confusion, headache, and hypogeusia/ageusia. However, long-term neurological consequences have received little interest compared to respiratory, cardiovascular, and renal manifestations. Several mechanisms have been proposed to explain the potential SARS-CoV-2 neurological injury that could lead to the development of neurodegenerative diseases, including Alzheimer's Disease (AD). A mutualistic relationship between AD and COVID-19 seems to exist. On the one hand, COVID-19 patients seem to be more prone to developing AD. On the other hand, AD patients could be more susceptible to severe COVID-19. In this review, we sought to provide an overview on the relationship between AD and COVID-19, focusing on the potential role of biomarkers, which could represent precious tool for early identification of COVID-19 patients at high risk of developing AD.

Keywords: AD; biomarkers; SARS-CoV-2; neuroinflammation; neurodegenerative nisease; nervous system

Citation: Ciaccio, M.; Lo Sasso, B.; Scazzone, C.; Gambino, C.M.; Ciaccio, A.M.; Bivona, G.; Piccoli, T.; Giglio, R.V.; Agnello, L. COVID-19 and Alzheimer's Disease. *Brain Sci.* **2021**, *11*, 305. https://doi.org/10.3390/brainsci11030305

Academic Editor: Fernando Aguado

Received: 5 February 2021
Accepted: 24 February 2021
Published: 27 February 2021

Publisher's Note: MDPI stays neutral with regard to jurisdictional claims in published maps and institutional affiliations.

1. Introduction

The Severe Acute Respiratory Syndrome Coronavirus 2 (SARS-CoV-2) is the pathogen responsible for COVID-19 disease, which is characterized by a wide spectrum of symptoms, from fever and cough to multiple organ dysfunctions [1]. Additionally, SARS-CoV-2 can induce, directly or indirectly, several complications involving different organs [2,3]. Nowadays, the clinical course of the infection is unpredictable and characterized by high inter-individual variability. However, more than 80% of COVID-19 patients present ageusia or anosmia, which occurs early during the infection and represent pathognomonic features of the disease [4].

SARS-CoV-2, as well as all members of the human coronaviruses (CoVs) family, is an opportunistic pathogen of the central nervous system (CNS) [5]. The neurological signs and symptoms associated with SARS-CoV-2 infection, such as confusion, headache, hypogeusia/ageusia, hyposmia/anosmia, dizziness, epilepsy, acute cerebrovascular disease [4], are caused by the direct invasion of the virus into the CNS, and the subsequent interaction between SARS-CoV-2 spike protein and the angiotensin-converting enzyme 2 (ACE2) [6–8]. Post-mortem studies revealed the presence of both SARS-CoV-2 antigen and RNA in the brain tissue of COVID-19 patients [9].

ACE-2 expression is a key determinant of viral tropism and COVID-19 pathogenesis. In the brain, ACE-2 is expressed both on neurons and glial cells as well as on endothelial and arterial smooth muscle cells. ACE-2 is also expressed on the temporal lobe and hippocampus, which represent cerebral regions involved in the pathogenesis of Alzheimer's Disease (AD) [6].

It has been hypothesized that SARS-CoV-2 could cause damage in the CNS by direct neurotoxicity or indirectly through the activation of the host immune response, which could lead to demyelination, neurodegeneration and cellular senescence. Thus, it could accelerate brain aging favoring the development of neurodegenerative diseases, including dementia [10]. However, after the acute recovery phase, the long-term consequences of SARS-CoV-2 infection on accelerated aging and age-related neurodegenerative disorders are actually unknown. Noteworthily, SARS-CoV-2 could potentially induce a worsening cognitive decline in AD patients.

On the other hand, dementia could represent an important risk factor for COVID-19 severity and mortality, as shown by preliminary reports [11,12]. Thus, a mutualistic relationship between SARS-CoV-2 infection and AD can be hypothesized. Figure 1 shows the possible association between AD and SARS-CoV-2 infection by summarizing the possible underlying mechanisms, which are described in the next paragraphs.

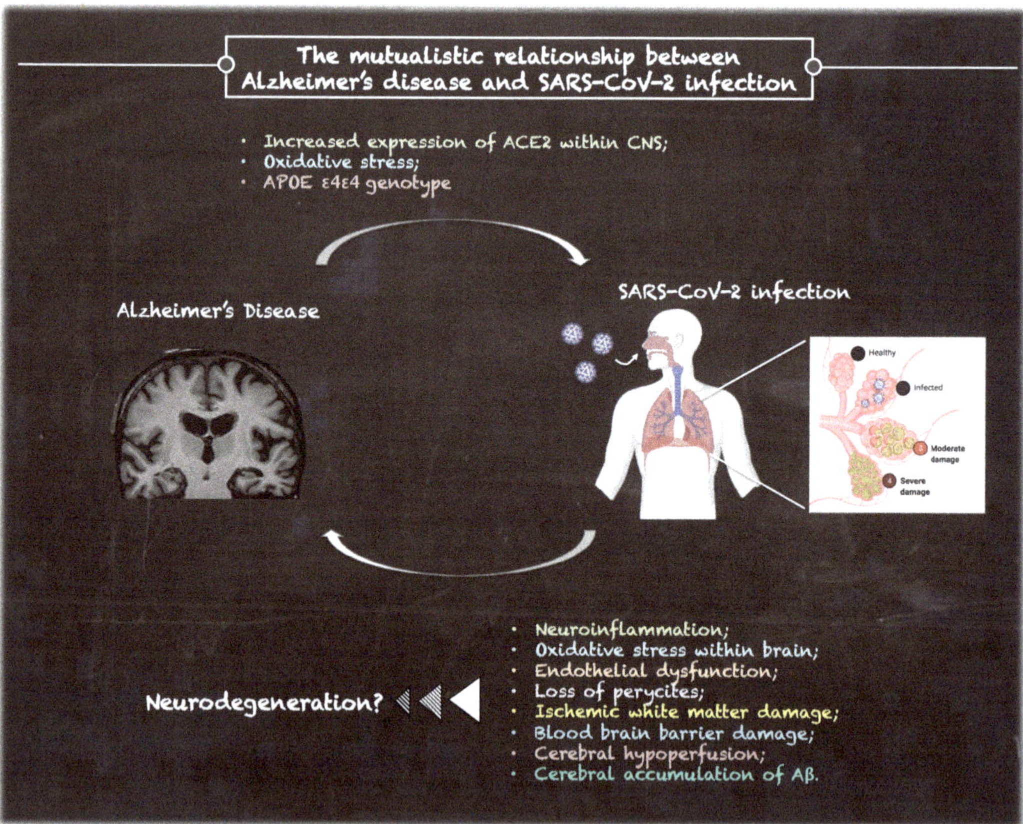

Figure 1. The complex relationship between Alzheimer's Disease and SARS-CoV-2 infection.

In this review, we sought to provide an overview on the relationship between AD and COVID-19, focusing on the potential role of biomarkers. This should represent a starting point for further investigations.

2. COVID-19 in Alzheimer's Disease

AD represents the most common form of dementia worldwide [13]. The term dementia refers to a wide spectrum of disorders characterized by global, chronic and generally irreversible cognitive deterioration, leading to the progressive alteration of several functions such as memory, the ability to orient oneself, and alterations of the personality and behavior, which compromise the autonomy of the subject in the daily life [13,14]. The incidence of dementia is increasing in the general population. Indeed, the World Health Organization and Alzheimer Disease International Report of 2016 defined it as a global public health priority [15]. Patients with dementia are frail, dependent on caregivers for daily living activities and need the support of several services resources, such as physical exercise and physiotherapy [16]. Thus, the measures introduced by government authorities during the current COVID-19 pandemic, including confinement and isolation, may exacerbate the cognitive decline. Additionally, patients with AD and mild dementia may either be unwilling or unable to follow recommendations from public health authorities such as sanitize their hands, cover their mouth and nose when coughing, maintain physical distance from others, in part due to the severity of their short-term memory loss and overall cognitive impairment [17].

The brain of AD patients is characterized by amyloid plaque deposition and the presence of neurofibrillary tangles, which induce neuronal damage and synapse loss as well as oligodendroglia degeneration and myelin impairment [18].

Post-mortem studies showed that ACE-2 expression is increased in the brain of AD patients in comparison to controls [19]. Additionally, genome-wide association studies (GWAS) showed that the expression of ACE-2 gene is elevated in the brain tissue of AD patients with increased levels in severe forms [20]. Thus, enhanced ACE-2 expression could represent a risk factor for COVID-19 transmission in AD patients. It has been postulated a direct link between AD and ACE-2 expression mediated by oxidative stress. Specifically, aging leads to the imbalance in the redox state, characterized by the generation of excess reactive oxygen species (ROS) or the dysfunction of the antioxidant system, leading to oxidative stress [21]. AD patients show a significant extent of intracerebral oxidative damage associated with the abnormal marked accumulation of Aβ and the deposition of neurofibrillary tangles [21]. Interestingly, ACE2 inhibitors have recently been suggested as potential treatment for neurodegenerative diseases, including AD [22].

Noteworthily, AD and COVID-19 share several risk factors and comorbidities, such as age, gender, hypertension, diabetes and APOE ε4 expression. Such evidence could in part explain the increased prevalence of SARS-CoV-2 infection in AD patients. However, further studies are mandatory in order to clarify the pathophysiological mechanisms linking AD and COVID-19.

3. Patients with COVID-19 Could Develop AD?

Overall, CoVs can enter the CNS via different routes, including retrograde axonal transport via the olfactory and enteric neurons or infected lymphocytes, which cross the disrupted blood-brain barrier (BBB) [23].

Aging is characterized by a gradual loss of the BBB integrity [24]. Thus, the elderly could be more susceptible to neuroinvasion during SARS-CoV-2 infection.

SARS CoV-2 infects the olfactory neurons and, through the neuro-epithelium of the olfactory mucosa, reaches the olfactory bulb in the hypothalamus [5,25]. The presence of SARS CoV-2 in the olfactory bulb leads to the activation of non-neuronal cells, such as mast cells, microglia, astrocytes, as well as to the tissue release of pro-inflammatory cytokines. SARS-CoV-2 uses the phospholipids of the infected cells to build its own envelope. The consequence is that the cells, in particular the innate immune cells, lose precursors for the synthesis of the autacoid local injury antagonist amides (ALIamides), which have a pivotal role for controlling the excessive reactivity [26]. Consequently, the resulting neuroinflammation could become uncontrollable, especially in the elderly, which have a less efficient immune system response [27,28]. Neuroinflammation, associated with intense oxidative stress, could induce neurodegeneration, potentially favoring the development of neurodegenerative diseases, such as AD [25,29]. COVID-19 patients with advanced age and comorbidities with an inflammatory basis, such as diabetes, atherosclerosis and sub-clinical dementia, could be at increased risk of developing AD.

Several pathological mechanisms seem to be involved in the potential increased risk of developing AD in COVID-19 patients.

A growing body of evidence suggested a role for neuroinflammation. Systemic inflammation induces the activation of microglia and astrocytes, which in turn secrete pro-inflammatory cytokines, including IL-1β, IL-6, IL-12, TNF-α. Such biomarkers could be involved in the synaptic dysfunction, inducing neurodegeneration, which could potentially lead to AD [30].

Hypoxic alterations and demyelinating lesions have been described in COVID-19 patients [31–33]. Neuroradiological studies showed alterations of functional brain integrity, especially in the hippocampus, in recovered COVID-19 patients at 3-month follow-up. The hippocampus is an area particularly vulnerable to respiratory viral infections, as shown in experimental studies [34]. Hippocampal atrophy is associated with cognitive decline and represents a common characteristic of AD patients [35,36]. Additionally, the altered BBB could allow the infiltration of immune cells, which may contribute to cognitive decline and dementia in COVID-19 patients. Moreover, endothelial dysfunction, which is a pathognomonic characteristic of COVID-19, and loss of pericytes could impair the clearance of cerebral metabolites, including Aβ peptides. The excess and accumulation of Aβ protein in senile plaques, especially in the hippocampus, represent the main pathophysiological mechanism underlying the AD. Some authors showed that severe COVID-19 presents ischemic white matter damage due to the reduced perfusion secondary to hypercoagulability and disseminated intravascular coagulation (DIC), which are common features of severe COVID-19. Neuroimaging and experimental studies showed that ischemic white matter damage occurs at a very early stage of AD, accelerates the progression of the disease and contributes to cognitive decline [37,38]. Moreover, cerebral hypoperfusion can increase the phosphorylation rate of tau [39].

In severe COVID-19, the systemic inflammation characterized by the so-called "cytokine storm" leads to the disruption of the blood–brain barrier and neural and glial cell damage that could be involved in long-term sequelae. Systemic inflammation is recognized as a pathophysiological mechanism underlying AD [40]. Also, pro-inflammatory cytokines alter the capacity of the microglial cells to phagocyte b-amyloid, promoting the accumulation of amyloid plaques [41]. The virus-induced systemic inflammatory storm, associated with a massive release of mediators able to access the CNS due to the increased permeability of the blood-brain barrier, could amplify neuroinflammation and contribute to the neurodegeneration process [42].

Another interesting piece of evidence suggests that the potential increase of AD risk in COVID-19 patients could be related to Aβ, which can act as an antimicrobial peptide. Thus, it could be postulated that the SARS-CoV-2 neuroinvasion could promote Aβ generation, as part of the immune response, and the b-amyloid cascade leading to b-amyloid deposition [43]. However, this is only a hypothesis that must be proved.

McLoughlin et al. showed that COVID-19 hospitalized patients who developed delirium during their hospitalization, after 1-month discharge had lower cognitive scores [44]. However, the difficulty in performing neuropsychological assessment leads to a poor understanding of the neurological impact of SARS-CoV-2 infection. Overall, ARDS is associated with a high prevalence of long-term cognitive impairment in critically ill patients [45]. Specifically, mechanical ventilation, which is a standard therapy to maintain adequate gas exchange during ARDS, also in severe COVID-19 patients, could contribute to long-term cognitive impairment [46–49]. Experimental studies showed that short-term mechanical ventilation triggers the neuropathology of AD by promoting cerebral accumulation of the Aβ peptide, systemic and neurologic inflammation, and blood–brain barrier dysfunction [50].

The long-term complications of COVID-19 would be expected in the next 10–15 years. Nowadays, it is not possible to assess them because the pandemic started last year. However, in the future, it will be pivotal to evaluate the risk of long-term COVID-19 neurological sequelae, especially in the elderly and patients who developed severe forms.

The potential mechanisms involved in cognitive impairment in COVID-19 patients can be summarized as follows: (i) direct SARS-CoV-2 infection in the CNS; (ii) systemic hyper inflammatory response to SARS-CoV-2; (iii) cerebrovascular ischemia due to endothelial dysfunction; (iv) severe coagulopathy; v) mechanical ventilation due to ARDS or severe disease; (vi) peripheral organ dysfunction.

Table 1 summarizes the potential mechanisms linking SARS-CoV-2 infection and the development of AD.

Table 1. Potential mechanisms involved in Alzheimer's Disease (AD) risk in COVID-19 patients.

Pathway	Mechanisms	References
Aβ deposition	Aβ is an antimicrobial peptide produced in response to neural infection as part of the innate immune innate response	[18,21,43]
APOEε4	<ul><li>APOEε4 represents a risk factor for both COVID-19 and AD;</li><li>APOEε4 enhance the disruption of BBB;</li><li>APOEε4 has an important role in neuroinflammation, which contributes to the pathogenic mechanism underlying AD.</li></ul>	[51,52]
Neuroinflammation	<ul><li>ACE-2 is expressed in the cells of the CNS;</li><li>SARS-CoV-2 can infect cells within CNS;</li><li>Pro-inflammatory cytokines can enter the CNS by crossing the altered BBB;</li><li>Inflammatory response within CNS can alter cells within CNS leading to cognitive decline.</li></ul>	[25,27–29,42]
Microglia activation	<ul><li>SARS-CoV-2 infection can induce microglial activation leading to neuronal loss;</li><li>Microglia activation promotes oxidative stress within brain;</li><li>Increased NO levels are neurotoxic and promote AD development.</li></ul>	[30,41]

ACE-2: Angiotensin Converting Enzyme 2; AD: Alzheimer's Disease; APOE: Apolipoprotein E; BBB: Blood Brain Barrier; COVID-19: Coronavirus Disease 2019; CSN: Central Nervous System; NO: Nitric Oxide; SARS-CoV-2: Severe Acute Respiratory Syndrome Coronavirus 2.

4. Biomarkers of Cognitive Decline in COVID-19 Patients

4.1. Neuronal Injury

Biomarkers of neurodegeneration in the cerebrospinal fluid (CSF), such as tau proteins, neurofilament light chain protein (NfL), and glial fibrillary acidic protein (GFAp), are increased in COVID-19 patients and associated both with neurological symptoms and disease severity [53–57].

T-tau is a biomarker of neuronal death. Its levels are increased in several neurodegenerative diseases, including AD. Specifically, the biochemical diagnosis of AD relies on the detection of a CSF biomarker profile characterized by the decrease of amyloid beta 1-42

(Aβ 1-42), the ratio Aβ 1-42/1-40 and the increase of t-Tau and p-Tau levels [13]. Some Authors found that COVID-19 patients have an increase of CSF t-Tau levels suggesting the presence of neuronal damage. However, to date, levels of amyloid beta have never been investigated in such patients.

Among intermediate filaments expressed in cerebral cells, GFAp and Neurofilmanents have been evaluated in COVID-19 patients.

GFAP is highly expressed in astrocytes and represents a biomarker of astrocytic activation/injury [58]. AD is characterized by amyloid plaques surrounded by reactive astrocytes, which show an increased expression of intermediate filaments, including GFAP [58]. To date, only two studies evaluated the role of GFAp in COVID-19 patients [53,54]. The authors showed that severe COVID-19 patients had higher plasma concentrations of GFAp than controls.

Neurofilaments are cytoskeletal proteins of neurons, particularly abundant in axons. Neurofilaments comprise three subunits: neurofilament light chain (NF-L), neurofilament medium (NF-M) and neurofilament heavy (NF-H). Among these, NF-Ls are the most abundant.

Following axonal damage, NFs are released into CSF. Thus, they represent a biomarker of axonal damage and neuronal death. CSF NFs levels are increased in several neurological disorders, including AD [59]. Increased levels of serum and CSF NF-L have been found in severe COVID-19 patients [53–57,60].

Only one study evaluated t-Tau in COVID-19 patients and reported increased levels of t-Tau in severe cases.

To date, a few authors evaluated the CSF biochemical profile of COVID-19 patients due to the difficulty of obtaining such biological fluid. However, preliminary literature evidence raises awareness for potential long-term neurologic sequelae following COVID-19. Although severe COVID-19 patients have CSF biochemical alterations indicative of neuronal and axonal damage, it is not possible to draw definitive conclusions on the cognitive impairment. Longitudinal studies are required to evaluate the potential neurological sequelae and the risk of developing AD.

4.2. Genetic Variants

The most important known predisposing risk factor for AD is the polymorphism APOE ε4, with the ε4ε4 (homozygous) genotype being associated with a 14-fold increase in AD risk. Specifically, APOE ε4 is correlated with low cerebral blood flow and subcortical ischaemic white matter damage, as well as neuroinflammation in AD patients [51]. Kuo et al. showed that individuals carrying APOE ε4 in homozygous had a higher prevalence of SARS-CoV-2 infection. Additionally, APOE ε4ε4 allele was associated with an increased risk of developing severe COVID-19, independently of dementia, and other comorbidities, including cardiovascular disease, and type-2 diabetes [52]. Thus, APOE ε4 represents a common risk factor for AD and SARS-CoV-2 infection. APOE ε4 could promote vulnerability to viral infection and neurodegeneration. Thus, it can be postulated that the SARS-CoV-2 infection could be a promoting factor for neurodegeneration in individuals with susceptible genetic variants [52].

However, the relationships between APOE ε4, COVID-19, and AD must be elucidated.

4.3. Inflammatory Biomarkers

Some inflammatory biomarkers, including IL-6, IL-1, and galectin-3 (Gal-3), have been proposed as a link between COVID-19 and AD.

IL-6 represents one of the most studied cytokines in COVID-19. Circulating increased levels of IL-6 are associated with a high risk of developing severe COVID-19 and mortality. Accordingly, it represents a reliable prognostic biomarker in SARS-CoV-2 infection [61]. IL-6 is also a prognostic biomarker of AD. Indeed, its increased levels are associated with the progression of the disease and worse cognitive performance [62]. Thus, IL-6 represents a common biomarker for COVID-19 and AD.

IL-6 exerts its biological effects by the interaction with IL-6R, which can be expressed on the membrane of immune, epithelial and liver cells or it can be present in soluble form. The latter represents an agonist of IL-6. The complex IL-6/IL-6R can activate intracellular pathways involved in the immunoinflammatory response [62,63].

Alterations in IL-6 and IL-6R genes could be involved in the onset and progression of several diseases, including infectious diseases, such as COVID-19, and neurodegenerative diseases, such as AD [64–66]. The "Disease and Function analysis" performed by Strafella et al. showed that IL-6 and IL-6R could be involved in neuroinflammation, synaptic damage, microglia activation and cognitive impairment in AD pathogenesis [63].

Similar to IL-6, IL-1 represents a prognostic biomarker of SARS-CoV-2 infection, with increased levels associated with worse prognosis [67,68]. IL-1 is a pro-inflammatory cytokine produced by several cell types, including glia and neurons. IL-1 levels have been found to increase in the brain of AD [69]. In vitro studies reported that IL-1 could induce neuronal death by the direct effect on neurons or indirectly by glial production of neurotoxic substances. Additionally, IL-1 is involved in the physiological regulation of hippocampal plasticity and memory processes. Literature evidence showed that alterations of IL-1 levels, both positively (increase) and negatively (decrease), are associated with impaired memory functioning. Thus, the increased levels of IL-1 found in COVID-19 patients could enhance cognitive decline, leading to the development of AD [70].

Gal-3 is a carbohydrate-binding protein belonging to the family of lectins. It has pleiotropic functions, with a key role in several physiological and pathological processes, including inflammation and fibrosis [71–73]. Increased levels of Gal-3 have been found in severe COVID-19 patients. It has been postulated that Gal-3 promotes COVID-19 progression by supporting the hyper-inflammation reaction and lung fibrosis, which is associated with the acute phase of diffuse alveolar damage, edema, and hypoxia [74]. Increased levels of Gal-3 have also been described in the serum of AD patients [75]. Studies on AD animal models showed that Gal-3 could be involved in the Aβ aggregation and amyloid plaque formation [76]. Thus, it can be hypothesised that increased levels of Gal-3 in COVID-19 patients could also be involved in the damage leading to the development of AD. However, further studies are mandatory to confirm such a hypothesis.

5. Conclusions

Nowadays, the question "Can SARS-CoV-2 infection increase the risk for development of Alzheimer's Disease?" actually remains unanswered. There is an urgent need for prospective studies to address such question.

Neurological sequelae, including the cognitive impairment leading to AD, could represent an important complication of COVID-19. Further detailed clinical, laboratory, and neuropathological studies will help to elucidate the underlying pathophysiological mechanisms of the COVID-19 neurological complications. A longitudinal follow-up of COVID-19 patients, especially older adults and severe cases, is required to detect the potential long-term neurological consequences of SARS-CoV-2 infection. In such scenario, biomarkers represent reliable tools for early monitoring of COVID-19 patients and early detection of those at high risk of developing neurological sequelae, such as AD. Currently, there is still little literature evidence to draw definitive conclusions. However, an important relationship between AD and COVID-19 seems to exist.

Funding: This research received no external funding.

Conflicts of Interest: The authors declare no conflict of interest.

References

1. Wang, C.; Horby, P.W.; Hayden, F.G.; Gao, G.F. A novel coronavirus outbreak of global health concern. *Lancet* **2020**, *395*, 470–473. [CrossRef]
2. Menter, T.; Haslbauer, J.D.; Nienhold, R.; Savic, S.; Hopfer, H.; Deigendesch, N.; Frank, S.; Turek, D.; Willi, N.; Pargger, H.; et al. Postmortem examination of COVID-19 patients reveals diffuse alveolar damage with severe capillary congestion and variegated findings in lungs and other organs suggesting vascular dysfunction. *Histopathology* **2020**, *77*, 198–209. [CrossRef]

3. Ciaccio, M.; Agnello, L. Biochemical biomarkers alterations in Coronavirus Disease 2019 (COVID-19). *Diagnosis* **2020**, *7*, 365–372. [CrossRef] [PubMed]
4. Jiménez-Ruiz, A.; García-Grimshaw, M.; Ruiz-Sandoval, J.L. Neurological manifestations of COVID-19. *Gac. Med. Mex.* **2020**, *156*, 4.
5. Desforges, M.; Le Coupanec, A.; Dubeau, P.; Bourgouin, A.; Lajoie, L.; Dubé, M.; Talbot, P.J. Human Coronaviruses and Other Respiratory Viruses: Underestimated Opportunistic Pathogens of the Central Nervous System? *Viruses* **2019**, *12*, 14. [CrossRef]
6. Dong, M.; Zhang, J.; Ma, X.; Tan, J.; Chen, L.; Liu, S.; Xin, Y.; Zhuang, L. ACE2, TMPRSS2 distribution and extrapulmonary organ injury in patients with COVID-19. *Biomed. Pharmacother.* **2020**, *131*, 110678. [CrossRef] [PubMed]
7. Berger, J.R. COVID-19 and the nervous system. *J. Neurovirol.* **2020**, *23*, 1–6. [CrossRef] [PubMed]
8. Lechien, J.R.; Chiesa-Estomba, C.M.; De Siati, D.R.; Horoi, M.; Le Bon, S.D.; Rodriguez, A.; Dequanter, D.; Blecic, S.; El Afia, F.; Distinguin, L.; et al. Olfactory and gustatory dysfunctions as a clinical presentation of mild-to-moderate forms of the coronavirus disease (COVID-19): A multicenter European study. *Eur. Arch. Otorhinolaryngol.* **2020**, *6*, 1–11. [CrossRef]
9. Matschke, J.; Lütgehetmann, M.; Hagel, C.; Sperhake, J.P.; Schröder, A.S.; Edler, C.; Mushumba, H.; Fitzek, A.; Allweiss, L.; Dandri, M.; et al. Neuropathology of patients with COVID-19 in Germany: A post-mortem case series. *Lancet Neurol.* **2020**, *19*, 919–929. [CrossRef]
10. De Erausquin, G.A.; Snyder, H.; Carrillo, M.; Hosseini, A.A.; Brugha, T.S.; Seshadri, S.; CNS SARS-CoV-2 Consortium. The chronic neuropsychiatric sequelae of COVID-19: The need for a prospective study of viral impact on brain functioning. *Alzheimers Dement.* **2021**. [CrossRef]
11. Atkins, J.L.; Masoli, J.A.H.; Delgado, J.; Pilling, L.C.; Kuo, C.L.C.; Kuchel, G.; Melzer, D. Preexisting comorbidities predicting severe COVID-19 in older adults in the UK biobank community cohort. *medRxiv* **2020**. Available online: http://medrxiv.org/content/early/2020/05/08/2020.05.06.20092700.abstract (accessed on 4 May 2020).
12. Docherty, A.B.; Harrison, E.M.; Green, C.A.; Hardwick, H.; Pius, R.; Norman, L.; Holden, K.A.; Read, J.M.; Dondelinger, F.; Carsona, G.; et al. Features of 16,749 hospitalised UK patients with COVID-19 using the ISARIC WHO clinical characterisation protocol. *medRxiv* **2020**. Available online: http://medrxiv.org/content/early/2020/04/28/2020.04.23.20076042.abstract (accessed on 25 February 2021).
13. Agnello, L.; Piccoli, T.; Vidali, M.; Cuffaro, L.; Lo Sasso, B.; Iacolino, G.; Giglio, V.R.; Lupo, F.; Alongi, P.; Bivona, G.; et al. Diagnostic accuracy of cerebrospinal fluid biomarkers measured by chemiluminescent enzyme immunoassay for Alzheimer disease diagnosis. *Scand. J. Clin. Lab. Investig.* **2020**, *80*, 313–317. [CrossRef]
14. Lo Sasso, B.; Agnello, L.; Bivona, G.; Bellia, C.; Ciaccio, M. Cerebrospinal Fluid Analysis in Multiple Sclerosis Diagnosis: An Update. *Medicina* **2019**, *55*, 245. [CrossRef]
15. Available online: https://www.alzint.org/u/WorldAlzheimerReport2016.pdf (accessed on 25 February 2021).
16. Cagnin, A.; Di Lorenzo, R.; Marra, C.; Bonanni, L.; Cupidi, C.; Laganà, V.; Rubino, E.; Vacca, A.; Provero, P.; Isella, V.; et al. Behavioral and Psychological Effects of Coronavirus Disease-19 Quarantine in Patients With Dementia. *Front. Psychiatry* **2020**, *11*, 578015. [CrossRef]
17. Brown, E.E.; Kumar, S.; Rajji, T.K.; Pollock, B.G.; Mulsant, B.H. Anticipating and Mitigating the Impact of the COVID-19 Pandemic on Alzheimer's Disease and Related Dementias. *Am. J. Geriatr. Psychiatry* **2020**, *28*, 712–721. [CrossRef]
18. Papuć, E.; Rejdak, K. The role of myelin damage in Alzheimer's disease pathology. *Arch. Med. Sci.* **2018**, *16*, 345–351. [CrossRef] [PubMed]
19. Ding, Q.; Shults, N.V.; Harris, B.T.; Suzuki, Y.J. Angiotensin-converting enzyme 2 (ACE2) is upregulated in Alzheimer's disease brain. *reprinted in bioRxiv* **2020**, *8*. [CrossRef]
20. Lim, K.H.; Yang, S.; Kim, S.H.; Joo, J.Y. Elevation of ACE2 as a SARS-CoV-2 entry receptor gene expression in Alzheimer's disease. *J. Infect.* **2020**, *81*, e33–e34. [CrossRef] [PubMed]
21. Huang, W.J.; Zhang, X.; Chen, W.W. Role of oxidative stress in Alzheimer's disease. *Biomed. Rep.* **2016**, *4*, 519–522. [CrossRef]
22. Kaur, P.; Muthuraman, A.; Kaur, M. The implications of angiotensin-converting enzymes and their modulators in neurodegenerative disorders: Current and future perspectives. *ACS Chem. Neurosci.* **2015**, *6*, 508–521. [CrossRef] [PubMed]
23. Hascup, E.R.; Hascup, K.N. Does SARS-CoV-2 infection cause chronic neurological complications? *Geroscience* **2020**, *42*, 1083–1087. [CrossRef] [PubMed]
24. Montagne, A.; Barnes, S.R.; Sweeney, M.D.; Halliday, M.R.; Sagare, A.P.; Zhao, Z.; Toga, A.W.; Jacobs, R.E.; Liu, C.Y.; Amezcua, L.; et al. Blood-brain barrier breakdown in the aging human hippocampus. *Neuron* **2015**, *85*, 296–302. [CrossRef] [PubMed]
25. Steardo, L., Jr.; Zorec, R.; Verkhratsky, A. Neuroinfection may contribute to pathophysiology and clinical manifestations of COVID-19. *Acta Physiol.* **2020**, *229*, e13473. [CrossRef] [PubMed]
26. Yan, B.; Freiwald, T.; Chauss, D.; Wang, L.; West, E.; Bibby, J. SARS-CoV2 drives JAK1/2-dependent local and systemic complement hyper-activation. *Res. Sq.* **2020**, *9*, rs.3.rs-33390.
27. Kritas, S.K.; Ronconi, G.; Caraffa, A.; Gallenga, C.E.; Ross, R.; Conti, P. Mast cells contribute to coronavirus-induced inflammation: New anti-inflammatory strategy. *J. Biol. Regul. Homeost. Agents* **2020**, *34*, 9–14. [PubMed]
28. Schirinzi, T.; Landi, D.; Liguori, C. COVID-19: Dealing with a potential risk factor for chronic neurological disorders. *J. Neurol.* **2020**, *27*, 1–8.

29. Okba, N.M.A.; Müller, M.A.; Li, W.; Wang, C.; GeurtsvanKessel, C.H.; Corman, V.M.; Lamers, M.M.; Sikkema, R.S.; de Bruin, E.; Chandler, F.D.; et al. Severe Acute Respiratory Syndrome Coronavirus 2-Specific Antibody Responses in Coronavirus Disease Patients. *Emerg. Infect. Dis.* **2020**, *26*, 1478–1488. [CrossRef]
30. Mohammadi, S.; Moosaie, F.; Aarabi, M.H. Understanding the Immunologic Characteristics of Neurologic Manifestations of SARS-CoV-2 and Potential Immunological Mechanisms. *Mol. Neurobiol.* **2020**, *57*, 5263–5275. [CrossRef] [PubMed]
31. Coolen, T.; Lolli, V.; Sadeghi, N.; Rovai, A.; Trotta, N.; Taccone, F.S. Early postmortem brain MRI findings in COVID-19 non-survivors. *Neurology* **2020**, *95*, e2016–e2027. [CrossRef] [PubMed]
32. Reichard, R.R.; Kashani, K.B.; Boire, N.A.; Constantopoulos, E.; Guo, Y.; Lucchinetti, C.F. Neuropathology of COVID-19: A spectrum of vascular and acute disseminated encephalomyelitis (ADEM)-like pathology. *Acta Neuropathol.* **2020**, *140*, 1–6. [CrossRef] [PubMed]
33. Solomon, I.H.; Normandin, E.; Bhattacharyya, S.; Mukerji, S.S.; Keller, K.; Ali, A.S.; Adams, G.; Hornick, J.L.; Padera, R.F., Jr.; Sabeti, P. Neuropathological features of Covid-19. *N. Engl. J. Med.* **2020**, *383*, 989–992. [CrossRef] [PubMed]
34. Jacomy, H.; Fragoso, G.; Almazan, G.; Mushynski, W.E.; Talbot, P.J. Human coronavirus OC43 infection induces chronic encephalitis leading to disabilities in BALB/C mice. *Virology* **2006**, *349*, 335–346. [CrossRef] [PubMed]
35. Lu, Y.; Li, X.; Geng, D.; Mei, N.; Wu, P.Y.; Huang, C.C.; Jia, T.; Zhao, Y.; Wang, D.; Xiao, A.; et al. Cerebral micro-structural changes in COVID-19 patients—an MRI—based 3-month follow-up study. *EClinicalMedicine* **2020**, *25*, 100484. [CrossRef] [PubMed]
36. Pereira, A. Long-term neurological threats of COVID-19: A call to update the thinking about the outcomes of the coronavirus pandemic. *Front. Neurol.* **2020**, *11*, 308. [CrossRef]
37. Lee, S.; Viqar, F.; Zimmerman, M.E.; Narkhede, A.; Tosto, G.; Benzinger, T.L.; Marcus, D.S.; Fagan, A.M.; Goate, A.; Fox, N.C.; et al. White matter hyperintensities are a core feature of Alzheimer's disease: Evidence from the dominantly inherited Alzheimer network. *Ann. Neurol.* **2016**, *79*, 929–939. [CrossRef]
38. Love, S.; Miners, J.S. Cerebrovascular disease in ageing and Alzheimer's disease. *Acta Neuropathol.* **2016**, *131*, 645–658. [CrossRef]
39. Wen, Y.; Yang, S.; Liu, R.; Simpkins, J.W. Transient cerebral ischemia induces site-specific hyperphosphorylation of tau protein. *Brain Res.* **2004**, *1022*, 30–38. [CrossRef] [PubMed]
40. Akiyama, H.; Barger, S.; Barnum, S.; Bradt, B.; Bauer, J.; Cole, G.M.; Cooper, N.R.; Eikelenboom, P.; Emmerling, M.; Fiebich, B.L.; et al. Inflammation and Alzheimer's disease. *Neurobiol. Aging.* **2020**, *21*, 383–421. [CrossRef]
41. Koenigsknecht-Talboo, J.; Landreth, G.E. Microglial phagocytosis induced by fibrillar beta-amyloid and IgGs are differentially regulated by proinflammatory cytokines. *J. Neurosci.* **2005**, *25*, 8240–8249. [CrossRef] [PubMed]
42. De Felice, F.G.; Tovar-Moll, F.; Moll, J.; Munoz, D.P.; Ferreira, S.T. Severe Acute Respiratory Syndrome Coronavirus 2 (SARS-CoV-2) and the Central Nervous System. *Trends Neurosci.* **2020**, *S0166-2236*, 30091–30096. [CrossRef] [PubMed]
43. Soscia, S.J.; Kirby, J.E.; Washicosky, K.J.; Tucker, S.M.; Ingelsson, M.; Hyman, B.; Burton, M.A.; Goldstein, L.E.; Duong, S.; Tanzi, R.E.; et al. The Alzheimer's disease-associated amyloid beta-protein is an antimicrobial peptide. *PLoS ONE* **2010**, *5*, e9505. [CrossRef] [PubMed]
44. McLoughlin, B.C.; Miles, A.; Webb, T.E.; Knopp, P.; Eyres, C.; Fabbri, A.; Humphries, F.; Davis, D. Functional and cognitive outcomes after COVID-19 delirium. *Eur. Geriatr. Med.* **2020**, *11*, 857–862. [CrossRef] [PubMed]
45. Sasannejad, C.; Ely, E.W.; Lahiri, S. Long-term cognitive impairment after acute respiratory distress syndrome: A review of clinical impact and pathophysiological mechanisms. *Crit. Care* **2019**, *23*, 352. [CrossRef] [PubMed]
46. Van den Boogaard, M.; Kox, M.; Quinn, K.L.; van Achterberg, T.; van der Hoeven, J.G.; Schoonhoven, L.; Pickkers, P. Biomarkers associated with delirium in critically ill patients and their relation with long-term subjective cognitive dysfunction; indications for different pathways governing delirium in inflamed and noninflamed patients. *Crit. Care* **2011**, *15*, R297. [CrossRef] [PubMed]
47. Sharshar, T.; Hopkinson, N.S.; Orlikowski, D.; Annane, D. Science review: The brain in sepsis-culprit and victim. *Crit. Care* **2005**, *9*, 37–44. [CrossRef]
48. Montagne, A.; Zhao, Z.; Zlokovic, B.V. Alzheimer's disease: A matter of blood brain barrier dysfunction? *J. Exp. Med.* **2017**, *214*, 3151–3169. [CrossRef]
49. Zlokovic, B.V. Clearing amyloid through the blood-brain barrier. *J. Neurochem.* **2004**, *89*, 807–811. [CrossRef]
50. Lahiri, S.; Regis, G.C.; Koronyo, Y.; Fuchs, D.T.; Sheyn, J.; Kim, E.H.; Mastali, M.; Van Eyk, J.E.; Rajput, P.S.; Lyden, P.D.; et al. Acute neuropathological consequences of short-term mechanical ventilation in wild-type and Alzheimer's disease mice. *Crit. Care* **2019**, *23*, 63. [CrossRef]
51. Kloske, C.M.; Wilcock, D.M. The important interface between apolipoprotein E and neuroinflammation in Alzheimer's disease. *Front. Immunol.* **2020**, *11*, 754. [CrossRef]
52. Kuo, C.L.; Pilling, L.C.; Atkins, J.L.; Masoli, J.; Delgado, J.; Kuchel, G.A.; Melzer, D. APOEe4 genotype predicts severe COVID-19 in the UK Biobank community cohort. *J. Gerontol. A Biol. Sci. Med. Sci.* **2020**, *75*, 2231–2232. [CrossRef]
53. Virhammar, J.; Nääs, A.; Fällmar, D.; Cunningham, J.L.; Klang, A.; Ashton, N.J.; Jackmann, S.; Westman, G.; Frithiof, R.; Blennow, K.; et al. Biomarkers for CNS injury in CSF are elevated in COVID-19 and associated with neurological symptoms and disease severity. *Eur. J. Neurol.* **2020**. [CrossRef]
54. Kanberg, N.; Ashton, N.J.; Andersson, L.M.; Yilmaz, A.; Lindh, M.; Nilsson, S.; Price, R.W.; Blennow, K.; Zetterberg, H.; Gisslén, M. Neurochemical evidence of astrocytic and neuronal injury commonly found in COVID-19. *Neurology* **2020**, *95*, e1754–e1759. [CrossRef]

55. Ameres, M.; Brandstetter, S.; Toncheva, A.A.; Kabesch, M.; Leppert, D.; Kuhle, J.; Wellmann, S. Association of neuronal injury blood marker neurofilament light chain with mild-to-moderate COVID-19. *J. Neurol.* **2020**, *267*, 3476–3478. [CrossRef] [PubMed]

56. Edén, A.; Kanberg, N.; Gostner, J.; Fuchs, D.; Hagberg, L.; Andersson, L.M.; Lindh, M.; Price, R.W.; Zetterberg, H.; Gisslén, M. CSF biomarkers in patients with COVID-19 and neurological symptoms: A case series. *Neurology* **2021**, *96*, e294–e300.

57. Espíndola, O.M.; Brandão, C.O.; Gomes, Y.C.P.; Siqueira, M.; Soares, C.N.; Lima, M.A.S.D.; Leite, A.C.C.B.; Torezani, G.; Araujo, A.Q.C.; Silva, M.T.T. Cerebrospinal fluid findings in neurological diseases associated with COVID-19 and insights into mechanisms of disease development. *Int. J. Infect. Dis.* **2021**, *102*, 155–162. [CrossRef]

58. Kamphuis, W.; Middeldorp, J.; Kooijman, L.; Sluijs, J.A.; Kooi, E.J.; Moeton, M.; Freriks, M.; Mizee, M.R.; Hol, E.M. Glial fibrillary acidic protein isoform expression in plaque related astrogliosis in Alzheimer's disease. *Neurobiol. Aging.* **2014**, *35*, 492–510. [CrossRef] [PubMed]

59. Jin, M.; Cao, L.; Dai, Y.P. Role of Neurofilament Light Chain as a Potential Biomarker for Alzheimer's Disease: A Correlative Meta-Analysis. *Front. Aging. Neurosci.* **2019**, *11*, 254. [CrossRef]

60. Sutter, R.; Hert, L.; De Marchis, G.M.; Twerenbold, R.; Kappos, L.; Naegelin, Y.; Kuster, G.M.; Benkert, P.; Jost, J.; Maceski, A.M.; et al. Serum Neurofilament Light Chain Levels in the Intensive Care Unit: Comparison between Severely Ill Patients with and without Coronavirus Disease 2019. *Ann. Neurol.* **2020**, *89*, 610–616. [CrossRef] [PubMed]

61. Chen, X.; Zhao, B.; Qu, Y.; Chen, Y.; Xiong, J.; Feng, Y.; Men, D.; Huang, Q.; Liu, Y.; Yang, B.; et al. Detectable serum SARS- CoV-2 viral load (RNAaemia) is closely correlated with drastically elevated interleukin 6 (IL-6) level in critically ill COVID-19 patients. *Clin. Infect. Dis.* **2020**, *71*, 1937–1942. [CrossRef] [PubMed]

62. Cojocaru, I.M.; Cojocaru, M.; Miu, G.; Sapira, V. Study of interleukin-6 production in Alzheimer's disease. *Rom. J. Intern. Med.* **2011**, *49*, 55–58. [PubMed]

63. Strafella, C.; Caputo, V.; Termine, A.; Barati, S.; Caltagirone, C.; Giardina, E.; Cascella, R. Investigation of Genetic Variations of IL6 and IL6R as Potential Prognostic and Pharmacogenetics Biomarkers: Implications for COVID-19 and Neuroinflammatory Disorders. *Life (Basel)* **2020**, *10*, 351.

64. Mun, M.J.; Kim, J.H.; Choi, J.Y.; Jang, W.C. Genetic polymorphisms of interleukin genes and the risk of Alzheimer's disease: An update meta-analysis. *Meta Gene* **2016**, *8*, 1–10. [CrossRef] [PubMed]

65. Woo, P.; Humphries, S.E. IL-6 polymorphisms: A useful genetic tool for inflammation research? *J. Clin. Investig.* **2013**, *123*, 1413–1414. [CrossRef] [PubMed]

66. Shah, T.; Zabaneh, D.; Gaunt, T.; Swerdlow, D.I.; Shah, S.; Talmud, P.J.; Day, I.N.; Whittaker, J.; Holmes, M.V.; Sofat, R.; et al. Gene-centric analysis identifies variants associated with interleukin-6 levels and shared pathways with other inflammation markers. *Circ. Cardiovasc. Genet.* **2013**, *6*, 163–170. [CrossRef]

67. Cauchois, R.; Koubi, M.; Delarbre, D.; Manet, C.; Carvelli, J.; Blasco, V.B.; Jean, R.; Fouche, L.; Bornet, C.; Pauly, V.; et al. Early IL-1 receptor blockade in severe inflammatory respiratory failure complicating COVID-19. *Proc. Natl. Acad. Sci. USA* **2020**, *117*, 18951–18953. [CrossRef]

68. Pugh, R.C.; Fleshner, M.; Watkins, L.R.; Maier, S.F.; Rudy, J.W. The immune system and memory consolidation: A role for the cytokine IL-1beta. *Neurosci. Biobehav. Rev.* **2001**, *25*, 29–41. [CrossRef]

69. Griffin, W.S.; Stanley, L.C.; Ling, C.; White, L. Brain interleukin 1 and S-100 immunoreactivity are elevated in Down syndrome and Alzheimer disease. *Proc. Natl. Acad. Sci. USA* **1989**, *86*, 7611–7615. [CrossRef] [PubMed]

70. Goshen, I.; Kreisel, T.; Ounallah-Saad, H.; Renbaum, P.; Zalzstein, Y.; Ben-Hur, T.; Levy-Lahad, E.; Yirmiya, R. A dual role for interleukin-1 in hippocampal-dependent memory processes. *Psychoneuroendocrinology* **2007**, *32*, 1106–1115. [CrossRef]

71. Agnello, L.; Bivona, G.; Lo Sasso, B.; Scazzone, C.; Bazan, V.; Bellia, C.; Ciaccio, M. Galectin-3 in acute coronary syndrome. *Clin. Biochem.* **2017**, *50*, 797–803. [CrossRef]

72. Bivona, G.; Bellia, C.; Lo Sasso, B.; Agnello, L.; Scazzone, C.; Novo, G.; Ciaccio, M. Short-term Changes in Gal 3 Circulating Levels After Acute Myocardial Infarction. *Arch. Med. Res.* **2016**, *47*, 521–525. [CrossRef] [PubMed]

73. Agnello, L.; Bellia, C.; Lo Sasso, B.; Pivetti, A.; Muratore, M.; Scazzone, C.; Bivona, G.; Lippi, G.; Ciaccio, M. Establishing the upper reference limit of Galectin-3 in healthy blood donors. *Biochem. Med. (Zagreb)* **2017**, *27*, 030709. [CrossRef]

74. Garcia-Revilla, J.; Deierborg, T.; Venero, J.L.; Boza-Serrano, A. Hyperinflammation and Fibrosis in Severe COVID-19 Patients: Galectin-3, a Target Molecule to Consider. *Front. Immunol.* **2020**, *11*, 2069. [CrossRef] [PubMed]

75. Wang, X.; Zhang, S.; Lin, F.; Chu, W.; Yue, S. Elevated Galectin-3 Levels in the Serum of Patients With Alzheimer's Disease. *Am. J. Alzheimers Dis. Other Demen.* **2015**, *30*, 729–732. [CrossRef] [PubMed]

76. Tao, C.C.; Cheng, K.M.; Ma, Y.L.; Hsu, W.L.; Chen, Y.C.; Fuh, J.L.; Lee, W.J.; Chao, C.C.; Lee, E.H.Y. Galectin-3 promotes Aβ oligomerization and Aβ toxicity in a mouse model of Alzheimer's disease. *Cell Death Differ.* **2020**, *27*, 192–209. [CrossRef] [PubMed]

brain sciences

Perspective

Might Fibroblasts from Patients with Alzheimer's Disease Reflect the Brain Pathology? A Focus on the Increased Phosphorylation of Amyloid Precursor Protein Tyr$_{682}$ Residue

Filomena Iannuzzi [1], Vincenza Frisardi [2], Lucio Annunziato [3] and Carmela Matrone [4,*]

[1] Department of Biomedicine, University of Aarhus, Bartholins Allé, 8000 Aarhus, Denmark; Filomena.iannuzzi@biomed.au.dk

[2] Geriatric and Neuro Rehabilitation Department, Clinical Center for Nutrition in the Elderly, AUSL-IRCCS Reggio Emilia, Giovanni Amendola Street, 42122 Reggio Emilia, Italy; Vincenza.Frisardi@ausl.re.it

[3] SDN Research Institute Diagnostics and Nuclear (IRCCS SDN), Gianturco, 80131 Naples, Italy; Iannunzi@unina.it

[4] Division of Pharmacology, Department of Neuroscience, School of Medicine, University of Naples Federico II, 80131 Naples, Italy

* Correspondence: carmela.matrone@unina.it

Abstract: Alzheimer's disease (AD) is a devastating neurodegenerative disorder with no cure and no effective diagnostic criteria. The greatest challenge in effectively treating AD is identifying biomarkers specific for each patient when neurodegenerative processes have not yet begun, an outcome that would allow the design of a personalised therapeutic approach for each patient and the monitoring of the therapeutic response during the treatment. We found that the excessive phosphorylation of the amyloid precursor protein (APP) Tyr$_{682}$ residue on the APP $_{682}$YENPTY$_{687}$ motif precedes amyloid β accumulation and leads to neuronal degeneration in AD neurons. We proved that Fyn tyrosine kinase elicits APP phosphorylation on Tyr$_{682}$ residue, and we reported increased levels of APP Tyr$_{682}$ and Fyn overactivation in AD neurons. Here, we want to contemplate the possibility of using fibroblasts as tools to assess APP Tyr$_{682}$ phosphorylation in AD patients, thus making the changes in APP Tyr$_{682}$ phosphorylation levels a potential diagnostic strategy to detect early pathological alterations present in the peripheral cells of AD patients' AD brains.

Keywords: Alzheimer's disease; amyloid precursor protein; Tyr682 residue; YENPTY motif; Fyn tyrosine kinase; amyloid beta

check for **updates**

Citation: Iannuzzi, F.; Frisardi, V.; Annunziato, L.; Matrone, C. Might Fibroblasts from Patients with Alzheimer's Disease Reflect the Brain Pathology? A Focus on the Increased Phosphorylation of Amyloid Precursor Protein Tyr$_{682}$ Residue. *Brain Sci.* **2021**, *11*, 103. https://doi.org/10.3390/brainsci11010103

Received: 12 December 2020
Accepted: 4 January 2021
Published: 14 January 2021

Publisher's Note: MDPI stays neutral with regard to jurisdictional claims in published maps and institutional affiliations.

1. Introduction

Alzheimer's disease (AD) is a prominent neurodegenerative disease which takes a major toll on the elderly and places an enormous burden on the healthcare system.

The amyloid precursor protein (APP) is one of most extensively studied molecules in AD, and its cleavage, mediated by β-site APP-cleaving enzyme 1 (BACE1 or β-secretase) and by γ-secretase, leads to the dysregulated production of Aβ peptides, which mostly accumulate at central synapses [1].

Our group have previously underlined the crucial role of APP Tyr$_{682}$ residue in the processes responsible for the generation of Aβ peptides in human neurons [2,3]. APP Tyr$_{682}$ residue is located in the highly conserved $_{682}$YENPTY$_{687}$ motif, which binds specific adaptor proteins depending on its phosphorylation/dephosphorylation state [4]. The APP interaction with these proteins acts as the major regulator of APP fate [5–7]. In particular, increased APP Tyr$_{682}$ phosphorylation prevents the APP from binding to the Clathrin and AP2 endocytic proteins [8,9], and affects APP endocytosis and trafficking inside neurons [8,9]. Consequently, APP accumulates in the acidic neuronal compartments, such as late endosomes and lysosomes, where it is preferentially cleaved to generate Aβ peptides [3]. Notably, we previously showed that high levels of phosphorylation of APP Tyr_{682} residue in neurons differentiated from neuronal stem cells of AD patients [2,3,10] preceding Aβ accumulation, and promoting neuronal degencration. We therefore suggested that this high levels of APP Tyr_{682} phosphorylation might be an early marker of neuronal degeneration [5,6]. Furthermore, we identified Fyn tyrosine kinase (TK) in AD neurons as being responsible for the increased levels of APP Tyr_{682} phosphorylation, that in turn triggers biochemical events downstream, finally resulting in Aβ production and neuronal death [2]. Consistently, the reduction in Fyn activity, using Fyn TK inhibitors (TKI) or Fyn mRNA silencing, led to the decrease in APP Tyr_{682} phosphorylation and Aβ release in AD neurons [2,3]. These findings opened up a promising perspective for the use of APP Tyr_{682} residue as a biomarker to identify patients with high levels of APP Tyr_{682} phosphorylation that may benefit from personalised treatments with Fyn TKI.

We found elevated APP Tyr phosphorylation levels and Fyn overactivation both in skin fibroblasts and cortical tissues of minipigs carrying PSEN1 M146I mutation [3]. The increased APP Tyr phosphorylation in fibroblasts of mutant minipigs was associated to the same defects in APP trafficking that we also reported in neurons from AD patients carrying the same PSEN1 M146L mutation [3]. Relevantly, both APP Tyr phosphorylation and the defects in APP trafficking were rescued by Fyn TKI [2,3]. Of note, minipigs carrying PSEN1 M146L mutation displayed Aβ plaques and tau tangles in brain [11]. This evidence suggested the existence of possible common molecular pathways between fibroblasts and AD neurons, and encouraged us to explore the possibility of using skin fibroblasts to assess the levels of APP Tyr phosphorylation in AD patients. Indeed, the need to identify changes in peripheral cells recapitulating genetic, epigenetic, or metabolic deficiencies recognised in the brains of AD patients, and to develop standardised procedures to analyse changes in these cellular populations, is crucial for biomarker discovery.

2. Materials and Methods

2.1. Human Fibroblasts

Human fibroblasts were purchased from the Coriell Institute (New Orleans, LA, USA) and cultured following the guidelines reported on the manufacturer's web site (https://catalog.coriell.org/). All fibroblast cultures derived from an ante mortem skin biopsy from the forearm, and had approximately the same biological age in culture (number of passages of cell lines and cell divisions). Fibroblast cell cultures grew very slow, but they all showed the same morphology and shape in culture. No differences were found in the extent of neuronal survival or proliferation. In addition, Aβ40 and Aβ42 were undetectable in media of fibroblast cultures, because the levels appeared below the sensitivity of a commercially available ELISA kit (data not shown).

Patients were classified according to the criteria reported on the Coriell website and are summarised in Table 1: AD, sporadic AD patients; fAD 1, patients carrying APP mutations or APP duplication on chromosome 21 [12]; and PSEN1, patients carrying Ala246Glu (A246E) PSEN1 mutation. As controls, based on age and sex of the donors, we included six samples from unaffected spouse and four from donors with a familiar history of AD (fAD 1) or one sibling (#AG06846) of a patient carrying the PSEN 1 mutation A246E (#AG06848). As further controls of our experiments, we included three patients with a clinical diagnosis of Parkinson's disease (PD) (#442, #446, #527), and one of epilepsy (#159). In particular, donor #527 (PD + AD) was initially diagnosed as PD because he exhibited difficulties in movements. However, one year after diagnosis he developed problems with attention span and memory. Autopsy confirmed AD. All clinically diagnosed AD and PD patients—according to clinical or neuroimaging features—received histopathological confirmation at autopsy.

Some more information about patient clinical features and specific characteristics of the fibroblast cultures are available on the Coriell online catalogue (https://catalog.coriell. org/), using the source number (#) reported in Table 1.

Table 1. Genotype, age, and sex of fibroblast donors. Individual clinical data (diagnostic criteria) and specific clinical features for the majority of donors are available on the Coriell website.

Healthy			
Genotype	**Age (years)**	**Sex**	**Source (#)**
Unaffected spouse	48	F	AG07865
	49	F	AG07871
	60	F	AG08379
	66	F	AG08517
	73	M	AG08509
	64	M	AG08125
fAD 1 Family members	69	F	AG07936
	47	F	AG07928
	42	M	AG08658
Family member with #AG06848	75	F	AG06846
Alzheimer's Patients			
fAD 1	41	F	AG08110
	38	F	AG08563
	75	M	AG08245
	41	M	AG08064
	43	M	AG08523
PSEN 1 A246E	49	F	AG06840
	53	M	AG07872
	52	M	AG06848 family member of AG06846
AD	61	F	AG04400 sibling of AG04401and AG04402
	53	F	AG04401 sibling of AG04400 and AG04402
	60	F	AG06869
	49	M	AG06844
Apo E3/E4	87	F	AG10788
Apo E4/E4	47	M	AG04402 sibling of AG04400and AG04401
Other Dementias			
Parkinson's	57	M	AG20446
	53	M	AG20442
Parkinson's + AD	60	M	AG08527
Epilepsy + AD	52	F	AG04159

2.2. Immunoprecipitation and Western Blot Assays

Fibroblasts were kept in culture until cell confluence was reached. Lysates were firstly processed for IP (pTyr) using mouse anti-p-Tyr magnetic beads, and subsequently analysed by WB. For pTyr protein enrichment, we used Phospho-Tyrosine Mouse mAb Magnetic Bead Conjugate (P-Tyr-100) (Cell Signaling, #8095). The samples were loaded on a 4–20% precast gel. Western blot analysis was performed using anti APP, clone Y188 (abcam, ab32136), rabbit anti-pan Fyn (Cell Signaling, BioNordika, Herlev, Denmark, #4023), rabbit anti-Src pTyr$_{416}$ (Cell Signaling Technology, #2101), mouse anti-p-Tyr (Sigma-Aldrich, #9416) and monoclonal anti-β-actin-peroxidase (Sigma-Aldrich, A3854, Søborg, Denmark) antibodies.

2.3. Statistical Analysis

Data are expressed as mean ± standard error of the means. All the experiments were performed at least three times. The appropriate statistical test was selected using GraphPad Prism software (version 9.0c) (GraphPad, San Diego, CA, USA); details are reported in the legend for each figure.

3. Results

In the study, we included fibroblasts from ante mortem skin biopsies of 14 AD patients, six unaffected spouses, and four healthy donors with a familiar history of AD. In addition, four patients with a diagnosis of dementia, three associated to Parkinson's disease (PD) and one to epilepsy were included. Post-mortem histopathology brain examinations were performed in all patients with mild cognitive impairment or dementia diagnosis to confirm AD.

The levels of phosphorylation of APP Tyr were assessed using the same procedure previously applied to neurons [2,3] and consisting of immunoprecipitating the total lysates with mouse anti pTyr conjugated beads and analysing samples by Western blot (WB) using anti APP antibody. Although this procedure is not selective for APP Tyr_{682} residue (APP has indeed other two Tyr(s) whose changes in the phosphorylation can also easily be detected by the same procedure), previous findings from our group proved that only APP Tyr_{682} residue phosphorylation is crucial in initiating APP amyloidogenic processes in AD neurons, whereas Tyr phosphorylation at different APP sites does not affect the extent of $A\beta$ production or accumulation [8,9]. The increase in APP Tyr phosphorylation was analysed with respect to the basal APP expression levels in the total lysate of each sample (Figures 1–4). In addition, the Fyn Tyr_{420} phosphorylation was also assessed as an indirect measure of Fyn TK activity [13], and the results were normalised on Fyn basal expression levels (Figures 1–4).

We found a significant increase in the level of APP Tyr phosphorylation in the majority of fibroblasts from AD patients when compared to healthy donors, as well as patients with other dementias (Figure 1). In particular, APP Tyr phosphorylation levels were significantly increased in all AD patients when compared to healthy unaffected spouse (Figure 2). When we analysed fibroblasts from four familiar AD-type 1 (fAD 1) patients, we noted that the increase in the phosphorylation of APP Tyr residue was significantly greater than that in either healthy donors with a familiar history of fAD 1 or sporadic AD (Figure 3). Interestingly, the healthy donor #AG06846 did not show changes in APP Tyr phosphorylation levels when compared to the family member carrying the PSEN1 A246E genotype (Figure 4). However, due to the very small sample size of the single panel, the power of these statistical analyses appears to be very speculative. Of interest, Tyr phosphorylation levels were increased in AD and PSEN1 A246E patients when compared to patients affected by other neurodegenerative diseases (Figure 4).

In regard to the Fyn TK activity, fibroblasts in which APP Tyr phosphorylation levels were increased also showed elevated Fyn Tyr_{420} phosphorylation levels (Figure 1). However, Fyn Tyr_{420} residue was also phosphorylated in fibroblasts from PD patients (Figure 4) as well as in some healthy donors (Figure 1). The findings that Fyn activation is not restricted to AD patients appeared consistent with previous evidence, pointing to the involvement of Fyn in a multitude of processes related either to cellular functions or dysfunctions [13] and strengthening the importance to confine any potential treatments with Fyn TKI to those patients in which APP Tyr_{682} phosphorylation levels are elevated.

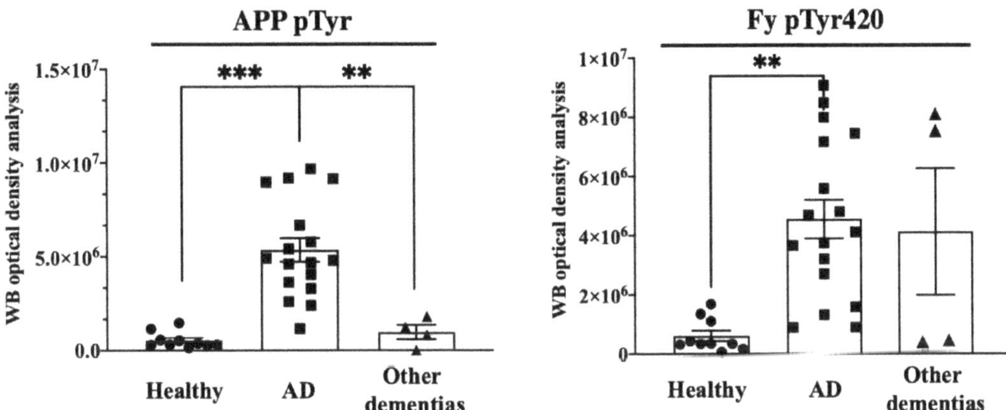

Figure 1. APP Tyr phosphorylation increases in fibroblasts from AD patients but not in patients with other forms of dementia or in healthy controls. The left panel reports the optical density analysis of APP pTyr bands, expressed as the mean optical density ratio between APP pTyr and APP basal levels from each sample in AD neurons (square) vs. healthy (circle) controls and other dementias (triangle). *** $p < 0.0001$ and ** $p < 0.001$, one-way ANOVA followed by Tukey's test. The right panel reports the densitometric analysis of the bands expressed as the mean optical density (OD) ratio of pFyn Tyr$_{420}$ relative to basal Fyn levels. ** $p < 0.001$, one-way ANOVA followed by Tukey's test.

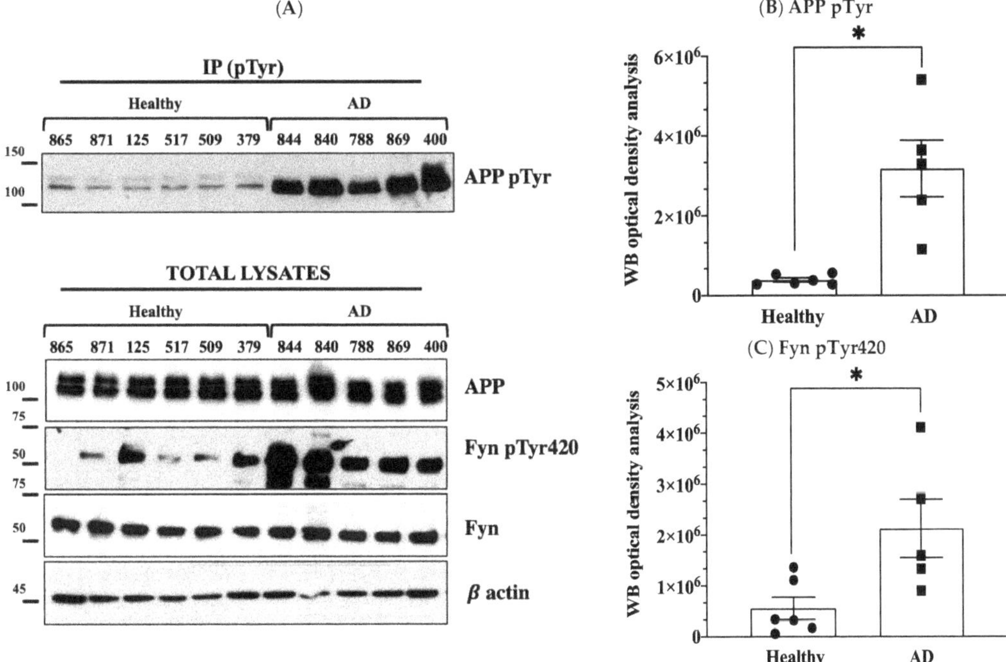

Figure 2. *Cont.*

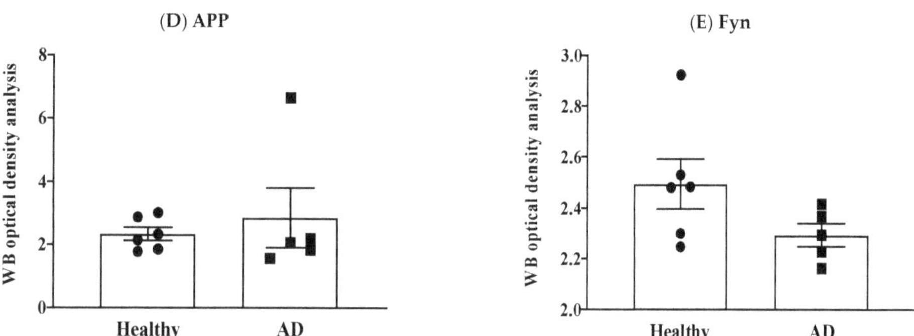

Figure 2. APP Tyr phosphorylation increases in fibroblasts from AD patients but not in healthy donors. Panel (**A**) reports data from six unaffected spouses (circle) and five AD cases (square). Panel (**B**) shows the optical density analysis of APP pTyr bands expressed as the mean optical density ratio between APP pTyr and APP basal levels from each sample in AD neurons vs. healthy controls. * $p < 0.05$; paired *t*-test. Panel (**C**) reports the ratio of pFyn Tyr$_{420}$ relative to basal Fyn levels. * $p < 0.05$; unpaired *t*-test. Western blot (WB) APP and Fyn expression levels are reported in panels (**D,E**). β-actin values were used as loading controls.

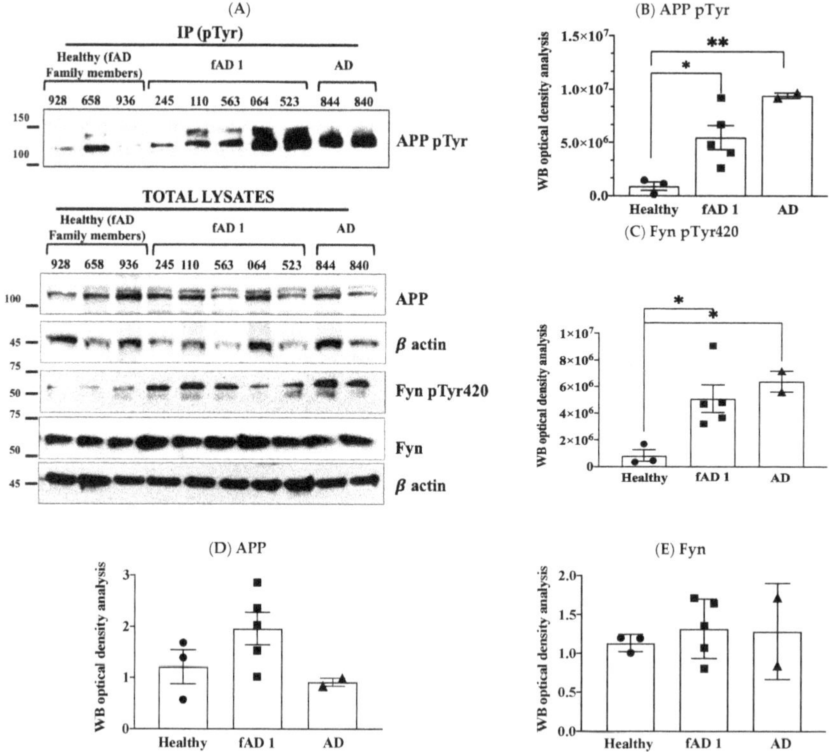

Figure 3. APP Tyr phosphorylation increases in fibroblasts from familial AD type 1 (fAD 1) and AD patients but not in healthy controls with a history of fAD 1. Panel (**A**) shows WB analyses of APP pTyr, Fyn pTyr$_{420}$, APP and Fyn in healthy donors (circle) patients with a diagnosis of fAD 1 (square) and AD (triangle). Panels (**B–E**) report the densitometric analyses of the bands shown in panels (**A**). In particular, panel (**B**) reports the optical density analysis of APP pTyr bands expressed as the mean optical density ratio between APP pTyr and APP basal levels from each sample. * $p < 0.05$; ** $p < 0.005$. Panel **C** reports the ratio of pFyn to Tyr$_{420}$ relative to basal Fyn levels. * $p < 0.05$, one-way ANOVA followed by Tukey's test. APP and Fyn expression levels, normalised on β-actin values, are reported in panels (**D,E**).

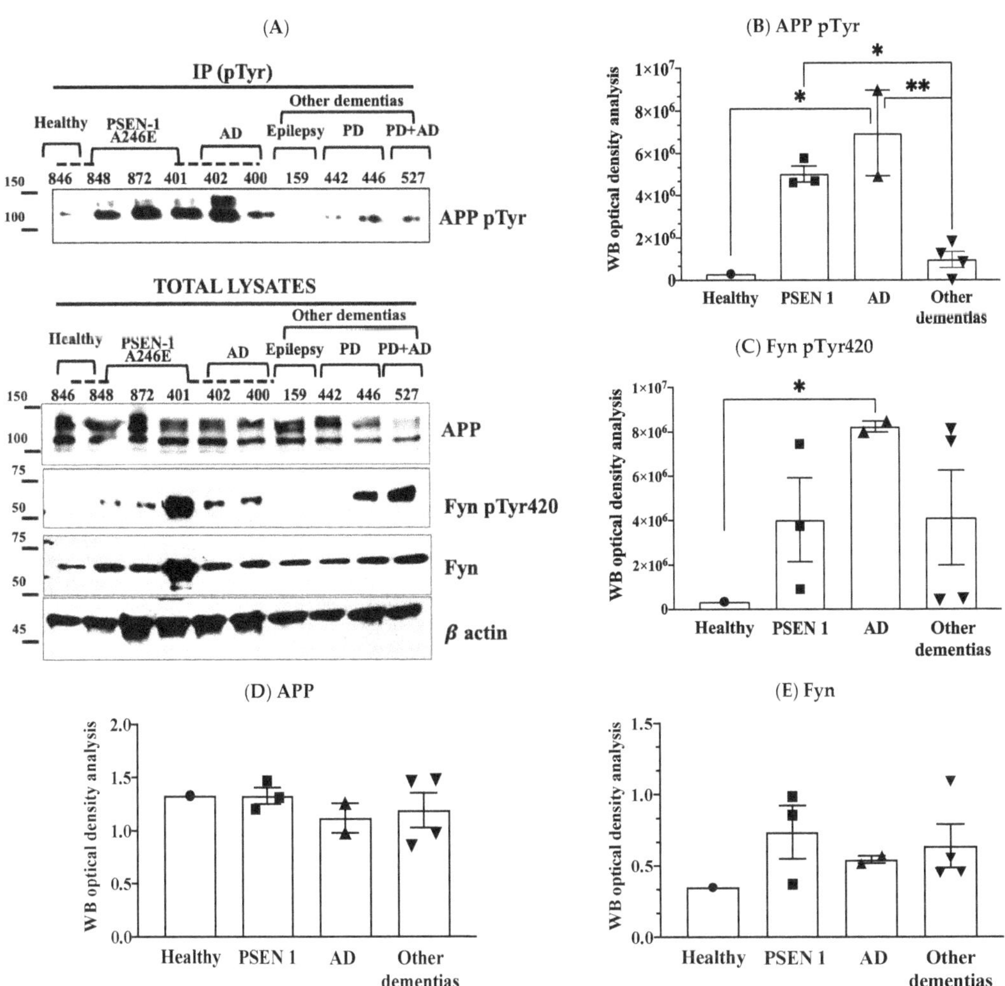

Figure 4. APP Tyr phosphorylation increases in fibroblasts from AD patients or in patients carrying PSEN 1 mutation, but not in patients with other forms of dementia or in healthy donors. Panel (**A**) shows WB analyses of APP pTyr, Fyn pTyr$_{420}$, basal APP and Fyn levels. Panel (**A**) in particular reports data from one healthy donor #846 with his sibling #848 who carried PSEN1 A246E mutation (indicated with a discontinued line); the PSEN1 A246E genotype was also present in patients #872 and #401; patients #400, #401 and #402 were siblings, and are indicated by the discontinued line; patients #159 had a diagnosis of epilepsy for their entire life, however post-mortem histologic analysis revealed an AD phenotype; patients #442 and #446 were PD, whereas #527 was clinically diagnosed with PD, however after death brain histology indicated an AD phenotype. Panel (**B**) reports the optical density analysis of APP pTyr bands expressed as the mean optical density ratio between APP pTyr and APP basal levels from each sample in fAD 1 and AD fibroblasts vs. healthy controls and other dementias. * $p < 0.005$; ** $p < 0.001$, one-way ANOVA followed by Tukey's test. Panel (**C**) reports the ratio of pFyn Tyr$_{420}$ relative to basal Fyn levels. The graph represents the densitometric analysis of the bands expressed as the mean optical density (OD). * $p < 0.005$, one-way ANOVA followed by Tukey's test. APP and Fyn expression levels are reported in panels (**D,E**). β actin values were used as loading controls.

4. Discussion

Despite the increasing efforts to understand the pathophysiology of AD, there are many unresolved questions regarding how to diagnose the disease early and the treatments available to ameliorate the condition in affected patients. To date, no efficient therapy has been found, and the majority of the pharmacologic approaches are only palliative because none of them delay the pathology [14]. This lack of success is partly explained by the complex aetiology of AD which makes this pathology heterogeneous and unpredictable.

We previously showed the potential role of APP Tyr_{682} phosphorylation as a biomarker for APP dysregulation in trafficking and processing, and as a predictive factor for the increased production of neurotoxic Aβ peptide and neurodegenerative processes [2,3,5–10,13,15]. Indeed, the dysregulation of kinase activity has been associated with the progression of other neurodegenerative diseases, and has been proposed as a potential biomarker for other major proteinopathies besides AD, including PD [16–18] and Huntington's disease [19,20].

Given the above scenario, in this study we explored the possibility of using peripheral cells, such as skin fibroblasts, to assess changes in APP Tyr_{682} phosphorylation levels. We found that APP Tyr_{682} phosphorylation increases in fibroblasts of AD patients. In addition, hyperactivated Fyn and elevated APP Tyr phosphorylation levels were detectable in the neurons from the same AD patients.

It is well established that Fyn can play a dual role in both the Aβ and tau pathologies observed in AD [21], exacerbating Aβ neurotoxicity through cellular prion protein (PrP^C) and interacting directly with tau [22] by phosphorylating it at Tyr_{18} [23]. We previously delineated a new possible mechanism in which Fyn initiated the amyloid cascade by phosphorylating the APP Tyr_{682} residue, which would then induce the amyloidogenic cleavage [2,13]. This cascade of events would amplify the neurotoxic effects of Aβ, tau hyperphosphorylation, neuronal tangles, and, consequently, neuronal damage. Consistently, we found here that Fyn was activated in #379 and #125 healthy donors and in PD patients (#446, #527), emphasising the involvement of Fyn in other processes besides AD [13]. This observation further strengthens the importance to restrict any potential treatments with Fyn TKI to those patients in which APP Tyr_{682} phosphorylation levels are elevated, thus preventing off-target effects, and improving the outcome of the disease.

Indeed, the identification of Fyn as an actor in APP Tyr_{682} phosphorylation in AD neurons opens the scenario to either test already commercially available Fyn TKIs or develop more specific compounds that can potentially control Fyn hyperactivity in AD patients and prevent $Aβ_{42}$ production in patients in whom APP Tyr_{682} phosphorylation is increased.

Overall, the observations reported here prospect the possibility that events related to APP Tyr phosphorylation in fibroblasts may reflect Aβ related abnormalities in the brain, and that, if correlated with changes in cognitive performance and with other biomarkers including neuroimaging approaches or to the levels of Aβ and phosphorylated tau in biofluids, may allow the identification of patients who could benefit from a personalised pharmacologic approach, by using Fyn TKI. Indeed, these results underline the need to extend this evidence to a larger number of samples/fibroblasts, as well as to develop a procedure selective and quantitative for the detection of the levels of APP Tyr_{682} phosphorylation.

In conclusion, this study proposes to use changes in APP Tyr_{682} phosphorylation levels in peripheric cells, such as fibroblasts, as a potential diagnostic approach to design personalised therapy in patients in which APP Tyr_{682} phosphorylation is increased. Indeed, further studies will be necessary to extend this analysis to a larger number of patients.

Author Contributions: F.I. performed experiments; F.I. and C.M. conceived and planned the experiments; V.F. and L.A. helped in data analyses and the critical revision of the paper. C.M. wrote the manuscript. All the authors discussed the results and contributed to the final manuscript. All authors have read and agreed to the published version of the manuscript.

Funding: This study was funded by the Lundbeck Foundation (Grant nos. R208–2015-3075 to Carmela Matrone), by PRIN 2017 (Grant nos. 2017T9JNLT) and by PON (PERMEDNET ARS 01_0126 to Lucio Annunziato).

Acknowledgments: The authors appreciate the valuable help of Henriette Gram Johanson (Department of Biomedicine, Aarhus University, Denmark) for her excellent technical assistance.

Conflicts of Interest: The authors declare no conflict of interest. The funders had no role in the design of the study; in the collection, analyses, or interpretation of data; in the writing of the manuscript, or in the decision to publish the results

References

1. Robakis, N.K. What Do Recent Clinical Trials Teach Us About the Etiology of AD. *Adv. Exp. Med. Biol.* **2020**, *1195*, 167. [CrossRef]
2. Iannuzzi, F.; Sirabella, R.; Canu, N.; Maier, T.J.; Annunziato, L.; Matrone, C. Fyn Tyrosine Kinase Elicits Amyloid Precursor Protein Tyr682 Phosphorylation in Neurons from Alzheimer's Disease Patients. *Cells* **2020**, *9*, 1807. [CrossRef]
3. Poulsen, E.T.; Iannuzzi, F.; Rasmussen, H.F.; Maier, T.J.; Enghild, J.J.; Jørgensen, A.L.; Matrone, C. An Aberrant Phosphorylation of Amyloid Precursor Protein Tyrosine Regulates Its Trafficking and the Binding to the Clathrin Endocytic Complex in Neural Stem Cells of Alzheimer's Disease Patients. *Front. Mol. Neurosci.* **2017**, *10*, 59. [CrossRef] [PubMed]
4. Klevanski, M.; Herrmann, U.; Weyer, S.W.; Fol, R.; Cartier, N.; Wolfer, D.P.; Caldwell, J.H.; Korte, M.; Müller, U.C. The APP Intracellular Domain Is Required for Normal Synaptic Morphology, Synaptic Plasticity, and Hippocampus-Dependent Behavior. *J. Neurosci.* **2015**, *35*, 16018–16033. [CrossRef] [PubMed]
5. Matrone, C. A new molecular explanation for age-related neurodegeneration: The Tyr682 residue of amyloid precursor protein. *Bioessays* **2013**, *35*, 847–852. [CrossRef] [PubMed]
6. Matrone, C.; Iannuzzi, F.; Annunziato, L. The $Y_{682}ENPTY_{687}$ motif of APP: Progress and insights toward a targeted therapy for Alzheimer's disease patients. *Ageing Res. Rev.* **2019**, *52*, 120–128. [CrossRef] [PubMed]
7. Matrone, C.; Luvisetto, S.; La Rosa, L.R.; Tamayev, R.; Pignataro, A.; Canu, N.; Yang, L.; Barbagallo, A.P.; Biundo, F.; Lombino, F.; et al. Tyr682 in the Abeta-precursor protein intracellular domain regulates synaptic connectivity, cholinergic function, and cognitive performance. *Aging Cell* **2012**, *11*, 1084–1093. [CrossRef]
8. Poulsen, E.T.; Larsen, A.; Zollo, A.; Jørgensen, A.L.; Sanggaard, K.W.; Enghild, J.J.; Matrone, C. New Insights to Clathrin and Adaptor Protein 2 for the Design and Development of Therapeutic Strategies. *Int. J. Mol. Sci.* **2015**, *16*, 29446–29453. [CrossRef] [PubMed]
9. La Rosa, L.R.; Perrone, L.; Nielsen, M.S.; Calissano, P.; Andersen, O.M.; Matrone, C. Y682G Mutation of Amyloid Precursor Protein Promotes Endo-Lysosomal Dysfunction by Disrupting APP-SorLA Interaction. *Front. Cell Neurosci.* **2015**, *9*, 109. [CrossRef]
10. Zollo, A.; Allen, Z.; Rasmussen, H.F.; Iannuzzi, F.; Shi, Y.; Larsen, A.; Maier, T.J.; Matrone, C. Sortilin-Related Receptor Expression in Human Neural Stem Cells Derived from Alzheimer's Disease Patients Carrying the APOE Epsilon 4 Allele. *Neural Plast.* **2017**, *2017*, 1892612. [CrossRef]
11. Jakobsen, J.E.; Johansen, M.G.; Schmidt, M.; Liu, Y.; Li, R.; Callesen, H.; Melnikova, M.; Habekost, M.; Matrone, C.; Bouter, Y.; et al. Expression of the Alzheimer's Disease Mutations AβPP695sw and PSEN1M146I in Double-Transgenic Göttingen Minipigs. *J. Alzheimers Dis.* **2016**, *53*, 1617–1630. [CrossRef]
12. St George-Hyslop, P.H.; Tanzi, R.E.; Polinsky, R.J.; Haines, J.L.; Nee, L.; Watkins, P.C.; Myers, R.H.; Feldman, R.G.; Pollen, D.; Drachman, D. The genetic defect causing familial Alzheimer's disease maps on chromosome 21. *Science* **1987**, *235*, 885–890. [CrossRef] [PubMed]
13. Matrone, C.; Petrillo, F.; Nasso, R.; Ferretti, G. Fyn Tyrosine Kinase as Harmonizing Factor in Neuronal Functions and Dysfunctions. *Int. J. Mol. Sci.* **2020**, *21*, 4444. [CrossRef] [PubMed]
14. Bachurin, S.O.; Bovina, E.V.; Ustyugov, A.A. Drugs in Clinical Trials for Alzheimer's Disease: The Major Trends. *Med. Res. Rev.* **2017**, *37*, 1186–1225. [CrossRef] [PubMed]
15. Matrone, C.; Barbagallo, A.P.; La Rosa, L.R.; Florenzano, F.; Ciotti, M.T.; Mercanti, D.; Chao, M.V.; Calissano, P.; D'Adamio, L. APP is phosphorylated by TrkA and regulates NGF/TrkA signaling. *J. Neurosci.* **2011**, *31*, 11756–11761. [CrossRef]
16. Mehdi, S.J.; Rosas-Hernandez, H.; Cuevas, E.; Lantz, S.M.; Barger, S.W.; Sarkar, S.; Paule, M.G.; Ali, S.F.; Imam, S.Z. Protein Kinases and Parkinson's Disease. *Int. J. Mol. Sci.* **2016**, *17*, 1585. [CrossRef]
17. Jan, A.; Jansonius, B.; Delaidelli, A.; Bhanshali, F.; An, Y.A.; Ferreira, N.; Smits, L.M.; Negri, G.L.; Schwamborn, J.C.; Jensen, P.H.; et al. Activity of translation regulator eukaryotic elongation factor-2 kinase is increased in Parkinson disease brain and its inhibition reduces alpha synuclein toxicity. *Acta Neuropathol. Commun.* **2018**, *6*, 54. [CrossRef]
18. Kofoed, R.H.; Betzer, C.; Ferreira, N.; Jensen, P.H. Glycogen synthase kinase 3 β activity is essential for Polo-like kinase 2- and Leucine-rich repeat kinase 2-mediated regulation of α-synuclein. *Neurobiol. Dis.* **2020**, *136*, 104720. [CrossRef]
19. Kim, J.; Amante, D.J.; Moody, J.P.; Edgerly, C.K.; Bordiuk, O.L.; Smith, K.; Matson, S.A.; Matson, W.R.; Scherzer, C.R.; Rosas, H.D.; et al. Reduced creatine kinase as a central and peripheral biomarker in Huntington's disease. *Biochim. Biophys. Acta* **2010**, *1802*, 673–681. [CrossRef]
20. Narayanan, K.L.; Chopra, V.; Rosas, H.D.; Malarick, K.; Hersch, S. Rho Kinase Pathway Alterations in the Brain and Leukocytes in Huntington's Disease. *Mol. Neurobiol.* **2016**, *53*, 2132–2140. [CrossRef]

21. Nygaard, H.B. Targeting Fyn Kinase in Alzheimer's Disease. *Biol. Psychiatry* **2018**, *83*, 369–376. [CrossRef] [PubMed]
22. Lee, G.; Newman, S.T.; Gard, D.L.; Band, H.; Panchamoorthy, G. Tau interacts with src-family non-receptor tyrosine kinases. *J. Cell Sci.* **1998**, *111*, 3167–3177. [PubMed]
23. Lee, G.; Thangavel, R.; Sharma, V.M.; Litersky, J.M.; Bhaskar, K.; Fang, S.M.; Do, L.H.; Andreadis, A.; Van Hoesen, G.; Ksiezak-Reding, H. Phosphorylation of Tau by Fyn: Implications for Alzheimer's Disease. *J. Neurosci.* **2004**, *24*, 2304–2312. [CrossRef] [PubMed]

Article

Transcranial Magnetic Resonance Imaging-Guided Focused Ultrasound with a 1.5 Tesla Scanner: A Prospective Intraindividual Comparison Study of Intraoperative Imaging

Cesare Gagliardo [1,*,†], Roberto Cannella [1,†], Costanza D'Angelo [1], Patrizia Toia [1], Giuseppe Salvaggio [1], Paola Feraco [2], Maurizio Marrale [3], Domenico Gerardo Iacopino [4], Marco D'Amelio [5], Giuseppe La Tona [1], Ludovico La Grutta [1] and Massimo Midiri [1]

[1] Section of Radiological Sciences, Department of Biomedicine, Neuroscience and Advanced Diagnostics, University of Palermo, 90127 Palermo, Italy; rob.cannella89@gmail.com (R.C.); costanza.dangelo@gmail.com (C.D.); patrizia.toia@unipa.it (P.T.); p.salvaggio@libero.it (G.S.); giuseppe.latona@unipa.it (G.L.T.); ludovico.lagrutta@unipa.it (L.L.G.); massimo.midiri@unipa.it (M.M.)

[2] Neuroradiology Unit, S. Chiara Hospital, 38122 Trento, Italy; paola.feraco@apss.tn.it

[3] Department of Physics and Chemistry, University of Palermo, 90133 Palermo, Italy; maurizio.marrale@unipa.it

[4] Section of Neurosurgery, Department of Biomedicine, Neuroscience and Advanced Diagnostics, University of Palermo, 90133 Palermo, Italy; gerardo.iacopino@unipa.it

[5] Section of Neurology, Department of Biomedicine, Neuroscience and Advanced Diagnostics, University of Palermo, 90133 Palermo, Italy; marco.damelio@unipa.it

* Correspondence: cesare.gagliardo@unipa.it

† These authors contributed equally to this work.

Citation: Gagliardo, C.; Cannella, R.; D'Angelo, C.; Toia, P.; Salvaggio, G.; Feraco, P.; Marrale, M.; Iacopino, D.G.; D'Amelio, M.; La Tona, G.; et al. Transcranial Magnetic Resonance Imaging-Guided Focused Ultrasound with a 1.5 Tesla Scanner: A Prospective Intraindividual Comparison Study of Intraoperative Imaging. *Brain Sci.* **2021**, *11*, 46. https://doi.org/10.3390/brainsci11010046

Received: 17 November 2020
Accepted: 28 December 2020
Published: 4 January 2021

Publisher's Note: MDPI stays neutral with regard to jurisdictional claims in published maps and institutional affiliations.

Abstract: Background: High-quality intraoperative imaging is needed for optimal monitoring of patients undergoing transcranial MR-guided Focused Ultrasound (tcMRgFUS) thalamotomy. In this paper, we compare the intraoperative imaging obtained with dedicated FUS-Head coil and standard body radiofrequency coil in tcMRgFUS thalamotomy using 1.5-T MR scanner. Methods: This prospective study included adult patients undergoing tcMRgFUS for treatment of essential tremor. Intraoperative T2-weighted FRFSE sequences were acquired after the last high-energy sonication using a dedicated two-channel FUS-Head (2ch-FUS) coil and body radiofrequency (body-RF) coil. Postoperative follow-ups were performed at 48 h using an eight-channel phased-array (8ch-HEAD) coil. Two readers independently assessed the signal-to-noise ratio (SNR) and evaluated the presence of concentric lesional zones (zone I, II and III). Intraindividual differences in SNR and lesional findings were compared using the Wilcoxon signed rank sum test and McNemar test. Results: Eight patients underwent tcMRgFUS thalamotomy. Intraoperative T2-weighted FRFSE images acquired using the 2ch-FUS coil demonstrated significantly higher SNR (R1 median SNR: 10.54; R2: 9.52) compared to the body-RF coil (R1: 2.96, $p < 0.001$; R2: 2.99, $p < 0.001$). The SNR was lower compared to the 48-h follow-up ($p < 0.001$ for both readers). Intraoperative zone I and zone II were more commonly visualized using the 2ch-FUS coil (R1, $p = 0.031$ and $p = 0.008$, R2, $p = 0.016$, $p = 0.008$), without significant differences with 48-h follow-up ($p \geq 0.063$). The inter-reader agreement was almost perfect for both SNR (ICC: 0.85) and lesional findings (k: 0.82 0.91). Conclusions. In the study population, the dedicated 2ch-FUS coil significantly improved the SNR and visualization of lesional zones on intraoperative imaging during tcMRgFUS performed with a 1.5-T MR scanner.

Keywords: focused ultrasound; MR-guided focused ultrasound; high-intensity focused ultrasound ablation; magnetic resonance imaging; image quality; stereotaxic techniques; essential tremor

1. Introduction

Transcranial Magnetic Resonance Imaging-guided Focused Ultrasound (tcMRgFUS) is an emerging incisionless stereotactic procedure based on the thermal ablation of a brain area using a high-intensity focused ultrasound (HI-FU) beam. Randomized-controlled clinical trials and several prior studies have demonstrated the clinical efficacy of tcMRgFUS thalamotomy for the treatment of essential tremor [1–3], idiopathic asymmetrical tremor-dominant Parkinson's disease [4,5], and neuropathic pain [6].

The treatment is conducted under constant Magnetic Resonance (MR) imaging monitoring. MR allows to acquire detailed anatomical images to calculate the optimal target coordinates and MR thermometry for real-time thermal monitoring during sonications [7]. Anatomical intraoperative MR images may also depict the typical neuroradiological findings of tcMRgFUS-placed lesions, which consist of three concentric lesional zones, as originally described by Wintermark et al. [8]. Optimal MR imaging quality is, therefore, of the upmost importance for the HI-FU thermal ablation and lesion monitoring. TcMRgFUS procedures have been initially performed using 3.0-T MR scanners, but this technology is also rapidly expanding on 1.5-T MR [7]. On 3.0-T MR scanners, the anatomical images and MR thermometry are typically acquired using standard body radiofrequency coil built in the MR system because the 30-cm-diameter hemispherical FUS helmet, stereotactic frame, and supporting equipment almost fill the whole scanner space and do not allow the placement of specific head coil [9]. When the tcMRgFUS equipment is integrated into a 1.5-T MR scanner, a dedicated coil is used to compensate the lower field strength. Initial studies have reported significant improvement in signal-to-noise ratio (SNR) using dedicated head coil on 1.5-T MR, but these evidences were only based on phantom evaluations [10]. Moreover, there are still very limited experiences on FUS thalamotomy performed with 1.5-T MR scanners [11–13]. Therefore, there is a significant gap in knowledge regarding the added value of dedicated coil on the quality of intraoperative imaging acquired in patients undergoing tcMRgFUS procedure with 1.5-T MR. We hypothesize that dedicated head coil significantly increases the image quality and intraoperative lesions detection in patients undergoing tcMRgFUS on 1.5-T MR, compared to the images acquired using the standard body radiofrequency coil.

The purpose of our study was to conduct a prospective intraindividual comparison between intraoperative sequences acquired using a dedicated head coil and a standard body radiofrequency coil in order to assess the signal-to-noise ratio and intraoperative visualization of neuroimaging findings in patients undergoing transcranial MR-guided Focused Ultrasound thalamotomy using 1.5-T MR.

2. Materials and Methods

The institutional review board approved this study ("Comitato Etico Palermo 1"—seduta del 11/04/2018 verbale n 04/2018). All subjects provided written informed consent before enrolling for tcMRgFUS treatment in accordance with the Declaration of Helsinki.

2.1. Patients

This prospective study included all the adult patients undergoing tcMRgFUS using a 1.5-T MR for the treatment of the movement disorders between September and December 2019. Candidates for tcMRgFUS were selected after accurate screening visits including extensive neurological evaluation and preprocedural CT and MR imaging for treatment planning. A total of 10 patients were treated during the study time. Two patients were excluded due to the onset of unbearable headache or nausea and vomiting during the treatment sonications, which made the patient unsuitable for performing additional sequences at the end of the procedure.

The following data were collected in the included patients: age, gender, diagnosis of movement disorder, skull density ratio (SDR) [14], and skull area obtained from preoperative evaluations. In addition, tcMRgFUS technical parameters were recorded from the dedicated treatment workstation (ExAblate, INSIGHTEC Ltd., Tirat Carmel, Haifa, Israel) after reviewing the procedures, including: total number of sonications, number of high-energy sonications (i.e., sonications reaching an average temperature > 50 °C), maximum and average temperatures, effective delivered energy (measured in Joules), sonication power (Watt), and duration (seconds) of each high-energy sonication.

2.2. Procedure Details

TcMRgFUS procedures were performed using a focused ultrasound (FUS) equipment (ExAblate 4000; InSightec Ltd., Tirat Carmel, Haifa, Israel) integrated with a 1.5-T MR unit (Signa HDxt; GE Medical Systems, Milwaukee, WI, USA). All treatments were performed by a single primary operator (C.G. with 15 years of experience in neuroradiology) who had the full control of the workstation, in collaboration with a dedicated transdisciplinary team.

Detailed descriptions of tcMRgFUS thalamotomy have been extensively described in prior reports [7,15]. Briefly, before the procedure, the patient's head was comparatively shaved and immobilized to the FUS helmet using a stereotactic frame. A flexible silicone membrane, which integrates the 2-channel FUS-Head (2ch-FUS) coil, was placed on the patient's head and the space between the head and the helmet was filled with cooled degassed water in order to allow the HI-FU transmission. Once the patient was positioned in the FUS table, 2D FRFSE T2-weighted sequences were acquired on coronal, sagittal, and axial planes according to the anterior commissure–posterior commissure (AC-PC) anatomical landmarks in order to calculate the optimal stereotactic coordinates. In our study, the target was placed in the nucleus ventralis intermedius (Vim) in the contralateral thalamus to the hand-dominant tremor side (25% of the AC-PC distance in front of PC, 2 mm above AC-PC line, and 11–12 mm lateral to the third ventricle wall).

The procedure began with the alignment stage (Stage I), which consisted of few low energy sonications (with a maximum temperature between 40 °C and 45 °C), to confirm the accuracy of thermal spot according to the frequency-encoding direction. During the verify stage (Stage II), multiple intermediate-energy sonications (with a maximum temperature of 50 °C which minimize the risk of a permanent lesion) were performed to evaluate the optimal treatment target based on real-time clinical assessment of transient tremor suppression or any type of adverse events. In case of poor tremor response, intraoperative anatomical images were used to redefine the target coordinates to achieve the optimal tremor suppression. The HI-FU ablation consisted of a few high-energy sonications, reaching an average temperature $\geq$ 51 °C (Stage III: treatment low) for permanent lesion and $\geq$55 °C (Stage IV: treatment high) for lesion consolidation. The number of high-energy sonications varied according to the patient's skull characteristics, tremor disappearance, and neuroradiological findings on intraoperative imaging. Particularly, in our clinical practice, further high-energy sonications were usually considered for lesion consolidation in case of intraoperative imaging showing a FUS-placed lesion lacking of concentric lesional zones (zone I and zone II, see below).

2.3. MRI Data Acquisition

Intraoperative imaging was performed using a dedicated two-channel FUS-Head coil (INSIGHTEC Ltd., Tirat Carmel, Haifa, Israel). The coil is composed of two silicon-coated rings embedded in the elastic membrane. The two rings of this coil are positioned to either side of the patient's head without additional mechanical constraints to the patient. For the purpose of this study, intraoperative axial 2D fast recovery fast spin echo (FRFSE) T2-weighed sequences were acquired immediately after the last high-energy sonication using the dedicated 2ch-FUS coil in all patients. Then, a second axial FRFSE T2-weighed sequence was scanned with the identical MR parameters and conditions, but it was acquired using the standard body radiofrequency (body-RF) coil integrated in the MR scanner. Both sequences were acquired with the same setup used during the procedure, and thus, before emptying the helmet from the coupling degassed cooled water. The same sequence, along with a specific MR brain protocol (axial 3D T1w BRAVO, sagittal 3D T2w FLAIR with fat saturation, axial 3D SWAN, axial 2D T2w FRFSE, axial 2D EPI-DWI, axial 2D T1 FSE; after i.v. contrast medium injection axial 3D T1w BRAVO and axial 2D T1 FSE), was repeated at the 48-h follow-ups using an eight-channel phased-array head (8ch-HEAD) coil (GE's standard product 8ch BRAIN HD coil). Acquisition parameters for the sequences included in this study are reported in Table 1.

Table 1. MRI acquisition parameters used for axial fast recovery fast spin echo (FRFSE) T2-weighed images acquired intraoperatively and at 48-h follow-up with 1.5-T MR scanner.

	Axial FRFSE T2-Weighed Images		
	2ch-FUS Coil	**Body-RF Coil**	**8ch-HEAD Coil**
Slice thickness (mm)	2.0	2.0	2.0
Slice gap	0	0	0
Number or slice	19	19	19
TR (ms)	4461	4461	4380
TE (ms)	103	103	108
Matrix	384 × 288	384 × 288	320 × 288
NEX	2	2	5
FOV (cm)	22 × 22	22 × 22	24 × 24
Acquisition time (min)	4:06	4:06	4:06

Abbreviations: TR: Repetition Time; TE: Echo Time; NEX: number of excitations; FOV: Field of View.

2.4. MRI Data Analysis

Two readers (R1 R.C. and R2 C.D., with 6 and 2 years of experience in neuroimaging) independently analyzed the intraoperative and postoperative axial FRFSE T2-weighed images in order to evaluate the SNR and presence of concentric lesional zones. All the images were analyzed using a dedicated workstation equipped with Horos (Annapolis, MD, USA) a free and open source code software program that is distributed free of charge under the Lesser General Public License (LGPL) at Horosproject.org. The sequences were anonymized and reviewed in random order.

The SNR was evaluated following the approach proposed by the National Electrical Manufacturers Association (NEMA) [16]. A standard region of interest (ROI) with an area of 50 mm^2 was placed on the axial FRFSE T2-weighed images at the level of the AP-PC plane (treatment plane) in each of the following locations, bilaterally (Figure 1): (a) white matter of the frontal lobe; (b) head of the caudate nucleus; (c) lentiform nucleus; (d) posterior aspect of the thalamus; and (e) white matter of the occipital lobe. ROIs were placed carefully avoiding lateral ventricles white matter and FUS-placed lesions and related imaging findings (i.e., vasogenic edema). Two additional ROIs were placed in the body of the lateral ventricles. The signal was recorded as the mean pixel value within the ROI, while the noise was defined as the variations (i.e., standard deviation) of pixel intensities. The SNR was then calculated using the following equation: SNR = $(\sqrt{(2 \times S)})/\sigma$; where S is the mean signal in a ROI, and σ is the standard deviation from the same ROI.

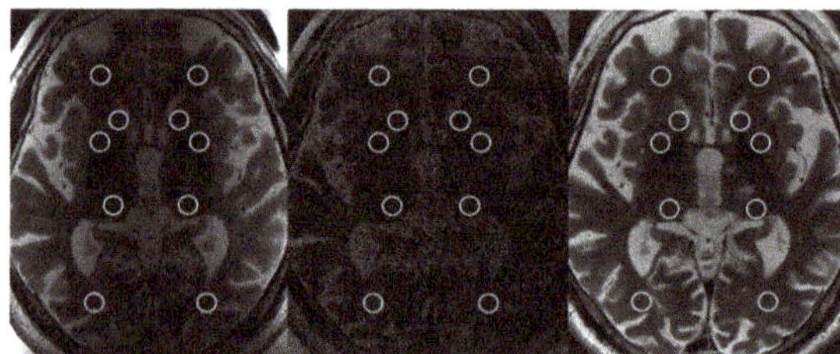

Figure 1. Axial T2-weighted FRFSE images acquired intraoperatively using the dedicated two-channel FUS-Head coil (**left image**); body radiofrequency coil (**central image**); and at 48-h using the eight-channel phased-array head coil (**right image**) at the level of the anterior commissure-posterior commissure plane, showing the placement of regions of interest for the evaluation of signal-to-noise ratio.

The readers also recorded the presence of the three concentric lesional zones in each sequence, as described in prior studies [8,12]. Zone I was defined as central spot markedly hypointense on T2-weighted images, and it represents the cavitating lesion. Zone II was considered as moderate-to-markedly hyperintense area on T2-weighted images, concentrically surrounding the zone I, and demarcated by a hypointense rim, which represents the cytotoxic edema. Zone III consisted of a peripheral slightly hyperintense area, which represents the perilesional vasogenic edema surrounding the ablation lesion.

2.5. Statistical Analysis

Data were summarized as continuous variables and expressed as mean and standard deviation (SD) or median and interquartile range (IQR), and categorical variable, expressed as numbers and percentages. The Shapiro–Wilk test was performed to assess the normality distribution of continuous variables. Intraindividual differences in SNR between sequences acquired with different coils were compared using the Wilcoxon signed rank sum test. Differences in qualitative imaging analysis were assessed using the McNemar Test. The intraclass correlation coefficient (ICC), with 95% confidence intervals (95% CI), was calculated to assess the inter-reader agreement for continuous variables (SNR), while the Cohen's kappa (*k*) test, with 95% confidence intervals (95% CI), was used for categorical variables. Agreement was categorized as poor (<0.00), slight (0.00–0.20), fair (0.21–0.40), moderate (0.41–0.60), substantial (0.61–0.80), or almost perfect (0.81–1.00). Statistical significance level was set at $p < 0.05$. Statistical analysis was conducted using SPSS software (Version 20.0. Armonk, NY, USA: IBM Corp).

3. Results

3.1. Patients

The characteristics of the final population and tcMRgFUS sonications parameters are summarized in Table 2. A total number of eight patients were protectively enrolled for the purpose of this study, including seven men and one woman with a mean age of 74.1 ± 5.4 years (range 65–81 years). All patients underwent tcMRgFUS for the treatment of essential tremor. All the TcMRgFUS thalamotomies were performed on the left Vim, according to the hand-dominant tremor side. The mean SDR was 0.48 ± 0.04 (range 0.42–0.56). The number of high-energy sonications reaching an average temperature greater than 50 °C ranged from two to five. The maximum temperatures in treatments sonications ranged from 51 to 62, while the average temperature ranged from 49 to 57.

3.2. Signal-to-Noise Ratio

A total number of 288 ROIs (12 ROIs in each sequence) were placed by each reader in order to assess the signal-to-noise ratio. The SNR on axial T2-weighted FRFSE acquired after the last high-energy sonication using different coils and at 48 h are reported in Table 3.

Intraoperative axial T2-weighted FRFSE images acquired using the dedicated 2ch-FUS coil demonstrated a significantly higher SNR (R1, median SNR: 10.54, IQR: 9.05, 12.61; R2, median SNR: 9.52, IQR: 7.74, 11.36) compared to the images acquired with the body-RF coil (R1, median SNR: 2.96, IQR: 2.77, 3.31, $p < 0.001$; R2, median SNR: 2.99, IQR: 2.83, 3.26, $p < 0.001$) (Figure 2). The dedicated 2ch-FUS coil allowed to increase the SNR on intraoperative images by an average of $254 \pm 74\%$ and $211 \pm 86\%$ measured by R1 and R2, respectively. However, when compared with the standard 8ch-HEAD coil, the dedicated 2ch-FUS coil achieved significantly lower SNR ($p < 0.001$ for both readers), with a loss of SNR on intraoperative images of $31 \pm 24\%$ for R1 and $15 \pm 46\%$ for R2, compared to the images acquired at 48-h follow-ups.

The inter-reader agreement for SNR measurements was almost perfect (ICC: 0.85, 95% CI: 0.78, 0.89).

Table 2. Characteristics of the final treated population.

Characteristics	Number
Patients	8
Sex	
Males	7 (87.5%)
Females	1 (12.5%)
Age (years)	
Mean ± SD (range)	74.1 ± 5.4 (65–81)
Thalamotomy side	
Left Vim	8 (100%)
Right Vim	0 (0%)
SDR	
Mean ± SD (range)	0.48 ± 0.04 (0.42–0.59)
Skull area	
Mean ± SD (range)	343.6 ± 18.2 (323–371)
Treatment elements	
Mean ± SD (range)	948 ± 48.1 (839–996)
Number of sonications	
Mean ± SD (range)	11.7 ± 2.1 (9–15)
Number of High-energy sonications (Stage IV)	
Mean ± SD (range)	4.0 ± 1.0 (2–5)
Energy (Joule)	
Mean ± SD (range)	11,696.5 ± 6646.4 (4536–28,062)
Power (Watt)	
Mean ± SD (range)	628.7 ± 97.6 (438–791)
Time (seconds)	
Mean ± SD (range)	19.5 ± 8.5 (11–41)
Maximum temperatures (°C)	
Mean ± SD (range)	55.5 ± 2.7 (51–62)
Average temperatures (°C)	
Mean ± SD (range)	52.7 ± 2.3 (49–57)

Continuous variables are expressed as mean ± standard deviation (SD), categorical variables are expressed as numbers and percentages. Abbreviation: SDR: skull-density ratio.

Table 3. Signal-to-nose ratio (SNR) differences between coils.

	2ch-FUS SNR	Body-RF SNR	8ch-HEAD SNR	p Value 2ch-FUS vs. body-RF	p Value 2ch-FUS vs. 8ch-HEAD	ICC (95% CI)
Reader 1	10.54 (9.05, 12.61)	2.96 (2.77, 3.31)	16.24 (13.10, 19.95)	<0.001	<0.001	0.85 (0.78, 0.89)
Reader 2	9.52 (7.74, 11.36)	2.99 (2.83, 3.26)	13.24 (10.67, 18.31)	<0.001	<0.001	

Variables are expressed as median and interquartile range (25th to 75th percentile). Variables were compared using the Wilcoxon signed rank sum test. Inter-reader agreement was assessed using the intraclass correlation coefficient (ICC), with 95% confidence intervals (95% CI). Statistically significant values ($p < 0.05$) are highlighted in bold. Abbreviations: 2ch-FUS: Two-channel FUS-Head Coil; Body-RF: Body Radiofrequency Coil; 8ch-HEAD: Eight-channel phased-array head Coil; SNR: Signal-to-Noise Ratio; ICC: Intraclass Correlation Coefficient.

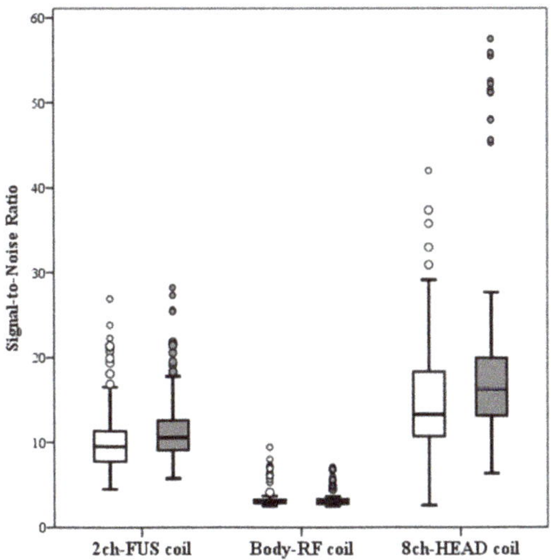

Figure 2. Plot box illustrates the distribution of signal-to-noise ratio in the images acquired using the dedicated two-channel FUS-Head (2ch-FUS) coil, body radiofrequency (body-RF) coil and eight-channel phased-array head (8ch-HEAD) coil by Reader 1 (white boxes) and Reader 2 (grey boxes).

3.3. Qualitative Imaging Findings

The visualization rate of concentric lesional zones at intraoperative and 48-h imaging are reported in Table 4. On intraoperative axial 2D FRFSE T2-weighted images, zone I was already visible in 75% of patients for R1 and 100% for R2 in the images acquired with the dedicated 2ch-FUS coil, while it was never visualized when the images were acquired with the body-RF coil (p = 0.031 and p = 0.008 for R1 and R2, respectively). Similarly, zone II was always observed by both readers in the images acquired with the dedicated 2ch-FUS coil, while it was recorded only in one (12.5%) case by R1 (p = 0.016) and in no case by R2 (p = 0.008) on the subsequent acquisition performed with body-RF coil. There was no difference in the visualization rate of zone III on intraoperative images acquired by both coils (Table 4).

Table 4. Comparison of visualization of the three concentric zones on intraoperative images obtained with head and body coils.

	2ch-FUS	Body-RF	8ch-HEAD	*p* Value 2ch-FUS vs Body-RF	*p* Value 2ch-FUS vs 8ch-HEAD	*k* Value (95% CI)
Zone I Present						
Reader 1	6 (75.0)	0 (0)	8 (100)	**0.031**	0.500	0.82
Reader 2	8 (100)	0 (0)	8 (100)	**0.008**	1.000	(0.59, 1.00)
Zone II Present						
Reader 1	8 (100)	1 (12.5)	8 (100)	**0.016**	1.00	0.90
Reader 2	8 (100)	0 (0)	8 (100)	**0.008**	1.00	(0.71, 1.00)
Zone III Present						
Reader 1	3 (37.5)	2 (25.0)	8 (100)	1.000	0.063	0.91
Reader 2	3 (37.5)	3 (37.5)	8 (100)	1.000	0.063	(0.75, 1.00)

Categorical variables (Zone I, II, III) are expressed as numbers and percentages in parenthesis and they were compared using the McNemar Test. Inter-reader agreement was assessed using the Cohen's kappa (k) test with 95% confidence intervals (95% CI). Statistically significant values (p < 0.05) are highlighted in bold. Abbreviations: 2ch-FUS: Two-channel FUS-Head Coil; Body-RF: Body Radiofrequency Coil; 8ch-HEAD: Eight-channel phased-array head Coil.

At 48-h follow-up, the three concentric zones were visualized in all patients by the two readers. There was no significant difference in the visualization of the three concentric zones between the intraoperative imaging acquired with dedicated 2ch-FUS coil and the 48 h follow-up, although zone III was more commonly visible at 48 h (37.5% vs. 100%, $p = 0.063$).

The inter-reader agreement was almost perfect for all three concentric zones (zone I, k: 0.82; zone II, k: 0.90; zone III, k: 0.91). An example of T2-weighted FRFSE intraoperative images acquired after the last sonication using both coils and at 48-h follow-up is reported in Figure 3.

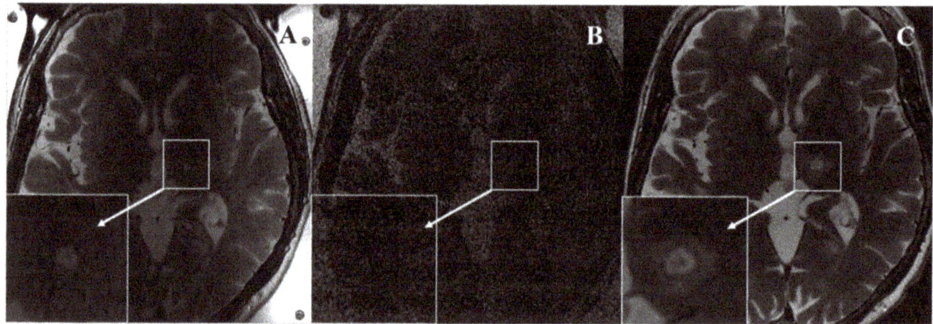

Figure 3. Seventy-eight-year-old woman with essential tremor. (**A**) Intraoperative axial 2D FRFSE T2-weighted sequence acquired after the last high-energy sonication using the dedicated two-channel FUS-Head coil well demonstrates the presence of zone I (hypointense) and zone II (hyperintense) in the left nucleus ventralis intermedius. (**B**) Intraoperative axial 2D FRFSE T2-weighted sequence acquired using body radiofrequency coil does not visualize any lesional zones. (**C**) Forty-eight-hour follow-up acquired with the eight-channel phased-array head coil demonstrates enlargement of zone I and zone II and development of peripheral vasogenic edema (zone III). Both readers agreed on the presence of these imaging findings.

4. Discussion and Future Directions

In this study, intraoperative tcMRgFUS anatomical MR sequences were prospectively acquired after the last high-energy sonication using a dedicated 2ch-FUS coil and the standard body-RF coil on a 1.5-T MR scanner. Our results demonstrate that the dedicated 2ch-FUS coil significantly increased the SNR ($p < 0.001$) compared to the identical images acquired with the body-RF coil. Overall, the SNR increased by an average of 254 $\pm$ 74% for R1 and 211 $\pm$ 86% for R2 on axial T2-weighed FRFSE sequence. To the best of our knowledge, this is the first study evaluating the improvements of intraoperative neuroimaging quality acquired in patients undergoing tcMRgFUS with 1.5-T MR for the treatment of essential tremor. A recent technical note [10] reported that the 2ch-FUS coil allowed to increase the SNR by 10 times using the 1.5-T MR. However, those results were based only on measurements performed using a gel phantom [10]. Werner et al. [17] also described a custom-built eight-channel phased head receive array coil integrated to 3.0-T scanner. Their experience reports an increase of SNR by a factor of 3.5 times in intraoperative imaging compared to the standard body coil [17]. However, this preliminary report did not assess the intraoperative visualization of concentric lesional zones.

High quality imaging during tcMRgFUS is particularly challenging due to the presence of FUS equipment composed of a 30-cm-diameter water-filled helmet which fills almost the entire space within the MR scanner. As a consequence, the image quality with body-RF coil results in being inferior to the current neuroradiological diagnostic standards in terms of SNR and image artifacts, even when using 3.0-T scanners [9,17]. Despite the use of dedicated 2ch-FUS coil, a similar trend was observed in our study using 1.5-T MR scanner. The intraoperative SNR remained significantly lower compared to the images acquired at

48-h using the 8ch-HEAD coil ($p < 0.001$). However, there was only a mean SNR decrease of $31 \pm 24\%$ for R1 and $15 \pm 46\%$ for R2. This lower SNR did not impede the visualization of typical lesional zones on intraoperative imaging when using the 2ch-FUS coil and allowed us to schedule the first MRI follow-up at 48-h post-treatment.

In our study, the intraoperative images acquired after the last high-energy sonication using the 2ch-FUS coil allowed the optimal visualization of ablation zones (zone I and zone II) in almost all patients (75% by R1 and 100% by R2). Despite the fact that body-RF coil sequence was acquired after being scanned using the 2ch-FUS coil in all patients, giving potential time for further lesion consolidation, zone I was not visualized in any patient by R1 and R2 ($p = 0.031$ and $p = 0.008$, respectively), while zone II was scored in only one case by R1 and in no patient by R2 ($p = 0.016$ and $p = 0.008$, respectively). In prior studies, neuroimaging lesional findings were most commonly described in the MR images acquired immediately after the procedure or at 24-h follow-ups [8]. Intraoperative high-resolution images that accurately detect the lesional findings, without emptying the water-filled FUS helmet or removing the stereotactic frame, may have a significant impact for the treatments monitoring and early detection of any possible adverse events [12]. In our clinical experience, intraoperative axial FRFSE T2-weighed images are acquired between the high-energy sonications according to the real-time clinical evaluation and patients' symptoms. Although the decision to stop the treatment is mainly based on the clinical efficacy and tremor suppression, the intraoperative visualization of the lesion is fundamental to document the correct location of the ablation zone and adequate lesion volume at the end of the treatment, without compromising the possibility of further sonications.

During follow-ups, the FUS-placed lesions typically enlarged in the first 48 h and then started to gradually decrease after 1 week [8,18,19]. In our study, all three concentric lesional zones were observed at 48 h in all the patients by both readers. As expected, zone III, representing the vasogenic edema, was more frequently observed at 48 h compared to intraoperative imaging (100% vs. 37.5%), although this difference did not reach the statistically significant level ($p = 0.063$ for both R1 and R2).

Our results may have significant implications for the further expansion of tcMRgFUS unique clinical and research applications. Optimization of high-quality intraoperative imaging may be necessary for making the tcMRgFUS a feasible image-guided intervention for the precise intracranial tumor ablation [20,21], target identification in psychiatric disorders [22–25], and reversible controlled blood-brain barrier opening to enhance therapeutic drugs delivery in specific brain regions [26–30], even using 1.5-T MR scanners. In this context, precise intraoperative imaging may provide accurate real-time delineation of pathological areas and direct visualization of the targeted spot, increasing the preciseness of the procedure and ensuring the preservation of nontarget tissues. All this will be even more interesting if other impulse sequences and/or weighting to be used in the intra-procedural phase will be implemented.

The most important limitation of our study is the small number of included patients. Performing additional time-consuming sequences (each with a scanning time of about four minutes) at the end of the procedure increases the total treatment time, which usually ranges from 2 to 3 h. However, despite the small population, there were already unequivocal differences in quantitative and qualitative analyses. Furthermore, our study did not compare the SNR and lesions visualization with the image quality achieved with 3.0-T MR unit using the standard body-RF coil. Further multicentric studies should be performed to compare the image quality and intraoperative neuroimaging findings in treatment performed with 1.5-T or 3.0-T MR scanners. Finally, we did not evaluate the correlations among intraoperative imaging findings, sonications parameters, patients' characteristics, and clinical outcome at subsequent follow-ups, since other studies [9,31–33] have already assessed these aspects in larger cohorts.

5. Conclusions

In conclusion, the dedicated two-channel FUS-Head coil significantly increases the SNR on intraoperative anatomical images when the tcMRgFUS treatment is performed on 1.5-T MR scanner. We anticipate that the use of a dedicated coil with 3-T integrated tcMRgFUS systems is desirable and would fill the gap with nowadays 1.5-T integrated scanners. Anatomical high-resolution intraoperative images allow the accurate visualization of concentrically lesional zones after the high-energy sonications and may have a significant role in guiding the tcMRgFUS procedures. Considering that more and more often it is possible to use imaging techniques for the identification of specific biomarkers or therapeutic targets, in consideration of the increasing advances in the field of trans-cranial focused ultrasound, high resolution intraoperative imaging could be the ideal companion for increasingly tailored therapies.

Author Contributions: Conceptualization, C.G. and R.C.; methodology, C.G. and R.C.; software, R.C.; validation, C.G., R.C. and C.D.; formal analysis, R.C. and C.D.; investigation, C.G., P.T., G.S., L.L.G., G.L.T., M.M. (Massimo Midiri), M.D., D.G.I.; resources, C.G. and M.M. (Massimo Midiri); data curation, C.G., R.C., P.F. and C.D.; writing—original draft preparation, C.G., R.C. and C.D.; writing—review and editing, all authors, supervision C.G., M.M. (Massimo Midiri); project administration, C.G., M.M. (Maurizio Marrale), M.M. (Massimo Midiri); funding acquisition, C.G. and M.M. (Massimo Midiri). All authors have read and agreed to the published version of the manuscript.

Funding: The installation of the tcMRgFUS equipment named in this paper was funded by the Italian Ministry of Education, University and Research (MIUR) within the project "Programma Operativo Nazionale 2007–3013" (PONa3_00011; project leader: Carlo Catalano). The research leading to these results has received funding from the Italian Ministry of Health "Ricerca Finalizzata 2016" call under grant agreement no. GR-2016-02364526; principal investigator: Cesare Gagliardo).

Institutional Review Board Statement: The institutional review board approved this study ("Comitato Etico Palermo 1"—seduta del 11/04/2018 verbale n 04/2018).

Informed Consent Statement: All subjects provided written informed consent before enrolling for tcMRgFUS treatment in accordance with the Declaration of Helsinki.

Data Availability Statement: The datasets generated during and/or analysed during the current study are available from the corresponding author on reasonable request.

Acknowledgments: The tcMRgFUS procedures performed at the University-Hospital of Palermo are the results of a trans-disciplinary collaboration lead by the Radiology sections of the Universities of Palermo and Rome, with the essential support of the Neurology and the Neurosurgery Units of the University-Hospital of Palermo and the Physicists from the Department of Chemistry and Physics of the University of Palermo. The authors feel very grateful for such a lively collaboration.

Conflicts of Interest: The authors declare no conflict of interest.

References

1. Elias, W.J.; Huss, D.; Voss, T.; Loomba, J.; Khaled, M.; Zadicario, E.; Frysinger, R.C.; Sperling, S.A.; Wylie, S.; Monteith, S.J.; et al. A pilot study of focused ultrasound thalamotomy for essential tremor. *N. Engl. J. Med.* **2013**, *369*, 640–648. [CrossRef]
2. Elias, W.J.; Lipsman, N.; Ondo, W.G.; Ghanouni, P.; Kim, Y.G.; Lee, W.; Schwartz, M.; Hynynen, K.; Lozano, A.M.; Shah, B.B.; et al. A Randomized Trial of Focused Ultrasound Thalamotomy for Essential Tremor. *N. Engl. J. Med.* **2016**, *375*, 730–739. [CrossRef]
3. Chang, J.W.; Park, C.K.; Lipsman, N.; Schwartz, M.L.; Ghanouni, P.; Henderson, J.M.; Gwinn, R.; Witt, J.; Tierney, T.S.; Cosgrove, G.R.; et al. A prospective trial of magnetic resonance-guided focused ultrasound thalamotomy for essential tremor: Results at the 2-year follow-up. *Ann. Neurol.* **2018**, *83*, 107–114. [CrossRef]
4. Magara, A.; Bühler, R.; Moser, D.; Kowalski, M.; Pourtehrani, P.; Jeanmonod, D. First experience with MR-guided focused ultrasound in the treatment of Parkinson's disease. *J. Ther. Ultrasound* **2014**, *2*, 11. [CrossRef]
5. Bond, A.E.; Shah, B.B.; Huss, D.S.; Dallapiazza, R.F.; Warren, A.; Harrison, M.B.; Sperling, S.A.; Wang, X.Q.; Gwinn, R.; Witt, J.; et al. Safety and Efficacy of Focused Ultrasound Thalamotomy for Patients with Medication-Refractory, Tremor-Dominant Parkinson Disease: A Randomized Clinical Trial. *JAMA Neurol.* **2017**, *74*, 1412–1418. [CrossRef]
6. Jeanmonod, D.; Werner, B.; Morel, A.; Michels, L.; Zadicario, E.; Schiff, G.; Martin, E. Transcranial magnetic resonance imaging-guided focused ultrasound: Noninvasive central lateral thalamotomy for chronic neuropathic pain. *Neurosurg. Focus* **2012**, *32*, E1. [CrossRef]

7.   Ghanouni, P.; Pauly, K.B.; Elias, W.J.; Henderson, J.; Sheehan, J.; Monteith, S.; Wintermark, M. Transcranial MRI-Guided Focused Ultrasound: A Review of the Technologic and Neurologic Applications. *AJR Am. J. Roentgenol.* **2015**, *205*, 150–159. [CrossRef] [PubMed]

8.   Wintermark, M.; Druzgal, J.; Huss, D.S.; Khaled, M.A.; Monteith, S.; Raghavan, P.; Huerta, T.; Schweickert, L.C.; Burkholder, B.; Loomba, J.J.; et al. Imaging findings in MR imaging-guided focused ultrasound treatment for patients with essential tremor. *AJNR Am. J. Neuroradiol.* **2014**, *35*, 891–896. [CrossRef] [PubMed]

9.   Federau, C.; Goubran, M.; Rosenberg, J.; Henderson, J.; Halpern, C.H.; Santini, V.; Wintermark, M.; Butts Pauly, K.; Ghanouni, P. Transcranial MRI-guided high-intensity focused ultrasound for treatment of essential tremor: A pilot study on the correlation between lesion size, lesion location, thermal dose, and clinical outcome. *J. Magn. Reson. Imaging* **2018**, *48*, 58–65. [CrossRef] [PubMed]

10. Gagliardo, C.; Midiri, M.; Cannella, R.; Napoli, A.; Wragg, P.; Collura, G.; Marrale, M.; Bartolotta, T.V.; Catalano, C.; Lagalla, R. Transcranial magnetic resonance-guided focused ultrasound surgery at 1.5T: A technical note. *Neuroradiol. J.* **2019**, *32*, 132–138. [CrossRef] [PubMed]

11. Iacopino, D.G.; Gagliardo, C.; Giugno, A.; Giammalva, G.R.; Napoli, A.; Maugeri, R.; Graziano, F.; Valentino, F.; Cosentino, G.; D'Amelio, M.; et al. Preliminary experience with a transcranial magnetic resonance-guided focused ultrasound surgery system integrated with a 1.5-T MRI unit in a series of patients with essential tremor and Parkinson's disease. *Neurosurg. Focus* **2018**, *44*, E7. [CrossRef]

12. Gagliardo, C.; Cannella, R.; Quarrella, C.; D'Amelio, M.; Napoli, A.; Bartolotta, T.V.; Catalano, C.; Midiri, M.; Lagalla, R. Intraoperative imaging findings in transcranial MR imaging-guided focused ultrasound treatment at 1.5T may accurately detect typical lesional findings correlated with sonication parameters. *Eur. Radiol.* **2020**, *30*, 5059–5070. [CrossRef] [PubMed]

13. Yang, A.I.; Chaibainou, H.; Wang, S.; Hitti, F.L.; McShane, B.J.; Tilden, D.; Korn, M.; Blanke, A.; Dayan, M.; Wolf, R.L.; et al. Focused Ultrasound Thalamotomy for Essential Tremor in the Setting of a Ventricular Shunt: Technical Report. *Oper. Neurosurg.* **2019**, *17*, 376–381. [CrossRef] [PubMed]

14. Chang, W.S.; Jung, H.H.; Zadicario, E.; Rachmilevitch, I.; Tlusty, T.; Vitek, S.; Chang, J.W. Factors associated with successful magnetic resonance-guided focused ultrasound treatment: Efficiency of acoustic energy delivery through the skull. *J. Neurosurg.* **2016**, *124*, 411–416. [CrossRef] [PubMed]

15. Wang, T.R.; Bond, A.E.; Dallapiazza, R.F.; Blanke, A.; Tilden, D.; Huerta, T.E.; Moosa, S.; Prada, F.U.; Elias, W.J. Transcranial magnetic resonance imaging-guided focused ultrasound thalamotomy for tremor: Technical note. *Neurosurg. Focus* **2018**, *44*, E3. [CrossRef]

16. National Electrical Manufacturers Association (NEMA). *Determination of Signal-to-Noise Ratio (SNR) in Diagnostic Magnetic Resonance Imaging. MS 1-2001*; NEMA Standards Publication: Rosslyn, WV, USA, 2001.

17. Werner, B.; Martin, E.; Bauer, R.; O'Gorman, R. Optimizing MR imaging-guided navigation for focused ultrasound interventions in the brain. *AIP Conf. Proc.* **2017**, *1821*, 120001.

18. Harary, M.; Essayed, W.I.; Valdes, P.A.; McDannold, N.; Cosgrove, G.R. Volumetric analysis of magnetic resonance-guided focused ultrasound thalamotomy lesions. *Neurosurg. Focus* **2018**, *44*, E6. [CrossRef]

19. Jung, H.H.; Chang, W.S.; Rachmilevitch, I.; Tlusty, T.; Zadicario, E.; Chang, J.W. Different magnetic resonance imaging patterns after transcranial magnetic resonance-guided focused ultrasound of the ventral intermediate nucleus of the thalamus and anterior limb of the internal capsule in patients with essential tremor or obsessive-compulsive disorder. *J. Neurosurg.* **2015**, *122*, 162–168.

20. Coluccia, D.; Fandino, J.; Schwyzer, L.; O'Gorman, R.; Remonda, L.; Anon, J.; Martin, E.; Werner, B. First noninvasive thermal ablation of a brain tumor with MR-guided focused ultrasound. *J. Ther. Ultrasound* **2014**, *2*, 17. [CrossRef]

21. Grasso, G.; Midiri, M.; Catalano, C.; Gagliardo, C. Transcranial Magnetic Resonance-Guided Focused Ultrasound Surgery for Brain Tumor Ablation: Are We Ready for This Challenging Treatment? *World Neurosurg.* **2018**, *119*, 438–440. [CrossRef]

22. Jung, H.H.; Kim, S.J.; Roh, D.; Chang, J.G.; Chang, W.S.; Kweon, E.J.; Kim, C.H.; Chang, J.W. Bilateral thermal capsulotomy with MR-guided focused ultrasound for patients with treatment-refractory obsessive-compulsive disorder: A proof-of-concept study. *Mol. Psychiatry* **2015**, *20*, 1205–1211. [CrossRef] [PubMed]

23. Kim, S.J.; Roh, D.; Jung, H.H.; Chang, W.S.; Kim, C.H.; Chang, J.W. A study of novel bilateral thermal capsulotomy with focused ultrasound for treatment-refractory obsessive-compulsive disorder: 2-year follow-up. *J. Psychiatry Neurosci.* **2018**, *43*, 327–337. [CrossRef] [PubMed]

24. Davidson, B.; Hamani, C.; Rabin, J.S.; Goubran, M.; Meng, Y.; Huang, Y.; Baskaran, A.; Sharma, S.; Ozzoude, M.; Richter, M.A.; et al. Magnetic resonance-guided focused ultrasound capsulotomy for refractory obsessive compulsive disorder and major depressive disorder: Clinical and imaging results from two phase I trials. *Mol. Psychiatry* **2020**, *25*, 1946–1957. [CrossRef] [PubMed]

25. Kinfe, T.; Stadlbauer, A.; Winder, K.; Hurlemann, R.; Buchfelder, M. Incisionless MR-guided focused ultrasound: Technical considerations and current therapeutic approaches in psychiatric disorders. *Expert Rev. Neurother.* **2020**, *20*, 687–696. [CrossRef]

26. Kinfe, T.; Stadlbauer, A.; Winder, K.; Hurlemann, R.; Buchfelder, M. Intracranial Applications of MR Imaging-Guided Focused Ultrasound. *AJNR Am. J. Neuroradiol.* **2017**, *38*, 426–431.

27. Lipsman, N.; Meng, Y.; Bethune, A.J.; Huang, Y.; Lam, B.; Masellis, M.; Herrmann, N.; Heyn, C.; Aubert, I.; Boutet, A.; et al. Blood-brain barrier opening in Alzheimer's disease using MR-guided focused ultrasound. *Nat. Commun.* **2018**, *9*, 2336. [CrossRef]

28. Abrahao, A.; Meng, Y.; Llinas, M.; Huang, Y.; Hamani, C.; Mainprize, T.; Aubert, I.; Heyn, C.; Black, S.E.; Hynynen, K.; et al. First-in-human trial of blood-brain barrier opening in amyotrophic lateral sclerosis using MR-guided focused ultrasound. *Nat. Commun.* **2019**, *10*, 4373. [CrossRef]

29. Mainprize, T.; Lipsman, N.; Huang, Y.; Meng, Y.; Bethune, A.; Ironside, S.; Heyn, C.; Alkins, R.; Trudeau, M.; Sahgal, A.; et al. Blood-Brain Barrier Opening in Primary Brain Tumors with Non-invasive MR-Guided Focused Ultrasound: A Clinical Safety and Feasibility Study. *Sci. Rep.* **2019**, *9*, 321. [CrossRef]
30. Lee, E.J.; Fomenko, A.; Lozano, A.M. Magnetic Resonance-Guided Focused Ultrasound: Current Status and Future Perspectives in Thermal Ablation and Blood-Brain Barrier Opening. *J. Korean Neurosurg. Soc.* **2019**, *62*, 10–26. [CrossRef]
31. Huang, Y.; Lipsman, N.; Schwartz, M.L.; Krishna, V.; Sammartino, F.; Lozano, A.M.; Hynynen, K. Predicting lesion size by accumulated thermal dose in MR-guided focused ultrasound for essential tremor. *Med. Phys.* **2018**, *45*, 4704–4710. [CrossRef]
32. Pineda-Pardo, J.A.; Urso, D.; Martínez-Fernández, R.; Rodríguez-Rojas, R.; Del-Alamo, M.; Millar Vernetti, P.; Máñez-Miró, J.U.; Hernández-Fernández, F.; de Luis-Pastor, E.; Vela-Desojo, L.; et al. Transcranial Magnetic Resonance-Guided Focused Ultrasound Thalamotomy in Essential Tremor: A Comprehensive Lesion Characterization. *Neurosurgery* **2020**, *87*, 256–265. [CrossRef] [PubMed]
33. Gagliardo, C.; Marrale, M.; D'Angelo, C.; Cannella, R.; Collura, G.; Iacopino, G.; D'Amelio, M.; Napoli, A.; Bartolotta, T.V.; Catalano, C.; et al. Transcranial Magnetic Resonance Imaging-Guided Focused Ultrasound Treatment at 1.5 T: A Retrospective Study on Treatmentand Patient-Related Parameters Obtained From 52 Procedures. *Front. Phys.* **2020**, *7*, 223. [CrossRef]

Review

Arylsulfatase A (ASA) in Parkinson's Disease: From Pathogenesis to Biomarker Potential

Efthalia Angelopoulou [1], Yam Nath Paudel [2], Chiara Villa [3,*] and Christina Piperi [1,*]

[1] Department of Biological Chemistry, Medical School, National and Kapodistrian University of Athens, 11527 Athens, Greece; angelthal@med.uoa.gr

[2] Neuropharmacology Research Laboratory, Jeffrey Cheah School of Medicine and Health Sciences, Monash University Malaysia, Bandar Sunway, Selangor 47500, Malaysia; yam.paudel@monash.edu

[3] School of Medicine and Surgery, University of Milano-Bicocca, 20900 Monza, Italy

* Correspondence: chiara.villa@unimib.it (C.V.); cpiperi@med.uoa.gr (C.P.); Tel.: +39-3497314439 (C.V.); +30-210-7462610 (C.P.)

Received: 7 September 2020; Accepted: 2 October 2020; Published: 7 October 2020

Abstract: Parkinson's disease (PD), the second most common neurodegenerative disorder after Alzheimer's disease, is a clinically heterogeneous disorder, with obscure etiology and no disease-modifying therapy to date. Currently, there is no available biomarker for PD endophenotypes or disease progression. Accumulating evidence suggests that mutations in genes related to lysosomal function or lysosomal storage disorders may affect the risk of PD development, such as *GBA1* gene mutations. In this context, recent studies have revealed the emerging role of arylsulfatase A (ASA), a lysosomal hydrolase encoded by the *ARSA* gene causing metachromatic leukodystrophy (MLD) in PD pathogenesis. In particular, altered ASA levels have been detected during disease progression, and reduced enzymatic activity of ASA has been associated with an atypical PD clinical phenotype, including early cognitive impairment and essential-like tremor. Clinical evidence further reveals that specific *ARSA* gene variants may act as genetic modifiers in PD. Recent in vitro and in vivo studies indicate that ASA may function as a molecular chaperone interacting with α-synuclein (SNCA) in the cytoplasm, preventing its aggregation, secretion and cell-to-cell propagation. In this review, we summarize the results of recent preclinical and clinical studies on the role of ASA in PD, aiming to shed more light on the potential implication of ASA in PD pathogenesis and highlight its biomarker potential.

Keywords: PD; Arylsulfatase A; lysosomes; GWAS; Gaucher's disease; prognostic biomarker

---

1. Introduction

Parkinson's disease (PD) is the most common neurodegenerative movement disorder, affecting approximately 1–2% of the population above the age of 60 years [1]. Idiopathic PD represents the most common cause of parkinsonism, which is a clinical syndrome encompassing a number of nosologic entities sharing mainly three cardinal motor features: bradykinesia, resting tremor and rigidity [2]. Less common parkinsonian disorders include other neurodegenerative diseases, such as multiple system atrophy (MSA) and progressive supranuclear palsy (PSP), drug-induced and vascular parkinsonism [2]. PD is a progressive disorder characterized by motor and non-motor symptoms, including cognitive impairment, depression and autonomic dysfunction [1]. Currently, there is no disease-modifying therapy against PD, and its treatment remains mainly symptomatic.

PD is not a single entity but rather a clinically heterogeneous disorder with diverse subtypes including tremor-dominant, postural instability-gait difficulty and akinetic-rigid forms of PD, and there is still no consensus for clinical PD sub-classification [3–5]. Most cases of PD are sporadic, while only

5-10% are caused by known inherited mutations in specific genes, including *SNCA*, leucine-rich repeat kinase 2 *(LRRK2), Parkin (PARK2)* and phosphatase and tensin homolog (PTEN)-induced putative kinase 1 *(PINK1)* [6]. Apart from *LRRK2* mutation carriers, who are mostly indistinguishable from idiopathic PD reflecting the majority of genetic PD cases, display atypical features [6]. In particular, PD patients carrying *SNCA* mutations have an earlier age at disease onset, a faster motor deterioration, earlier cognitive decline and prominent multimodal hallucinations (visual, olfactory, auditory) [7]. *Parkin* mutation carriers display an earlier onset age of the disease, more frequently associated with dystonia as an initial manifestation, with dementia being less common despite the long disease course [8]. Notably, the pathophysiological mechanisms underlying this clinical diversity are still unclear, and there is still no available biomarker that can effectively distinguish between PD endophenotypes or reflect disease progression [4,5].

The neuropathological hallmarks of PD include the degeneration of dopaminergic neurons in the substantia nigra pars compacta (SNpc) and the accumulation of Lewy bodies and Lewy neurites mainly consisting of α-synuclein [9,10]. Although the etiology of PD remains obscure, mitochondrial dysfunction, abnormal α-synuclein aggregation, excessive neuroinflammation, lysosomal impairment and dysregulation of lipid metabolism contribute to its pathogenesis. Of note, α-synuclein is degraded via both proteasomal and autophagic-lysosomal pathways, and it may itself impair lysosomal activity [11]. Furthermore, lysosomal dysfunction has been shown to contribute to cell-to-cell spreading of α-synuclein aggregates in the brain, a process highly implicated in PD progression [12].

The core of Lewy bodies contains large amount of lipids coated with high local concentrations of non-fibrillar α-synuclein [13]. α-synuclein is able to interact with fatty acids and phospholipids [14], and the imbalance of α-synuclein-lipid interaction has been proposed to affect its aggregation [15]. Genome wide association studies (GWAS) have revealed that several genetic loci associated with PD risk are implicated in lipid metabolism, including *GBA1*, diacylglycerol kinase (*DGKQ*) and the phospholipase *PLA2G6* genes [16].

Among the causative genes of PD, *SNCA, LRRK2, PARK2, PINK1* and *ATP13A2 (PARK9),* are highly implicated in lysosomal function. In particular, α-synuclein aggregates have been demonstrated to impair autophagic-lysosomal pathways, either via direct disruption of lysosomal components or indirectly by suppressing lysosomal trafficking [17]. LRRK2 protein can modulate lysosomal vesicular trafficking by phosphorylating Rab GTPases [17], while *PARK2* and *PINK1* mutations have been associated with endo-lysosomal defects [18]. Moreover, the *ATP13A2* gene encodes a lysosomal P-type ATPase that is highly implicated in cation homeostasis, while *ATP13A2* mutations have been associated with mitochondrial and lysosomal dysfunction [19].

GWAS have revealed at least 24 genetic loci that may alter PD risk, and many of them, such as *SLC17A5, ASAH1* and *CTSD* have been implicated in the autophagy-lysosomal pathways [20–22]. Notably, heterozygous mutations in the *GBA1* gene encoding glucocerebrosidase, whose homozygous mutations cause Gaucher's disease, constitute the most common genetic risk factor for idiopathic PD [23,24]. The presence of *GBA1* mutations has been also shown to affect the clinical phenotype of PD with a more rapid disease progression and cognitive impairment [25]. In addition, mutations in the NPC intracellular cholesterol transporter 1 *(NPC1)* gene, which cause the lysosomal disorder Niemann-Pick type C, have been associated with PD [26]. A recent study demonstrated that the majority of patients carry at least one potentially damaging variant in lysosomal storage disorder-related genes, and approximately one fifth of them possess multiple alleles [22]. The *ATP13A2* gene mutations, involved in a rare form of juvenile-onset parkinsonism and dementia, have been associated with the lysosomal storage disorder neuronal ceroid lipofuscinosis [27]. Hence, lysosomal impairment is suggested to play a crucial role in PD development, either as a causative or a contributing factor, with gene mutations causing other lysosomal storage disorders possibly affecting PD susceptibility.

Lysosomal storage disorders belong to the Mendelian-inherited metabolic diseases that are characterized by defects in the activity of lysosomal enzymes and the abnormal accumulation of undegraded substrates in the lysosomes of various tissues, including the heart, skin and brain [22,28].

Neurological manifestations, such as mental retardation, epilepsy and parkinsonism have been described in more than two-thirds of lysosomal storage disorders [28], including gangliosidosis, Niemann-Pick disease and Fabry–Anderson disease [29]. Furthermore, higher levels of α-synuclein oligomers have been detected in the plasma of patients with lysosomal storage disorders including Gaucher's disease, Niemann-Pick type C, Krabbe disease and Wolman disease compared to controls, while this difference is absent in patients with Gaucher's disease after enzyme replacement therapy [30]. These findings further strengthen the hypothesis that impairment of lysosomal enzymes may play a significant role in the pathogenesis of PD.

Arylsulfatase A (ASA) is a lysosomal enzyme that mainly hydrolyzes sulfatide (sulfogalactosylceramide), a glycolipid of myelin into galactosylceramide [31,32]. ASA is encoded by the *ARSA* gene, which is located on the chromosome locus 22q13.33. Homozygosity for *ARSA* mutations leads to severe ASA deficiency (<10%) that causes metachromatic leukodystrophy (MLD), a lysosomal storage disease that is inherited in an autosomal recessive manner with a reported frequency of 1.40000 [32,33]. MLD is characterized by abnormal accumulation of sulfatide primarily in the central nervous system (CNS), resulting in demyelination accompanied by motor and cognitive impairment [34,35].

Apart from sulfatide, ASA can also hydrolyze the sulfated glycolipids seminolipid and lactosylceramide sulfate [36]. Seminolipid exists only in small amounts in rat and mouse brain, whereas lactosylceramide sulfate was not detected in mammalian brain [36]. In addition to glial cells, sulfatide was present in neurons of *ARSA*-deficient mice [36], and neuronal sulfatide accumulation was associated with degeneration of Purkinje cells, axonal degeneration and cortical hyperexcitability [37]. Neuronal sulfatide storage is most prominent in the nuclei of medulla oblongata, pons and midbrain of *ARSA*-deficient mice [38], implying the potential role of ASA in neuronal functions unrelated to myelination. Notably, reduced levels of sulfatides by approximately 30% have been observed in the superior frontal and cerebellar gray matter of incidental PD human cases (with no complains of neurological deficits, although PD-related lesions were found in the neuropathological postmortem study) [39], suggesting that dysregulation of sulfatide metabolism may be implicated in the pathogenesis of PD.

ASA deficiency may be observed in healthy individuals (pseudodeficiency) displaying about 10–20% of normal enzymatic activity [33,34]. Partial ASA deficiency may result from homozygosity for the pseudodeficient *ARSA* allele, heterozygosity for a disease-causing *ARSA* allele or compound heterozygosity pseudodeficiency/deficiency. It still remains unclear whether partial ASA deficiency is completely benign [40]. Moreover, complete or partial ASA deficiency has been associated with parkinsonism and other movement disorders, including chorea, athetosis, dystonia, as well as several neurological conditions [29,41,42]. These findings suggest that ASA may also play a role in PD pathology and has led to an increasing number of preclinical and clinical studies investigating this relationship.

Although the potential implication of lysosomal ceramide metabolism disorders in PD pathogenesis has been already discussed in the literature [33], there is no recent review focusing specifically on the role of ASA in PD. Herein, we summarize recent emerging clinical and preclinical evidence on the possible role of ASA in PD and its biomarker potential, aiming to shed more light on the underlying mechanisms and their clinical impact.

2. Clinical Evidence on the Emerging Role of ASA in PD

2.1. ASA Levels and Activity as a Potential PD Biomarker

Reduced GBA activity has been detected in the cerebrospinal fluid of PD patients in comparison to controls [43] and in the blood of PD patients with and without *GBA1* mutations [44]. The activity of β-galactosidase and β-hexosaminidase, two other lysosomal enzymes, has been detected elevated in the CSF or blood of PD patients [43,45,46]. Therefore, it has been hypothesized that other lysosomal enzymes may be differentially expressed or display altered activity in PD patients in comparison to controls, independently of the presence of *ARSA* gene mutations.

Indeed, enzymatic activity of ASA in blood leukocytes has been shown significantly lower in patients with movement disorders, including essential tremor or parkinsonism, as compared to healthy controls or neurological patients without any movement disorder [47]. In particular, patients with atypical clinical features displayed reduced enzymatic ASA activity, while patients with typical essential tremor or parkinsonism showed normal values of ASA activity [47]. Atypical manifestations included pyramidal signs, orthostatic and continuous tremor (postural and resting tremor with bilateral onset of the same large amplitude), focal dystonia and/or no treatment response to levodopa, primidone or phenobarbitone [47]. Conventional neuroimaging of these patients with CT brain scans did not reveal any significant structural abnormalities [47]. Notably, a positive family history for movement disorders was mainly reported in the subgroup of patients with atypical symptoms [47], highlighting the potential contribution of *ARSA* gene polymorphisms or other inheritable factors affecting ASA activity in this clinical variability. A retrospective pedigree analysis of three of the patients with parkinsonism, positive family history and partial ASA deficiency demonstrated that the affected relatives displayed also partial ASA deficiency [40]. These patients exhibited atypical clinical features, such as head tremor, early postural or kinetic tremor, orthostatic tremor, cognitive impairment, and mild to moderate levodopa response [40]. Therefore, ASA deficiency may be possibly associated with an atypical PD clinical phenotype, which may also reflect a different pathophysiological background.

Postural tremor was the initial symptom in almost all cases of the study described above, and the only manifestation in a young patient [40]. It has been suggested that essential tremor may represent a potential "risk factor" for PD development, although the underlying pathophysiological mechanism remains unclear [48]. Given the results of the study reported above, it can be speculated that ASA deficiency might partially underlie this observed clinical association in some cases.

Interestingly, early cognitive impairment has been associated with *GBA1*-related PD [23,24,49], postural and action tremor has been reported in asymptomatic carriers of *LRRK2* mutations [50], and PD patients with *SNCA* mutations display early dementia [51]. These clinical observations suggest that mixed tremor and early cognitive decline may characterize PD cases related to genes associated with lysosomal function.

A recent study has demonstrated that plasma ASA levels were lower in PD patients with dementia compared to controls, whereas PD patients without dementia displayed higher plasma levels [31]. Moreover, plasma ASA levels were positively correlated with the scores of global cognitive performance, total Mini-Mental State Examination (MMSE) score and each cognitive domain separately, except for visuospatial function [31]. In PD patients, the clinical Dementia Rating-Sum of Boxes (CDR-SOB) scores were also negatively correlated with plasma ASA levels [31]. The different ASA levels between PD patients and controls, as well as PD patients with and without dementia, further highlight the contribution of ASA to PD pathogenesis and clinical endophenotypes.

Notably, a very recent study has demonstrated an interesting relationship between ASA plasma levels and PD progression [52]. In particular, the early PD subgroup characterized by shorter disease duration and lower UPDRS motor scores had higher ASA plasma levels, as compared to the late PD subgroup and healthy controls, independent of age, gender and MMSE scores [52]. Plasma ASA levels were graphically represented as an inverted U-shape in regard to the duration of the disease, reaching a peak at approximately 2 years of disease duration [52]. In the early PD subgroup, plasma ASA levels were also positively correlated with UPDRS motor scores and striatal dopamine depletion as evaluated by DATscan [52]. Other demographic or clinical parameters, such as age at disease onset, education years or MMSE scores, were not associated with plasma ASA levels [52]. These findings partially agree with the results of the abovementioned study which showed that PD patients without dementia displayed increased ASA levels, compared to PD with dementia or healthy controls [31]. Plasma ASA levels may increase at the early stages of PD in relation to nigrostriatal degeneration possibly reflecting an initial compensatory mechanism in response to accumulation of aggregated α-synuclein in neurons [52]. In agreement with this evidence, the concentration of plasma lysosomal enzymes differed in AD patients in regard to disease progression [53]. Hence, the plasma concentration of ASA may not be constant but rather change dynamically throughout disease course, implying that it could be used as a PD biomarker of disease severity and/or duration. However, larger longitudinal studies with serial measurements of plasma ASA levels in PD patients are required to validate their actual alterations during disease progression [52].

Apart from the brain, ASA is also expressed in the spinal cord, blood leukocytes and other peripheral tissues, and can be secreted by the cells [54]. It has been demonstrated that some serum exosomes may also contain lysosomal enzymes [55]. Nevertheless, the exact source of plasma ASA in both healthy controls and PD patients, as well as the proportional contribution of each source to plasma ASA concentration, remains to be elucidated [31]. Furthermore, it is largely unknown whether plasma ASA levels or activity correlate with the respective ASA levels or activity in the brain of PD patients and healthy controls.

It has been proposed that partial ASA deficiency alone or in combination with other yet unknown endogenous or exogenous factors may lead to dysfunction of specific neuronal cell populations that are particularly susceptible to metabolic alterations [56]. This hypothesis might at least partially explain the diverse and atypical parkinsonian clinical phenotype of patients with ASA deficiency caused by impairment of specific functional neuronal networks. Reduced GBA enzymatic activity has been reported in the brain of PD patients carrying *GBA1* mutations [57]. Decreased activity of GBA has been found in the SN and frontal cortex of patients with PD and LBD compared to controls [58]. Different enzymatic activity of GBA has been demonstrated in the substantia nigra and cerebellum of the brain of PD patients [57], highlighting that GBA activity is rather brain region-specific. Hence, further post-mortem investigation of ASA activity in the different brain structures of PD patients may explain the potentially ASA-related atypical parkinsonian clinical features.

Importantly, it has been demonstrated that GBA activity in healthy subjects decreases in an age-dependent manner in the brain regions mostly affected by PD, and at the 7th to 8th decade of life, GBA activity in the SN and putamen is reduced to the same extent as in patients with sporadic PD [59]. Therefore, GBA activity seems to be age-dependent even in the brain, and this hypothesis should be also taken into account and investigated in future studies exploring the role of ASA in PD.

In summary, ASA deficiency may be at least partially responsible for some PD cases with atypical symptoms, and represent a potential biomarker for clinical PD endophenotypes, mainly including patients with early cognitive dysfunction and early postural or mixed tremor. In addition, ASA activity may also aid in the differential diagnosis between PD and essential tremor, especially in cases with atypical presentation. However, given the small sample size of the abovementioned studies, further larger validation is needed to test these hypotheses.

2.2. ASA Localization in the Brain of PD Patients

Deposition of α-synuclein has been demonstrated in brain cells of patients with Gaucher's disease [60], and α-synuclein accumulation has been shown to be diffuse in the axons of neurons and glial cells of the cerebral white matter and brainstem of MLD patients [61]. α-synuclein-immunoreactivity colocalized with abnormal storage products including lipids in the brain of MLD patients, suggesting that impaired lipid metabolism may affect α-synuclein deposition in these cases [61]. Additionally, a granular pattern of α-synuclein appeared in the cytoplasm without fibrils in the aforementioned study [61]. In vivo evidence has indicated that α-synuclein can interact with polyunsaturated fatty acids resulting in the production of α-synuclein soluble oligomers, a process that precedes the formation of aggregates associated with neurodegeneration [62]. Hence, it seems that in MLD, α-synuclein accumulates in the cytosol without fibril formation, through potential binding to fatty acids.

α-synuclein was also found to accumulate in the neurons and glial cells of MLD cases, including astrocytes and microglia [61] while it can act directly on microglia, initiating a neuroinflammatory response and affecting neuronal survival [63]. In parallel, microglia are also involved in clearing of the extracellular α-synuclein aggregates by internalization and degradation, thus avoiding its accumulation [63]. Given the importance of neuroinflammation in PD pathogenesis, the role of ASA in neuron-to-microglia communication should be further explored.

Collectively, these findings suggest that α-synuclein accumulation may accompany MLD-related neuropathology, strengthening the hypothesis that ASA might be possibly implicated in PD-related α-synuclein deposition in the human brain.

In this context, post-mortem investigation of brain tissues derived from the anterior cingulate cortex of PD patients and age-matched healthy controls has indicated that ASA was present in neuronal and glial cells, as well as in cells of blood vessels in PD patients and controls [31]. Interestingly, ASA had a puncta (particle-like) appearance, and it was localized throughout the cytoplasm of neurons, surrounding or sometimes co-localized with Lewy bodies, particularly in cases with longer disease duration [31]. However, no significant differences were detected in either the neuronal intensity of ASA fluorescence signal or the number of ASA-stained neurons between PD patients and controls in this study [31]. Although interesting, these findings only demonstrate that ASA and Lewy bodies share a cytosolic localization, and therefore, more evidence is needed before suggesting that ASA may contribute to α-synuclein-containing Lewy bodies in PD.

2.3. Possible Association between ARSA Gene Variants and PD

The potential role of *ARSA* gene mutations in PD has been illustrated in a recent study that genetically analyzed a female patient of Japanese descent with adult-onset MLD and a positive family history of PD with mild cognitive dysfunction and probable essential tremor [31]. In particular, the MLD patient bears the compound heterozygous missense mutations, L300S (c.899T>C, rs199476389) and C174Y (c.521G>A, rs199476381) in the *ARSA* gene, while a heterozygous L300S mutation was found only in her family members with PD—her father and paternal uncle—but not in those without PD [31]. These findings suggest that the heterozygous L300S mutation in *ARSA* gene may be a risk factor for PD development, and that *ARSA*-related PD may be associated with early cognitive decline and essential tremor. However, the L300S *ARSA* variant was not detected in a case-control study in the Chinese population in either sporadic PD patients or healthy controls [64], implying that this variant might be pathogenic but rarely detected in PD. Moreover, the L300S and C174Y *ARSA* variants were not detected in more than 200,000 alleles in public databases, implying that these variants are very rare and indicating the need for larger multicenter studies in patients with familial and sporadic PD [31].

The abovementioned study has also genetically analyzed *ARSA* variants in a cohort of 92 patients with autosomal dominant familial PD and revealed a non-synonymous N352S variant (c.1055A>G, rs2071421) in both homozygous and heterozygous state in 4 and 11 patients, respectively, while the frequency of this variant was significantly higher in healthy controls within the Integrative Japanese Genome Variation Database [31]. Further analysis in another cohort of 92 patients with sporadic PD revealed that the frequency of the N352S variant was similar to that of the cohort of patients with autosomal dominant familial PD [31]. It has been demonstrated that the L300S *ARSA* mutation results in complete loss of ASA activity, while the N352S variant has no effect on ASA activity but rather leads to the loss of the N-glycosylation site of the enzyme [65]. Although the sample size of this study is relatively small, these results suggest that the N352S *ARSA* variant may represent a protective genetic factor against PD development.

On the contrary, a very large GWAS among patients with PD or other α-synucleinopathies (MSA, LBD and REM-sleep behavior disorder, RBD) and controls of European ancestry has found no significant associations between the N352S *ARSA* variant and these diseases [66]. In accordance, a large Japanese PD GWAS did not reveal any association between this genetic locus and PD [67]. The N352S *ARSA* variant is a frequent genetic polymorphism [66], and its potential association with PD would have been identified with relative certainty on a genome-wide level [66]. In accordance, a recent study of 407 sporadic PD patients and 471 healthy controls in the Chinese population did not demonstrate any association between the *N352S ARSA* variant and PD development [64]. However, given the fact that its frequency is highly variable among European and East Asian populations [66], a potential population-specific effect cannot be ruled out.

In conclusion, data supporting a possible genetic association between these *ARSA* variants and PD are still rather weak. Hence, additional larger case-control studies among both familial and sporadic PD cases and/or in other ethnic groups are required to determine whether N352S or additional *ARSA* variants are related to PD pathogenesis.

3. Preclinical Evidence on the Potential Role of ASA in PD Pathogenesis

Glucocerebrosidase deficiency has been shown to lead to lysosomal impairment, thus promoting the aggregation and propagation of α-synuclein [12]. Given the implication of ASA in lysosomal sphingolipid metabolism [31], it has been speculated that ASA deficiency and subsequent sulfatide accumulation may also contribute to α-synuclein aggregation and cell-to-cell propagation.

In this context, a very recent study has revealed that ASA might be able to inhibit α-synuclein aggregation and propagation by acting as a non-lysosome-related molecular chaperone that could interact with α-synuclein in the cytosol [31]. In particular, *ARSA* knockout in SH-SY5Y human neuroblastoma cell lines, which resulted in almost complete loss of ASA enzymatic activity and increased sulfatide levels, was associated with elevated intracellular and extracellular soluble and insoluble fractions of α-synuclein aggregates, in comparison to the wild-type cells [31]. Notably, cell-to-cell transfer of aggregated α-synuclein was also enhanced in *ARSA* knockout cells, implying that *ARSA* deficiency may promote α-synuclein aggregates generation, secretion and propagation in vitro [31]. ASA could also protect against α-synuclein fibrillation in a dose-dependent manner in vitro, further strengthening this hypothesis. The enhanced *ARSA* depletion-induced α-synuclein propagation was also confirmed in vivo, in *Caenorhabditis elegans* models [31]. These findings suggest that ASA deficiency may promote α-synuclein aggregation and cell-to-cell transmission, potentially contributing to PD pathogenesis and/or progression.

Further analyses aiming to identify the underlying molecular mechanisms demonstrated that the effects of ASA on α-synuclein aggregation and propagation were independent of lysosomes or its enzymatic activity, but they might be mediated by its function as a molecular chaperone [31]. More specifically, ASA was able to directly interact with α-synuclein in the cytosol of differentiated SH-SY5Y cells where it displayed no enzymatic activity as well as in the brain of A53T mutant α-synuclein transgenic and non-transgenic mice [31]. Interestingly, the protein encoded by the pathogenic variant L300S *ARSA* interacted weakly with α-synuclein, while the protein encoded by the protective variant N352S *ARSA* displayed a stronger binding affinity with α-synuclein in vitro, compared to the wild-type ASA [31]. Furthermore, the protein encoded by the wild-type or N352S *ARSA* variant significantly reduced α-synuclein aggregation and propagation, in comparison to the L300S *ARSA* variant in vitro [31]. In α-synuclein transgenic fly lines, motor impairment as assessed by the climbing activity was rescued by the expression of the wild-type or the protective N352S *ARSA* variants, while L300S *ARSA* expression was not able to reverse the motor deficits of these insects [31]. It was also indicated that L300S *ARSA* expression was associated with increased mRNA but reduced protein levels, implying a potential decreased translation rate [31].

Collectively, these results suggest that the wild-type and N352S *ARSA* variant may protect against α-synuclein aggregation and spreading by acting as a molecular chaperone of α-synuclein, whereas the L300S *ARSA* variant may exert opposing effects, thus enhancing α-synuclein aggregation and transmission. These mechanisms may underlie the protective and pathogenic role of N352S and L300S *ARSA* variants in the risk of PD development respectively, as indicated by the clinical evidence described above.

4. Discussion and Future Perspectives

In most cases, PD is considered a multifactorial disorder, since a complex interplay between several genetic and environmental factors may contribute to its development. Apart from the possibility of yet unknown genetic risk loci, epigenetic modifications may also account for familial forms of PD [22]. Potential interactions among rare or common alleles at various genetic loci may also be responsible for PD development [22], affecting the rate of disease progression or the clinical phenotype. Therefore, it is possible that specific *ARSA* gene variants may act in a "multi-hit" manner in combination with other yet unknown genetic or environmental factors, resulting in lysosomal impairment, α-synuclein aggregation and subsequent increased risk of PD (Figure 1).

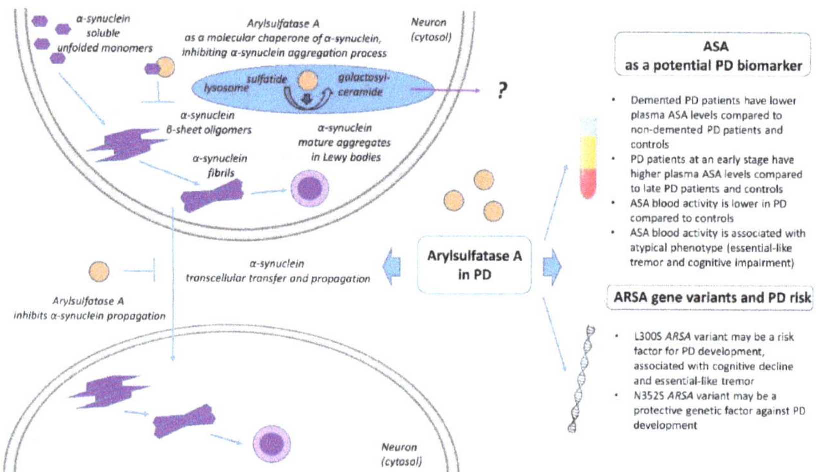

Figure 1. Implication of arylsulfatase A (ASA) in Parkinson's disease (PD). In vitro and in vivo evidence has demonstrated that ASA may act as a molecular chaperone interacting with α-synuclein (SNCA) in the cytoplasm, thereby preventing its aggregation, secretion, and cell-to-cell propagation. Clinical evidence supports the role of ASA blood levels and activity as a potential biomarker for PD pathology, progression and PD phenotype. ARSA gene variants have been also shown to act as genetic modifiers of PD development.

Currently, there is no available biomarker for determining PD progression or distinguishing between different clinical endophenotypes. In this context, ASA levels and/or its enzymatic activity represent a promising candidate towards this direction (Figure 1). However, larger studies, investigating a combination of potential clinical, neuroimaging or other biochemical biomarkers may prove as a more appropriate approach [68].

In regard to the molecular mechanisms underlying the effects of ASA in PD, there is limited information up to date. ASA has been shown to act as a molecular chaperone of α-synuclein, thus inhibiting its aggregation. Of note, neuronal loss does not always correlate with sulfatide storage, as for instance Purkinje cells in the cerebellum degenerate in old ASA-deficient mice without any evidence for lipid accumulation [69]. However, this may rather not be the single underlying mechanism, given the multifaceted role of lipid metabolism in PD pathogenesis and the interaction of α-synuclein with fatty acids [15]. Although sulfatide is primarily found in oligodendrocytes, low amounts have been also detected in astrocytes and neurons [70]. It has been proposed that ApoE-containing lipoproteins secreted by astrocytes might take up sulfatide and get endocytosed by neurons via LDL receptor or LDL receptor-related proteins via an unknown mechanism [70]. This hypothesis could be further explored in the case of ASA deficiency also in PD. Notably, the amphipathic N-terminal domain of α-synuclein displays lipid binding properties and the missense mutations A53T and A30P of α-synuclein reside in the N-terminal domain [71]. Hence, the potential effects of ASA on mutated-α-synuclein-lipid interaction would further aid in the elucidation of its role in α-synuclein aggregation in PD.

Ambroxol, a pharmacological chaperone that binds to glucocerebrosidase, has been associated with motor improvement in animal models of PD [33,72]. A recent clinical trial has demonstrated that oral administration of Ambroxol was safe, well-tolerated and led to motor improvement in PD patients carrying or not *GBA1* mutations [72], paving the way for development of future therapeutic strategies based on lysosomic enzymes against PD. Another rational therapeutic approach is enzyme-replacement therapy with recombinant ASA, although crossing of blood–brain barrier represents a significant obstacle that would need to be overcome. In this context, a clinical trial has demonstrated that intrathecal delivery of recombinant human arylsulfatase A in children with MLD is well-tolerated [73]. Gene therapy via adeno-associated viral vectors for delivery of engineered DNA into cells is another promising treatment strategy that is currently under investigation for *GBA1*-related PD [74]. Finally, small molecules acting as enzyme enhancers, capable to cross the blood-brain barrier may also prove effective. Other small molecules targeting specific downstream pathways implicated in ASA function represent another candidate therapeutic target [75]. The appropriate choice of patients that are more likely to respond to a specific therapy is of paramount importance, since both carriers of *ARSA* variants as well as patients with sporadic idiopathic PD may benefit from ASA-related treatment approaches.

5. Conclusions

Emerging preclinical and clinical evidence supports the role of ASA in PD pathogenesis (Figure 1). ASA has been associated with an atypical PD clinical phenotype, including early cognitive impairment and essential tremor, as well as disease progression. Specific *ARSA* variants may act as genetic modifiers in PD, whereas there is also evidence that does not confirm this relationship. In vitro and in vivo evidence has also revealed that ASA may act as a molecular chaperone interacting with α-synuclein in the cytoplasm, thus preventing its aggregation, secretion and cell-to-cell propagation. Further larger, longitudinal, case-control studies in different ethnic populations are required for the clarification of the role of ASA in PD, given the advances on lysosome-targeted therapeutic strategies.

Author Contributions: E.A. carried out the literature review, conceptualized and prepared the initial draft. Y.N.P. and C.V. edited and contributed to the final manuscript. C.P. provided critical inputs, edited and contributed to the final version of the manuscript. All authors read and approved the final manuscript.

Funding: This research has not received any specific grant from funding agencies in the public, commercial or not-for-profit sectors.

Acknowledgments: Y.N.P. would like to acknowledge Monash University Malaysia for supporting with HDR Scholarship.

Conflicts of Interest: The authors have no conflicts of interest to declare.

Abbreviations

ASA: Arylsulfatase A; PD, Parkinson's disease; SNCA, α-synuclein; SNpc, Substantia nigra pars compacta; LRRK2, *Leucine-rich repeat kinase 2*; *GWAS*, Genome wide association studies; MLD, Metachromatic leukodystrophy; CNS, Central nervous system; MMSE, Mini-Mental State Examination; CDR-SOB, Clinical Dementia Rating-Sum of Boxes; MSA, Multiple system atrophy; RBD, REM-sleep behavior disorder; LBD, Lewy body dementia.

References

1. Rizek, P.; Kumar, N.; Jog, M. An update on the diagnosis and treatment of Parkinson disease. *Can. Med. Assoc. J.* **2016**, *188*, 1157–1165. [CrossRef]
2. Armstrong, M.J.; Okun, M.S. Diagnosis and Treatment of Parkinson Disease. *JAMA* **2020**, *323*, 548–560. [CrossRef] [PubMed]
3. Angelopoulou, E.; Paudel, Y.N.; Piperi, C. miR-124 and Parkinson's disease: A biomarker with therapeutic potential. *Pharmacol. Res.* **2019**, *150*, 104515. [CrossRef] [PubMed]
4. Thenganatt, M.A.; Jankovic, J. Parkinson Disease Subtypes. *JAMA Neurol.* **2014**, *71*, 499. [CrossRef] [PubMed]

5.   Angelopoulou, E.; Bozi, M.; Simitsi, A.M.; Koros, C.; Antonelou, R.; Papagiannakis, N.; Michalopoulos, I. The relationship between environmental factors and different Parkinson's disease subtypes in Greece: Data analysis of the Hellenic Biobank of Parkinson's disease. *Parkinsonism Relat. Disord.* **2019**, *67*, 105–112. [CrossRef]

6.   Cheon, S.-M.; Chan, L.; Chan, D.K.Y.; Kim, J.W. Genetics of Parkinson's Disease—A Clinical Perspective. *J. Mov. Disord.* **2012**, *5*, 33–41. [CrossRef]

7.   Kasten, M.; Klein, C. The many faces of alpha-synuclein mutations. *Mov. Disord.* **2013**, *28*, 697–701. [CrossRef]

8.   Doherty, K.M.; Silveira-Moriyama, L.; Parkkinen, L.; Healy, D.G.; Farrell, M.; Mencacci, N.E.; Ahmed, Z.; Brett, F.M.; Hardy, J.; Quinn, N.; et al. Parkin Disease–A Clinicopathological Entity? *J. Neurol. Neurosurg. Psychiatry* **2013**, *84*. [CrossRef]

9.   Angelopoulou, E.; Paudel, Y.N.; Shaikh, M.F.; Piperi, C. Fractalkine (CX3CL1) signaling and neuroinflammation in Parkinson's disease: Potential clinical and therapeutic implications. *Pharmacol. Res.* **2020**, *158*, 104930. [CrossRef]

10.   Angelopoulou, E.; Paudel, Y.N.; Villa, C.; Shaikh, M.F.; Piperi, C. Lymphocyte-Activation Gene 3 (LAG3) Protein as a Possible Therapeutic Target for Parkinson's Disease: Molecular Mechanisms Connecting Neuroinflammation to α-Synuclein Spreading Pathology. *Biology* **2020**, *9*, 86. [CrossRef]

11.   Lin, K.-J.; Chen, S.-D.; Liou, C.-W.; Chuang, Y.-C.; Lin, H.-Y.; Lin, T.-K.; Lin, K.-L. The Overcrowded Crossroads: Mitochondria, Alpha-Synuclein, and the Endo-Lysosomal System Interaction in Parkinson's Disease. *Int. J. Mol. Sci.* **2019**, *20*, 5312. [CrossRef] [PubMed]

12.   Bae, E.-J.; Yang, N.-Y.; Song, M.; Lee, C.S.; Lee, J.S.; Jung, B.C.; Lee, H.-J.; Kim, S.; Masliah, E.; Sardi, S.P.; et al. Glucocerebrosidase depletion enhances cell-to-cell transmission of α-synuclein. *Nat. Commun.* **2014**, *5*, 1–11. [CrossRef] [PubMed]

13.   Shahmoradian, S.H.; Lewis, A.J.; Genoud, C.; Hench, J.; Moors, T.E.; Navarro, P.P.; Castaño-Díez, D.; Schweighauser, G.; Graff-Meyer, A.; Goldie, K.N.; et al. Lewy pathology in Parkinson's disease consists of crowded organelles and lipid membranes. *Nat. Neurosci.* **2019**, *22*, 1099–1109. [CrossRef] [PubMed]

14.   Fanning, S.; Selkoe, D.J.; Dettmer, U. Vesicle trafficking and lipid metabolism in synucleinopathy. *Acta Neuropathol.* **2020**, 1–20. [CrossRef]

15.   Mori, A.; Imai, Y.; Hattori, N. Lipids: Key Players That Modulate α-Synuclein Toxicity and Neurodegeneration in Parkinson's Disease. *Int. J. Mol. Sci.* **2020**, *21*, 3301. [CrossRef]

16.   Klemann, C.J.; Martens, G.J.; Sharma, M.; Martens, M.B.; Isacson, O.; Gasser, T.; Poelmans, G. Integrated molecular landscape of Parkinson's disease. *NPJ Parkinsons Dis.* **2017**, *3*, 14. [CrossRef]

17.   Klein, A.; Mazzulli, J.R. Is Parkinson's disease a lysosomal disorder? *Brain* **2018**, *141*, 2255–2262. [CrossRef]

18.   Plotegher, N.; Duchen, M.R. Crosstalk between Lysosomes and Mitochondria in Parkinson's Disease. *Front. Cell Dev. Biol.* **2017**, *5*, 110. [CrossRef]

19.   Gusdon, A.M.; Zhu, J.; Van Houten, B.; Chu, C.T. ATP13A2 regulates mitochondrial bioenergetics through macroautophagy. *Neurobiol. Dis.* **2012**, *45*, 962–972. [CrossRef]

20.   Nalls, M.A.; Pankratz, N.; Lill, C.M.; Do, C.B.; Hernandez, D.G.; Saad, M.; DeStefano, A.L.; Kara, E.; Bras, J.; Sharma, M.; et al. Large-scale meta-analysis of genome-wide association data identifies six new risk loci for Parkinson's disease. *Nat. Genet.* **2014**, *46*, 989–993. [CrossRef]

21.   Abeliovich, A.; Gitler, A.D. Defects in trafficking bridge Parkinson's disease pathology and genetics. *Nature* **2016**, *539*, 207–216. [CrossRef] [PubMed]

22.   Robak, L.A.; Jansen, I.E.; Van Rooij, J.; Kraaij, R.; Jankovic, J.; Shulman, J.M.; Nalls, M.A.; Plagnol, V.; Hernandez, D.G.; Sharma, M.; et al. Excessive burden of lysosomal storage disorder gene variants in Parkinson's disease. *Brain* **2017**, *140*, 3191–3203. [CrossRef] [PubMed]

23.   Simitsi, A.; Koros, C.; Moraitou, M.; Papagiannakis, N.; Antonellou, R.; Bozi, M.; Angelopoulou, E.; Stamelou, M.; Michelakakis, H.; Stefanis, L. Phenotypic Characteristics in GBA-Associated Parkinson's Disease: A Study in a Greek Population. *J. Park. Dis.* **2018**, *8*, 101–105. [CrossRef] [PubMed]

24.   Petrucci, S.; Ginevrino, M.; Trezzi, I.; Monfrini, E.; Ricciardi, L.; Albanese, A.; Bove, F. GBA-Related Parkinson's Disease: Dissection of Genotype-Phenotype Correlates in a Large Italian Cohort. *Mov. Disord.* **2020**. [CrossRef] [PubMed]

25.   Brockmann, K.; Srulijes, K.; Pflederer, S.; Hauser, A.K.; Schulte, C.; Maetzler, W.; Berg, D. GBA-associated Parkinson's disease: Reduced survival and more rapid progression in a prospective longitudinal study. *Mov. Disord.* **2015**, *30*, 407–411. [CrossRef]

26. Klüenemann, H.H.; Nutt, J.G.; Davis, M.Y.; Bird, T.D. Parkinsonism syndrome in heterozygotes for Niemann-Pick C1. *J. Neurol. Sci.* **2013**, *335*, 219–220. [CrossRef]

27. Brás, J.; Guerreiro, R.; Hardy, J. SnapShot: Genetics of Parkinson's Disease. *Cell* **2015**, *160*, 570. [CrossRef]

28. Cox, T.M.; Cachón-González, M.B. The cellular pathology of lysosomal diseases. *J. Pathol.* **2011**, *226*, 241–254. [CrossRef]

29. Shachar, T.; Bianco, C.L.; Recchia, A.; Wiessner, C.; Raas-Rothschild, A.; Futerman, A.H. Lysosomal storage disorders and Parkinson's disease: Gaucher disease and beyond. *Mov. Disord.* **2011**, *26*, 1593–1604. [CrossRef]

30. Pchelina, S.N.; Nuzhnyi, E.; Emelyanov, A.; Boukina, T.; Usenko, T.; Nikolaev, M.; Salogub, G.; Yakimovskii, A.; Zakharova, E.Y. Increased plasma oligomeric alpha-synuclein in patients with lysosomal storage diseases. *Neurosci. Lett.* **2014**, *583*, 188–193. [CrossRef]

31. Lee, J.S.; Kanai, K.; Suzuki, M.; Kim, W.S.; Yoo, H.S.; Fu, Y.; Kim, D.-K.; Jung, B.C.; Choi, M.; Oh, K.W.; et al. Arylsulfatase A, a genetic modifier of Parkinson's disease, is an α-synuclein chaperone. *Brain* **2019**, *142*, 2845–2859. [CrossRef] [PubMed]

32. Gieselmann, V.; Krägeloh-Mann, I. Metachromatic Leukodystrophy—An Update. *Neuropediatrics* **2010**, *41*, 1–6. [CrossRef] [PubMed]

33. Paciotti, S.; Albi, E.; Parnetti, L.; Beccari, T. Lysosomal Ceramide Metabolism Disorders: Implications in Parkinson's Disease. *J. Clin. Med.* **2020**, *9*, 594. [CrossRef] [PubMed]

34. Bindu, P.S.; Mahadevan, A.; Taly, A.B.; Christopher, R.; Gayathri, N.; Shankar, S.K. Peripheral neuropathy in metachromatic leucodystrophy. A study of 40 cases from south India. *J. Neurol. Neurosurg. Psychiatry* **2005**, *76*, 1698–1701. [CrossRef]

35. Troy, S.; Wasilewski, M.; Beusmans, J.; Godfrey, C. Pharmacokinetic Modeling of Intrathecally Administered Recombinant Human Arylsulfatase A (TAK-611) in Children with Metachromatic Leukodystrophy. *Clin. Pharmacol. Ther.* **2020**, *107*, 1394–1404. [CrossRef]

36. Molander-Melin, M.; Pernber, Z.; Franken, S.; Gieselmann, V.; Månsson, J.E.; Fredman, P. Accumulation of sulfatide in neuronal and glial cells of arylsulfatase A deficient mice. *J. Neurocytol.* **2004**, *33*, 417–427. [CrossRef]

37. Eckhardt, M. The Role and Metabolism of Sulfatide in the Nervous System. *Mol. Neurobiol.* **2008**, *37*, 93–103. [CrossRef]

38. Wittke, D.; Hartmann, D.; Gieselmann, V. Lysosomal sulfatide storage in the brain of arylsulfatase A-deficient mice: Cellular alterations and topographic distribution. *Acta Neuropathol.* **2004**, *108*, 261–271. [CrossRef]

39. Fabelo, N.; Martín, M.V.; Santpere, G.; Marín, R.; Torrent, L.; Ferrer, I.; Díaz, M. Severe Alterations in Lipid Composition of Frontal Cortex Lipid Rafts from Parkinson's Disease and Incidental Parkinson's Disease. *Mol. Med.* **2011**, *17*, 1107–1118. [CrossRef]

40. Antelmi, E.; Rizzo, G.; Fabbri, M.; Capellari, S.; Scaglione, C.; Martinelli, P. Arylsulphatase A activity in familial parkinsonism: A pathogenetic role? *J. Neurol.* **2014**, *261*, 1803–1809. [CrossRef]

41. Sangiorgi, S.; Ferlini, A.; Zanetti, A.; Mochi, M. Reduced activity of arylsulfatase A and predisposition to neurological disorders: Analysis of 140 pediatric patients. *Am. J. Med. Genet.* **1991**, *40*, 365–369. [CrossRef] [PubMed]

42. Kappler, J.; Watts, R.W.E.; Conzelmann, E.; Gibbs, D.A.; Propping, P.; Gieselmann, V. Low arylsulphatase A activity and choreoathetotic syndrome in three siblings: Differentiation of pseudodeficiency from metachromatic leukodystrophy. *Eur. J. Nucl. Med. Mol. Imaging* **1991**, *150*, 287–290. [CrossRef] [PubMed]

43. Parnetti, L.; Chiasserini, D.; Persichetti, E.; Eusebi, P.; Varghese, S.; Qureshi, M.M.; Dardis, A.; Deganuto, M.; De Carlo, C.; Castrioto, A.; et al. Cerebrospinal fluid lysosomal enzymes and alpha-synuclein in Parkinson's disease. *Mov. Disord.* **2014**, *29*, 1019–1027. [CrossRef] [PubMed]

44. Alcalay, R.N.; Levy, O.A.; Waters, C.C.; Fahn, S.; Ford, B.; Kuo, S.-H.; Mazzoni, P.; Pauciulo, M.W.; Nichols, W.C.; Gan-Or, Z.; et al. Glucocerebrosidase activity in Parkinson's disease with and without GBA mutations. *Brain* **2015**, *138*, 2648–2658. [CrossRef]

45. Niimi, Y.; Ito, S.; Mizutani, Y.; Murate, K.; Shima, S.; Ueda, A.; Mutoh, T. Altered regulation of serum lysosomal acid hydrolase activities in Parkinson's disease: A potential peripheral biomarker? *Parkinsonism Relat. Disord.* **2019**, *61*, 132–137. [CrossRef]

46. Van Dijk, K.D.; Persichetti, E.; Chiasserini, D.; Eusebi, P.; Beccari, T.; Calabresi, P.; van de Berg, W.D. Changes in endolysosomal enzyme activities in cerebrospinal fluid of patients with Parkinson's disease. *Mov. Disord.* **2013**, *28*, 747–754. [CrossRef]

47. Martinelli, P.; Ippoliti, M.; Montanari, M.; Martinelli, A.; Mochi, M.; Giuliani, S.; Sangiorgi, S. Arylsulphatase A (ASA) activity in parkinsonism and symptomatic essential tremor. *Acta Neurol. Scand.* **1994**, *89*, 171–174. [CrossRef]

48. Thenganatt, M.A.; Jankovic, J. The relationship between essential tremor and Parkinson's disease. *Parkinsonism Relat. Disord.* **2016**, *22* (Suppl. 1), S162–S165. [CrossRef]

49. Jiang, Z.; Huang, Y.; Zhang, P.; Han, C.; Lu, Y.; Mo, Z.; Zhang, Z.; Li, X.; Zhao, S.; Cai, F.; et al. Characterization of a pathogenic variant in GBA for Parkinson's disease with mild cognitive impairment patients. *Mol. Brain* **2020**, *13*, 1–10. [CrossRef]

50. Marras, C.; Schüle, B.; Schuele, B.; Munhoz, R.P.; Rogaeva, E.; Langston, J.W.; Kasten, M.; Meaney, C.; Klein, C.; Wadia, P.M.; et al. Phenotype in parkinsonian and nonparkinsonian LRRK2 G2019S mutation carriers. *Neurology* **2011**, *77*, 325–333. [CrossRef]

51. Konno, T.; Ross, O.A.; Puschmann, A.; Dickson, D.W.; Wszolek, Z.K. Autosomal dominant Parkinson's disease caused by SNCA duplications. *Parkinsonism Relat. Disord.* **2016**, *22* (Suppl. 1), S1–S6. [CrossRef] [PubMed]

52. Yoo, H.S.; Lee, J.S.; Chung, S.J.; Ye, B.S.; Sohn, Y.H.; Lee, S.-J.; Lee, P.H. Changes in plasma arylsulfatase A level as a compensatory biomarker of early Parkinson's disease. *Sci. Rep.* **2020**, *10*, 1–6. [CrossRef] [PubMed]

53. Morena, F.; Argentati, C.; Trotta, R.; Crispoltoni, L.; Stabile, A.; Pistilli, A.; Di Baldassarre, A.; Calafiore, R.; Montanucci, P.; Basta, G.; et al. A Comparison of Lysosomal Enzymes Expression Levels in Peripheral Blood of Mild- and Severe-Alzheimer's Disease and MCI Patients: Implications for Regenerative Medicine Approaches. *Int. J. Mol. Sci.* **2017**, *18*, 1806. [CrossRef] [PubMed]

54. Doerr, J.; Böckenhoff, A.; Ewald, B.; Ladewig, J.; Eckhardt, M.; Gieselmann, V.; Matzner, U.; Brüstle, O.; Koch, P. Arylsulfatase A Overexpressing Human iPSC-derived Neural Cells Reduce CNS Sulfatide Storage in a Mouse Model of Metachromatic Leukodystrophy. *Mol. Ther.* **2015**, *23*, 1519–1531. [CrossRef] [PubMed]

55. Goetzl, E.J.; Boxer, A.; Schwartz, J.B.; Abner, E.L.; Petersen, R.C.; Miller, B.L.; Kapogiannis, D. Altered lysosomal proteins in neural-derived plasma exosomes in preclinical Alzheimer disease. *Neurology* **2015**, *85*, 40–47. [CrossRef] [PubMed]

56. Kappler, J.; Pötter, W.; Gieselmann, V.; Kiessling, W.; Friedl, W.; Propping, P. Phenotypic consequences of low arylsulfatase A genotypes (ASAp/ASAp and ASA-/ASAp): Does there exist an association with multiple sclerosis? *Dev. Neurosci.* **1991**, *13*, 228–231. [CrossRef]

57. Gegg, M.E.; Burke, D.; Heales, S.J.R.; Cooper, J.M.; Hardy, J.; Wood, N.W.; Schapira, A.H.V. Glucocerebrosidase deficiency in substantia nigra of parkinson disease brains. *Ann. Neurol.* **2012**, *72*, 455–463. [CrossRef]

58. Moors, T.E.; Paciotti, S.; Ingrassia, A.; Quadri, M.; Breedveld, G.; Tasegian, A.; Chiasserini, D.; Eusebi, P.; Duran-Pacheco, G.; Kremer, T.; et al. Characterization of Brain Lysosomal Activities in GBA-Related and Sporadic Parkinson's Disease and Dementia with Lewy Bodies. *Mol. Neurobiol.* **2018**, *56*, 1344–1355. [CrossRef]

59. Rocha, E.M.; Smith, G.A.; Park, E.; Cao, H.; Brown, E.; Hallett, P.; Isacson, O. Progressive decline of glucocerebrosidase in aging and Parkinson's disease. *Ann. Clin. Transl. Neurol.* **2015**, *2*, 433–438. [CrossRef]

60. Wong, K.; Sidransky, E.; Verma, A.; Mixon, T.; Sandberg, G.D.; Wakefield, L.K.; Morrison, A.; Lwin, A.; Colegial, C.; Allman, J.M.; et al. Neuropathology provides clues to the pathophysiology of Gaucher disease. *Mol. Genet. Metab.* **2004**, *82*, 192–207. [CrossRef]

61. Suzuki, K.; Iseki, E.; Togo, T.; Yamaguchi, A.; Katsuse, O.; Katsuyama, K.; Kanzaki, S.; Shiozaki, K.; Kawanishi, C.; Yamashita, S.; et al. Neuronal and glial accumulation of α- and β-synucleins in human lipidoses. *Acta Neuropathol.* **2007**, *114*, 481–489. [CrossRef] [PubMed]

62. Sharon, R.; Bar-Joseph, I.; Frosch, M.P.; Walsh, D.M.; Hamilton, J.A.; Selkoe, D.J. The formation of highly soluble oligomers of alpha-synuclein is regulated by fatty acids and enhanced in Parkinson's disease. *Neuron* **2003**, *37*, 583–595. [CrossRef]

63. Ferreira, S.A.; Romero-Ramos, M. Microglia Response During Parkinson's Disease: Alpha-Synuclein Intervention. *Front. Cell. Neurosci.* **2018**, *12*, 247. [CrossRef] [PubMed]

64. Ruan, Y.; Zheng, R.; Lin, Z.-H.; Gao, T.; Xue, N.-J.; Cao, J.; Tian, J.; Zhang, B.; Pu, J. Genetic analysis of arylsulfatase A (ARSA) in Chinese patients with Parkinson's disease. *Neurosci. Lett.* **2020**, *734*, 135094. [CrossRef] [PubMed]

65. Gieselmann, V.; Polten, A.; Kreysing, J.; Von Figura, K. Arylsulfatase A pseudodeficiency: Loss of a polyadenylylation signal and N-glycosylation site. *Proc. Natl. Acad. Sci. USA* **1989**, *86*, 9436–9440. [CrossRef]

66. Makarious, M.B.; Diez-Fairen, M.; Krohn, L.; Blauwendraat, C.; Bandres-Ciga, S.; Ding, J.; Pihlstrøm, L.; Houlden, H.; Scholz, S.W.; Gan-Or, Z. ARSA variants in α-synucleinopathies. *Brain* **2019**, *142*, e70. [CrossRef]

67. Satake, W.; Nakabayashi, Y.; Mizuta, I.; Hirota, Y.; Ito, C.; Kubo, M.; Tomiyama, H. Genome-wide association study identifies common variants at four loci as genetic risk factors for Parkinson's disease. *Nat. Genet.* **2009**, *41*, 1303–1307. [CrossRef]

68. He, R.; Yan, X.; Guo, J.; Xu, Q.; Tang, B.; Sun, Q. Recent Advances in Biomarkers for Parkinson's Disease. *Front. Aging Neurosci.* **2018**, *10*, 305. [CrossRef]

69. D'Hooge, R.; Hartmann, D.; Manil, J.; Colin, F.; Gieselmann, V.; De Deyn, P. Neuromotor alterations and cerebellar deficits in aged arylsulfatase A-deficient transgenic mice. *Neurosci. Lett.* **1999**, *273*, 93–96. [CrossRef]

70. Han, X. Potential mechanisms contributing to sulfatide depletion at the earliest clinically recognizable stage of Alzheimer's disease: A tale of shotgun lipidomics. *J. Neurochem.* **2007**, *103*, 171–179. [CrossRef]

71. Alza, N.P.; González, P.A.I.; Conde, M.A.; Uranga, R.M.; Salvador, G.A. Lipids at the Crossroad of α-Synuclein Function and Dysfunction: Biological and Pathological Implications. *Front. Cell. Neurosci.* **2019**, *13*, 175. [CrossRef] [PubMed]

72. Mullin, S.; Smith, L.; Lee, K.; D'Souza, G.; Woodgate, P.; Elflein, J.; Hällqvist, J.; Toffoli, M.; Streeter, A.; Hosking, J.; et al. Ambroxol for the Treatment of Patients with Parkinson Disease with and Without Glucocerebrosidase Gene Mutations. *JAMA Neurol.* **2020**, *77*, 427. [CrossRef] [PubMed]

73. Í Dali, C.; Sevin, C.; Krägeloh-Mann, I.; Giugliani, R.; Sakai, N.; Wu, J.; Wasilewski, M. Safety of intrathecal delivery of recombinant human arylsulfatase A in children with metachromatic leukodystrophy: Results from a phase 1/2 clinical trial. *Mol. Genet. Metab.* **2020**, in press.

74. Schneider, S.A.; Alcalay, R.N. Precision medicine in Parkinson's disease: Emerging treatments for genetic Parkinson's disease. *J. Neurol.* **2020**, *267*, 860–869. [CrossRef]

75. Maegawa, G.; A Patil, S. Developing therapeutic approaches for metachromatic leukodystrophy. *Drug Des. Dev. Ther.* **2013**, *7*, 729–745. [CrossRef] [PubMed]

Review

Recent Advances on the Role of GSK3β in the Pathogenesis of Amyotrophic Lateral Sclerosis

Hyun-Jun Choi [1], Sun Joo Cha [2], Jang-Won Lee [3], Hyung-Jun Kim [4] and Kiyoung Kim [2,5,*]

[1] Department of Integrated Biomedical Science, Soonchunhyang University, Cheonan 31151, Korea; chj5913@sch.ac.kr
[2] Department of Medical Science, Soonchunhyang University, Asan 31538, Korea; cktjswn92@sch.ac.kr
[3] Department of Integrated Bio-industry, Sejong University, Seoul 05006, Korea; wintrelove@sejong.ac.kr
[4] Dementia Research Group, Korea Brain Research Institute (KBRI), Daegu 41068, Korea; kijang1@kbri.re.kr
[5] Department of Medical Biotechnology, Soonchunhyang University, Asan 31538, Korea
* Correspondence: kiyoung2@sch.ac.kr; Tel.: +82-41-413-5024; Fax: +82-41-413-5006

Received: 20 August 2020; Accepted: 25 September 2020; Published: 26 September 2020

Abstract: Amyotrophic lateral sclerosis (ALS) is a common neurodegenerative disease characterized by progressive motor neuron degeneration. Although several studies on genes involved in ALS have substantially expanded and improved our understanding of ALS pathogenesis, the exact molecular mechanisms underlying this disease remain poorly understood. Glycogen synthase kinase 3 (GSK3) is a multifunctional serine/threonine-protein kinase that plays a critical role in the regulation of various cellular signaling pathways. Dysregulation of GSK3β activity in neuronal cells has been implicated in the pathogenesis of neurodegenerative diseases. Previous research indicates that GSK3β inactivation plays a neuroprotective role in ALS pathogenesis. GSK3β activity shows an increase in various ALS models and patients. Furthermore, GSK3β inhibition can suppress the defective phenotypes caused by SOD, TDP-43, and FUS expression in various models. This review focuses on the most recent studies related to the therapeutic effect of GSK3β in ALS and provides an overview of how the dysfunction of GSK3β activity contributes to ALS pathogenesis.

Keywords: amyotrophic lateral sclerosis; GSK3β; neurodegenerative disease

---

1. Introduction

Amyotrophic lateral sclerosis (ALS) is a fatal neurodegenerative disease characterized by progressive and selective degeneration of the upper and lower motor neurons in the brain and spinal cord [1–3]. The majority of reported ALS cases are classified into two categories: sporadic ALS (sALS) and familial ALS (fALS). Approximately 90% of patients suffer from sALS and do not have a family history of the disease. The remaining 10% of patients suffer from fALS, which is strongly associated with the presence of a family history and genetic cause of the disease [1]. Since the initial discovery of *copper–zinc superoxide dismutase (SOD1)* as a causal gene for ALS, numerous genes such as *TAR DNA-binding protein 43 (TDP-43)*, *fused in sarcoma (FUS)*, *TATA-binding protein-associated factor 15 (TAF15)*, and *chromosome 9 open reading frame 72 (C9orf72)*, have been found to be associated with fALS and sALS [4–10]. Various RNA/DNA-binding proteins share structural and functional properties that contribute to the pathogenesis of ALS. Although several studies on ALS-causing genes have substantially expanded and improved our understanding of ALS pathogenesis, the underlying molecular mechanisms of the disease remain poorly understood. Recent studies have discovered various genetic susceptibility factors, using several in vivo and in vitro models, that could explain the underlying mechanisms involved in ALS pathogenesis. Understanding these mechanisms would be immensely helpful for developing new drug targets for ALS treatment.

Glycogen synthase kinase 3 (GSK3) is a cellular serine/threonine protein kinase that is involved in glycogen metabolism. GSK3 is also a phosphorylating and an inactivating agent of glycogen synthase [11,12]. GSK3 is encoded by two paralogous genes, *GSK3α* and *GSK3β*. GSK3β is the more studied and better characterized GSK3 isoform because of its predominant expression in the majority of cells and tissues. The specific function of GSK3α is less known. GSK3β is expressed ubiquitously, but the only cells currently known to express GSK3α predominantly, compared to GSK3β, are spermatozoa [13]. Conditional knockout of *GSK3α* in developing testicular germ cells in mice results in male infertility [14]. Other studies reported a role for GSK3α in central nervous system functioning and possible involvement in the development of psychiatric disorders [15]. GSK3β participates in a variety of critical cellular processes, such as glycogen metabolism, gene transcription, apoptosis, and microtubule stability [16]. Dysfunction of GSK3β is implicated in a variety of diseases, including type 2 diabetes and cancer. Recent studies have also suggested a possible role of GSK3β in neurodegenerative diseases, such as Parkinson's disease (PD) and Alzheimer's disease (AD) [16,17]. As it inhibits several common pathogenic pathways in neurodegenerative diseases, GSK3β could be a potential target for the development of novel therapeutics for neurodegenerative diseases. Recent studies suggest that GSK3β may have a definitive role in the pathogenesis of ALS. In this review, we present the evidence from these GSK3β studies in ALS and summarize the data into two categories: in vitro and in vivo models.

2. Role of GSK3β Signaling in Neurons

There are two types of GSK3 isoforms, GSK3α and GSK3β, which have an overall 85% sequence identity and 95% homology in the kinase domains [18]. Moreover, it has been reported that both GSK3 isoforms are highly expressed in the brain and spinal cord [19]. There are two splice variants of GSK3β in rodents and humans, short-form (GSK3β1) and long-form (GSK3β2). GSK3β contains a 13 amino acid insert in the catalytic domain due to alternative splicing [19]. In contrast to ubiquitously expressed human GSK3β1, GSK3β2 is specifically expressed in the developing nervous system [19]; human GSK3β2 is expressed only in neurons during the differentiation stage and not in glial cells, whereas human GSK3β1 is expressed in glial cells [20]. Furthermore, recent studies suggest that the upregulation of human GSK3β2 in PC12 cells induces nerve growth factor to differentiate into a neuronal phenotype, playing a specific role in neuronal morphogenesis [20–22]. These findings suggest that the GSK3β isoforms may have a significant role in the development of neurons in the nervous system. Neuronal progenitor proliferation and differentiation are regulated by multiple extracellular and intracellular signaling pathways that are closely associated with GSK3β [23–25]. Previous studies revealed that GSK3 signaling is an essential mediator of homeostatic controls that regulate neural progenitors during mammalian brain development. Inactivation of GSK3β in mouse neural progenitors resulted in the hyperproliferation of neural progenitors. Moreover, the generation of both intermediate neural progenitors and postmitotic neurons was markedly suppressed [24]. GSK3β regulates the stability of various proteins through the ubiquitin–proteasome system [26,27]. GSK3β controls progenitor proliferation or differentiation by regulating the levels of transcription regulators involved in neurogenesis, such as β-catenin in Wnt signaling, Gli in Sonic Hedgehog (Shh) signaling, and c-Myc in fibroblast growth factor signaling in the nervous system [24,28,29]. In addition, GSK3β is an essential regulator of microtubule–cytoskeleton reorganization, neuronal polarity, and neuronal migration by phosphorylating key microtubule-associated proteins [30–32], such as microtubule plus-end-tracking proteins (+TIPs) [33], collapsing response mediator protein-2 (CRMP-2) [34], adenomatous polyposis coli (APC) [35], cytoplasmic linker associated proteins (CLASPs) [36], microtubule-associated protein 1B (MAP1B) [37], and tau [38]. APC and CLASP promote microtubule stability. However, phosphorylation of APC and CLASPs by GSK3β induces decreased activity and leads to the destabilization of microtubules in neurons [36,39]. Therefore, polarized deposition of polarity proteins underlies asymmetric cell division, which is necessary for the neurogenic division of neural progenitors. Indeed, the polarized apical concentrations of markers, including APC, cadherin, and end-binding 1 (EB1) were found to be significantly reduced in the cortex of GSK3β-deleted mice [24].

In addition, GSK3β is associated with the formation of neuronal morphology, including axonal growth, dendritic branching, and the development of synapses [40]. Inhibition of GSK3β activity impairs axon formation and disturbs polarity development by leading to the formation of multiple axon-like processes in neurons [41,42]. Furthermore, the genetic activation of GSK3β activity results in a shrunken form of dendrites, whereas inhibition of GSK3β activity promotes dendritic growth in vivo [43].

GSK3β activity affects neuroplasticity and neurotransmission. Glutamatergic synapses are the main excitatory synapses in the brain and consist of glutamate localized inside the presynaptic vesicles and glutamate receptors on the postsynaptic membrane. GSK3β regulates the interaction between two major forms of synaptic plasticity at glutamatergic synapses, in an N-methyl-D-aspartate (NMDA)-dependent long-term potentiation (LTP) and a long-term depression (LTD) manner. During LTP, the activation of NMDA receptors causes the inhibition of GSK3β activity through the phosphatidylinositol 3-kinase (PI3-K)/Akt pathway. The inhibition of GSK3β is necessary for the induction of LTP, the process underlying new memory formation, whereas the action of protein phosphatase 1 (PP1) in LTD causes an increase in GSK3β activity, which is related to the downsizing of synapses and decreased excitability of neurons. Thus, GSK3β is crucial for the initiation of NMDA-induced LTD in neurons [44,45]. Furthermore, the inhibition of GSK3β-mediated phosphorylation of gephyrin-Ser270, which is a scaffolding protein, induces GABAergic synaptogenesis [43,46].

In another function, GSK3β plays a key role in regulating metabolic proteins, intermediary metabolism, and mitochondrial function. AMP-activated protein kinase (AMPK) is a major factor that regulates cellular energy homeostasis by inhibiting anabolic processes and activating catabolic ones [47]. GSK3β constitutively interacts with the AMPK complex through the β subunit and phosphorylates the α subunit of AMPK. This phosphorylation enhances the accessibility of the activation loop of the subunit to phosphatases, thereby inhibiting the AMPK kinase activity [48]. Furthermore, the PI3-K/Akt pathway, which is a major anabolic signaling pathway, enhances GSK3β-dependent phosphorylation of the α subunit. This suggests that GSK3β-dependent AMPK inhibition is critical for cells to enter an anabolic state [48]. GSK3β is enriched in many neurons that critically depend on mitochondrial function. GSK3β also acts as a negative regulator of mitochondrial energy metabolism by inhibiting the activity of pyruvate dehydrogenase, which attenuates mitochondrial activity [49]. Mitochondrial dynamics are important for the maintenance of mitochondrial homeostasis [50,51]. GSK3β regulates mitochondrial dynamics via phosphorylation of dynamin-related protein 1 (Drp1), which regulates mitochondrial fission [52]. Mitochondrial fusion or fission is regulated by the phosphorylation site of Drp1, which is mediated by GSK3β [53]. GSK3β also regulates mitochondrial metabolism through PGC-1α, a transcriptional coactivator and master regulator of mitochondrial function. GSK3β regulates PGC-1α stability through phosphorylation, which can be recognized by the E3 ubiquitin ligase [54]. Several studies have shown that GSK3β regulates mitochondrial energy metabolism in the neurons and glia [55]. Moreover, GSK3β inhibition increases mitochondrial respiration and membrane potential, and alters NAD(P)H metabolism in the neurons. Moreover, GSK3β inhibition alters PGC-1α protein stability, localization, and activity. These studies support the idea that GSK3β may be important in neuronal metabolic integrity [55].

Thus, functional studies of GSK3β have shown that GSK3β plays key roles in many fundamental processes in the neurons during neurodevelopment, including neuronal progenitor homeostasis, neuronal migration, neuronal morphology, synaptic development, and neurotransmission. Furthermore, it plays a crucial role in the energy metabolism of neurons (Figure 1).

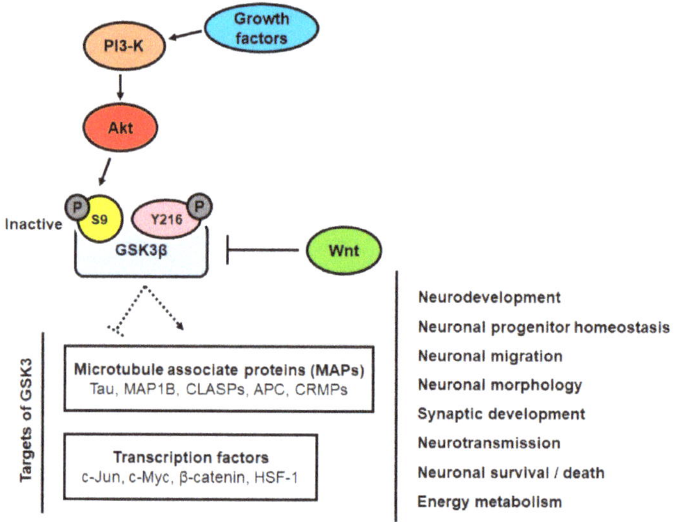

Figure 1. Role and regulation of glycogen synthase kinase 3 (GSK3) in the neuron. The Wnt and phosphatidylinositol 3-kinase (PI3-K)/Akt signaling pathways induce the inhibition of GSK3 activity by phosphorylating the serine 9 residue. This leads to subsequent regulation of its target proteins. GSK3 and its target proteins regulate a variety of biological processes, including neurodevelopment, neuronal migration, neurotransmission, neuronal cell death, and energy metabolism in the neurons.

3. Role of GSK3β in Neurodegenerative Diseases

GSK3 is a serine/threonine-protein kinase involved in glycogen metabolism and is also a phosphorylating and an inactivating agent of glycogen synthase [11,12]. It has two isoforms, GSK3α and GSK3β. GSK3α is predominantly present in the nucleus, and GSK3β is present in the cytoplasm. GSK3β plays an important role in the brain and participates in a variety of cellular processes, such as glycogen metabolism, gene transcription, apoptosis, and microtubule stability [16]. GSK3β is classified as a predominantly cytoplasmic enzyme but also localizes to three cellular compartments: the cytosol, the nucleus, and mitochondria [56]. The enzymatic activity of GSK3β is regulated by phosphorylation and depends on the phosphorylation of certain sites. This occurs through phosphorylation of the tyrosine 216 residue (Tyr216) located in the kinase region and is inactivated via phosphorylation of the amino-terminal serine 9 residue (Ser9) [16]. GSK3β is also involved in diverse cell signaling pathways. GSK3β activity is inhibited by the Wnt signaling pathway and is also negatively regulated by the PI3-K/Akt pathway (Figure 1). Additionally, activated GSK3β inhibits heat shock transcription factor-1 (HSF-1), affects the mitochondrial death pathway, and releases cytochrome *c* from the mitochondria. Released cytochrome *c* activates caspase-3 and induces apoptosis [57]. Dysfunction of GSK3β has been implicated in a variety of neurodegenerative diseases, such as PD and AD [16,17]. Many studies have shown that GSK3β is activated by amyloid-beta (Aβ) in AD, which eventually induces neuronal cell death and axonal transport defects by hyperphosphorylation of tau [58].

GSK3β also plays a critical role in the pathogenic mechanisms of PD and AD. α-synuclein is the most important factor in the pathogenesis of PD. Studies have shown that α-synuclein can also activate GSK3β. GSK3β inactivation decreases the protein level of α-synuclein, which in turn decreases α-synuclein-induced cell death in a cellular model of PD. Furthermore, activated GSK3β mediates the neurotoxic effects of tau hyperphosphorylation in AD [59]. Together, these findings suggest that inactivation of the GSK3 pathway may be an important therapeutic strategy for the treatment of PD and AD. In patients who have suffered a stroke, treatment with GSK3β inhibitors promotes neurovascular remodeling and improves ischemia. Recent studies have shown that GSK3β is closely associated with neuronal diseases [60]. Although the role of GSK3β in the pathogenesis of ALS remains poorly understood, recent evidence implicates GSK3β dysfunction in ALS models. Further studies are needed to determine the exact molecular mechanisms through which GSK3β affects ALS pathogenesis.

4. Studies Exploring the Role of GSK3β in In Vivo Models of ALS

Recent studies have reported that GSK3β is involved in the pathogenesis of ALS. Hu et al. reported that GSK3β expression and cytosolic levels of phospho-Y216 GSK3β increased in the spinal cord and frontal and temporal cortices of ALS patients [61]. In addition, Yang et al. demonstrated that the expression of GSK3β and phospho-Y216 GSK3β increased in patients with ALS compared to that in the controls, and that GSK3β and phospho-GSK3β immunoreactive neurons were mainly located in the frontal cortex and hippocampus of ALS patients [62]. These results show that GSK3β plays a critical role in ALS pathogenesis. Further studies are needed to determine the molecular mechanism of GSK3β in ALS pathogenesis. To investigate the molecular mechanism of GSK3β in ALS, many studies have been conducted using animal models. A genetic mutation in *SOD1* is one of the most common and important causes of ALS, accounting for 23% of fALS causes and approximately 7% of sALS causes worldwide [63]. A number of animal models of ALS are based on this gene.

PI3-K and its main downstream effector, Akt/protein kinase B, have been shown to play a central role in neuronal survival against apoptosis [64–67]. Interestingly, in the spinal cord of a SOD1^{G93A} mouse model of ALS, PI3-K and Akt expression decreased in a time-dependent manner [68]. Akt can inhibit GSK3β through the phosphorylation of GSK3β, indicating that altered GSK3β upstream signaling in SOD1^{G93A} mice may affect GSK3β activity (Figure 2). Semaphorin-3A (Sema3A), a member of the class 3 semaphorins, regulates the guidance of axonal and dendritic growth in the nervous system [69]. GSK3β regulates anterograde and retrograde dynein-dependent axonal transport in *Drosophila* and rats [70,71] and mediates Sema3A-induced axonal transport through the phosphorylation of Axin-1 [72]. It is also involved in Sema3A signaling through the motor neuron neuropilin-1 (NRP1) receptor to trigger distal axonopathy and muscle denervation in SOD1^{G93A} mice. Inhibition of Sema3A signaling via anti-NRP1 antibody restored the life span and rotarod motor function, reduced neuromuscular junction (NMJ) denervation, and attenuated ventral root pathology in the SOD1^{G93A} mouse model of ALS [73]. Insulin-like growth factor 2 (IGF-2), an activator of the PI3-K/Akt pathway, was maintained in oculomotor neurons in ALS and thus could play a role in oculomotor resistance in this disease. IGF-2 prolonged the survival of SOD1^{G93A} mice by preserving motor neurons and inducing nerve regeneration [74].

Recent studies have shown that chronic traumatic encephalopathy (CTE)-ALS is characterized by the presence of all six tau isoforms in both soluble and insoluble tau isolates. Activated GSK3β, pThr^{175} tau, pThr^{231} tau, and oligomerized tau protein expression were observed in hippocampal neurons and spinal motor neurons [75]. Other groups identified that genetic mutations in TDP-43 are associated with sALS and fALS [76,77]. A TDP-43-associated *Drosophila* ALS model, created by overexpression of wild-type human TDP-43 in *Drosophila* motor neurons, exhibited motor dysfunction and a dramatic reduction in life span [78,79]. The leg of an adult *Drosophila* is a valuable new tool for studying adult motor neuron phenotypes in vivo. Overexpression of TDP-43^{Q331K} in *Drosophila* motor neurons caused the progressive degeneration of adult motor axons and NMJs. Forward genetic studies have identified three genes that suppress TDP-43 toxicity, including *GSK3β/Shaggy*. Loss of *GSK3β/Shaggy* suppresses TDP-43^{Q331K}-mediated axons and NMJ degeneration [80]. These results suggest that GSK3β plays a pivotal role in axonal degeneration in ALS. They also indicate that GSK3β is both directly and indirectly involved in the pathogenesis of ALS and could be a pharmacologically beneficial target for developing novel ALS treatments. However, the therapeutic potential of GSK3β-targeted drugs in patients with ALS remains uncertain, and further research is needed.

To identify the therapeutic potential of GSK3β-targeted drugs in ALS treatment, many studies using in vivo models have been conducted. These studies have shown that GSK3β inhibitors can attenuate ALS disease progression. Lithium was the first natural GSK3β inhibitor to be identified, and it significantly increases the level of phosphorylated GSK3β serine 9, an inhibitory phosphorylation site [81–84]. Furthermore, lithium has been shown to have a neuroprotective role in neurodegenerative diseases [85]. Lithium treatment inhibits the Fas-mediated apoptosis signaling pathway, restores motor function defects, and inhibits disease progression in SOD1^{G93A} mice [86]. Valproic acid (VPA) is a well-known mood stabilizer, and its therapeutic effects in bipolar and affective disorders have been well studied. It also indirectly inhibits GSK3β via the regulation of Akt [87]. VPA acts as a neuroprotective agent for motor neurons, delays disease progression, and extends life span in SOD1^{G93A} mice [88]. Treatment with a combination of lithium and VPA exhibited greater and more consistent rescue effects on motor dysfunction and disease progression in the SOD1^{G93A} mouse model through the upregulation of phosphorylated GSK3β serine 9 levels compared to treatment with lithium or VPA alone [89]. Some studies have also shown that lithium treatment triggers autophagy and exhibits beneficial effects in ALS. For example, lithium treatment effects were concomitant with the activation of autophagy, increased the number of mitochondria in motor neurons of SOD1^{G93A} mice, and suppressed reactive astrogliosis [90]. However, the exact effect on autophagy activation by lithium remains unclear and the underlying mechanism has yet to be elucidated. GSK3 inhibitor VIII treatment markedly delayed the onset of symptoms, extended life span, and inhibited GSK3β activity in SOD1^{G93A} mice [91]. Furthermore, the inhibition of GSK3β activates HSF-1, which is associated with cell survival, reduction of cytochrome c release, caspase-3 activation, poly (ADP-ribose) polymerase (PARP) cleavage, and reduction of inflammation-related signals including cyclooxygenase-2 (COX-2) and intercellular adhesion molecule-2 (ICAM-2). Taken together, these findings suggest that GSK3β could be a potential target for developing treatments against ALS (Figure 2).

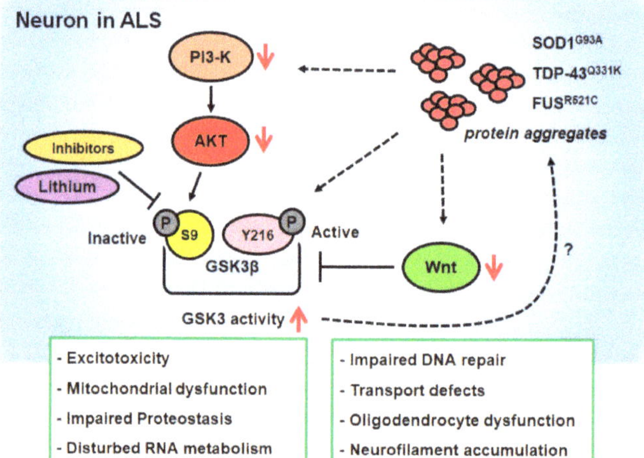

Figure 2. Abnormal regulation of the GSK3 activity in amyotrophic lateral sclerosis (ALS). A number of studies have found the inhibition of the PI3K/Akt and Wnt pathway in injured motor neurons of ALS pathological condition. This leads to abnormal activation of the GSK3 signaling in ALS neurons. The final outcome of disease phenotypes cause by GSK3 activation is further modulated by other mechanisms, which are currently incompletely understood. Further studies are needed to examine how ALS-causing genes such as *SOD*, *TDP-43*, and *FUS* increase GSK3 activity in neurons and how GSK3 activation is involved in the cytoplasmic aggregate formation of toxic proteins.

5. Studies Exploring the Role of GSK3β in In Vitro Models of ALS

Even with the availability of several excellent in vivo models of motor neuron degeneration for human ALS, in vitro models for the disease are still used in many studies, as they facilitate rapid screening of candidates for treatment. Several studies have investigated the effect of ALS on endoplasmic reticulum (ER)-mitochondrial signaling. Recent studies have shown that overexpression of TDP-43 or FUS leads to the activation of GSK3β, which regulates the endoplasmic reticulum (ER)–mitochondria association (Figure 2). It also perturbs ER–mitochondria interactions and cellular Ca^{2+} homeostasis in TDP-43- or FUS-overexpressed ALS models [92,93]. Other studies have shown that ALS with cognitive impairment (ALSci) is associated with tau phosphorylation at Thr^{175}, which leads to the activation of GSK3β. This induces phosphorylation at tau Thr^{231} in HEK293T and Neuro2A cells [94]. In addition, IGF-2 induces Akt phosphorylation, GSK3β phosphorylation, and β-catenin levels while protecting ALS patient motor neurons [74]. Numerous studies have been conducted using in vivo and in vitro models to establish the therapeutic effect of GSK3β in ALS. Yang et al. found that kenpaullone had the ability to prolong the survival of motor neurons via the inhibition of GSK3β through small-molecule survival screening, using both wild-type and $SOD1^{G93A}$ mouse embryonic stem cells [95]. Furthermore, kenpaullone also greatly improved the survival of human motor neurons derived from ALS-patient-induced pluripotent stem cells [95].

Vascular endothelial growth factor (VEGF) directly acts at the neuronal level as a potent neuroprotective agent against hypoxia and excitotoxicity [96–99]. VEGF treatment activates the PI3-K/Akt pathway and restores motor neuron cell death in $SOD1^{G93A}$-transfected NSC34 cells [100]. VPA treatment also shows a protective effect in spinal motor neurons and protects them against death from glutamate toxicity [88]. Epigallocatechin gallate (EGCG), a major constituent of green tea polyphenols, is known to have protective effects against neurodegenerative diseases [101]. Multiple studies have suggested that EGCG affects numerous cell signaling pathways, including PI3-K/Akt, GSK3β, and caspase-3 [102]. EGCG treatment restores viability in oxidative-stress-induced cell death $SOD1^{G93A}$-mutant cells by activating PI3-K/Akt and inhibiting GSK3β [103]. Studies have also shown that treatment with the GSK3β inhibitor 2-thio(3-iodobenzyl)-5-(1-pyridyl)-[1,3,4]-oxadiazole treatment in $SOD1^{G93A}$-transfected VSC4.1 motor neuron cells increased their viability by activating HSF-1 and reducing cytochrome c release, caspase-3 activation, and PARP cleavage [104]. Taken together, these findings highlight the importance of GSK3β activity and function in the prevention of ALS.

6. Inhibitors of GSK3β in ALS Clinical Trials

GSK3β inhibitors are divided into five types depending on the mechanism. First, there are magnesium-competitive inhibitors that compete with magnesium, a cofactor of GSK3β, including lithium and zinc. Lithium inhibits GSK3β directly by competing with magnesium ions, and indirectly inhibits GSK3β by activating Akt [82,105]. The second are ATP-competitive inhibitors, such as indirubin [106] and meridianins [107], which act competitively on adenosine triphosphate (ATP), to prevent GSK3β from obtaining the phosphate it requires to phosphorylate target substrates. The third type are substrate-competitive inhibitors such as thiazolidinone [108], which binds to GSK3β instead of a substrate to prevent the phosphorylation of GSK3β target substrates. The disadvantage for substrate-competitive inhibitors is that they cannot pass through the blood–brain barrier (BBB) due to their high molecular weight. Fourth are modulators of GSK3β Ser9 phosphorylation that act by inhibiting enzymes, such as Akt, PKC, and Rsk1, to inhibit GSK3β Ser9 phosphorylation. For example, tamoxifen, an inhibitor of PKC, induces phosphorylation of GSK3β Ser9 to inhibit enzyme activity [109]. The final type are inhibitors that regulate protein binding in GSK3β, such as tamoxifen [110], which combines with GSK3β to control enzyme activity.

Of the many developed GSK3 inhibitors, only a few have reached clinical trials targeting human subjects and have been attempted in several neurodegenerative studies. For example, several clinical trials have evaluated the therapeutic effect of lithium on AD [111–114]. However, conflicting results were reported. Some studies reported no effects, whereas others reported mild therapeutic effects. Tideglusib, also known as NP-12 and NP031112, is a potent, selective small-molecule non-ATP-competitive GSK3β inhibitor. Preclinical studies have shown that tideglusib treatment results in reduced tau phosphorylation, neuronal loss, and rescued spatial memory deficits in transgenic mice [115]. Furthermore, in a phase IIa clinical trial in 30 patients with mild-moderate AD (NCT00948259), improvement in mini-mental status examinations and cognitive function was observed with tideglusib treatment [116]. In a subsequent phase IIb clinical trial of 308 patients with mild to moderate AD (NCT01350362), tideglusib proved safe in the trial, but development was discontinued to the lack of clinical benefits [117]. In the case of ALS, the first clinical study using lithium in ALS was reported in 2008 [90]. Lithium treatment delayed disease progression in ALS. This study included 44 patients. Remarkably, all patients treated with lithium were alive at the end of the study, whereas in the group receiving riluzole only, 29% of the subjects did not survive. These data indicate that patients with ALS, receiving lithium, progressed very slowly in the disease [90]. To date, many preclinical studies are actively underway, and these findings suggest that GSK3β can be a pharmacologically significant therapeutic target for the treatment of ALS. However, there are limitations on the use of GSK3β inhibitors. Further studies are needed to determine the precise role and molecular function of GSK3β in ALS pathogenesis.

7. Discussion

In the nervous system, GSK3 signaling plays critical roles in neurodevelopment and energy metabolism (Figure 1). Dysregulation of the GSK3 signaling pathway has been reported to be involved in most of the pathogenic mechanisms described for various neurodegenerative diseases, including AD, PD, and ALS. GSK3 has been demonstrated to promote several pathological phenotypes, such as Ab production and tau phosphorylation in AD. Growing evidence has been reported suggesting that GSK3 inhibition is effective in PD, AD, and ALS models. However, despite several experimental results, the contribution and physiological role of GSK3 in relation to disease pathology are still unclear. The GSK3 signaling pathway is a complex process influenced not only by cellular type but also by cellular conditions. In addition, GSK3 has been shown to be a kinase that is recruited to phosphorylate numerous substrates, which participate in various cellular processes. Therefore, a critical challenge in the field of neurodegenerative disease research is to understand how GSK3 participates in various signaling pathways and cellular processes. Therefore, it will be necessary to identify its genetic/physical partners or substrates and to understand their relative contribution to neuronal function, and as a consequence, their contribution to the pathogenesis of neurodegenerative diseases.

Abnormal activation of the GSK3 signaling pathway may promote pathological processes in various types of ALS models (Figure 2). In other words, suppressing high GSK3 activation may be sufficient to be beneficial for neuronal function. This raises the possibility that GSK3 inhibition may have an important effect on ALS pathogenesis. However, our knowledge of the exact biological and molecular functions of GSK3 in the pathogenesis of ALS caused by several RNA/DNA-binding proteins is still unclear, and further experiments are still necessary.

8. Conclusions

Data from several studies indicate that protein aggregation, proteasomal dysfunction, mitochondrial defects, neuroinflammation, and oxidative stress are involved in ALS pathogenesis. Therefore, understanding cellular signaling pathways, such as GSK3, which regulate neuronal dysfunctions linked to ALS pathogenesis, is important for the development of new strategies for ALS treatment.

GSK3β activity is increased in multiple ALS models and patients and has been associated with neuronal cell death in ALS. Furthermore, several studies have shown that GSK3β inhibition can rescue defective phenotypes of ALS in various models (Table 1). Recent studies have also revealed that RNA/DNA-binding proteins, such as TDP-43 and FUS, activate GSK3β, and that GSK3β inactivation suppresses TDP-43-induced neuronal toxicity. However, the following questions still need to be addressed in order to understand the exact neuroprotective mechanisms of GSK3β activity in the pathogenesis of ALS:

(1) How do ALS-causing genes, such as *SOD*, *TDP-43*, and *FUS*, increase the GSK3β activity in neurons?

(2) Which factors are critical for GSK3β activation in ALS?

(3) What is the effect of GSK3β inactivation on proteasomal dysfunction, neuronal toxicity, and mitochondrial dysfunction in ALS?

(4) Is GSK3β activation involved in the cytoplasmic aggregate formation of toxic proteins?

Table 1. In vitro and in vivo studies exploring the role of GSK3β in ALS.

Type of Study	Model System	Key Findings	Reference
Ex Vivo	ALS patient sample (spinal cord, frontal and temporal cortices)	Increased GSK3β expression and cytosolic phospho-Y216 GSK3β in ALS patient sample	[61]
Ex Vivo	ALS patient sample (frontal and temporal cortices)	Increased GSK3β expression and cytosolic phospho-Y216 GSK3β in ALS patient sample	[62]
Ex Vivo	CTE-ALS patient sample (hippocampus and spinal cord)	Activated GSK3β, pThr175 tau, pThr231 tau, and oligomerized tau protein expression	[75]
In Vivo	SOD1^{G93A} transgenic mouse	Increased GSK3β activity via decreasing the PI3-K/Akt expression in an age-dependent manner	[68]
In Vivo	SOD1^{G93A} transgenic mouse	Inhibition of Sema3A/NRP1 signaling restored life span, motor function, and NMJ denervation	[73]
In Vivo	SOD1^{G93A} transgenic mouse	IGF-2 prolonged survival by preserving motor neurons and inducing nerve regeneration	[74]
In Vivo	TDP-43^{Q331K} transgenic fly	Loss of GSK3β/Shaggy suppressed TDP-43^{Q331K}-mediated axon and NMJ degeneration	[80]
In Vivo	SOD1^{G93A} transgenic mouse	Treatment with lithium (a GSK3β inhibitor) improved neuron survival, motor function, and mortality	[86]
In Vivo	SOD1^{G93A} transgenic mouse	Treatment with VPA (a GSK3β inhibitor) rescued motor neuronal defects and delayed the disease progression	[88]
In Vivo	SOD1^{G93A} transgenic mouse	Treatment with a combination of lithium and VPA strongly rescued the motor dysfunction and disease progression via upregulation of phospho-S9 GSK3β	[89]
In Vivo	SOD1^{G93A} transgenic mouse	GSK3β inhibitor VIII treatment prolonged the life span via inhibition of GSK3β activity, preserved survival signals, and attenuated death and inflammatory signals	[91]
In Vitro	FUS/TDP43-transfected HEK293, NSC34 cells	ALS associated FUS, TDP-43 activated GSK3β to disrupt the VAPB–PTPIP51 interaction and ER–mitochondria associations	[92,93]
In Vitro	Tau-transfected HEK293T and Neuro2A cells	ALSci is associated with tau phosphorylation at Thr175 and leads to the activation of GSK3β, which induces phosphorylation at tau Thr231.	[94]
In Vitro	ALS-patients iPSC-derived motor neuron	IGF-2 induced Akt phosphorylation, GSK3β phosphorylation, and β-catenin levels while protecting ALS patient motor neurons	[74]
In Vitro	SOD1^{G93A} mESCs, ALS-patients iPSC-derived motor neuron	Kenpaullone treatment improved the survival of human motor neurons via inhibition of GSK3β	[95]
In Vitro	SOD1^{G93A}-transfected NSC34 cells	GSK3β inhibitor: VEGF treatment activated PI3-K/Akt signaling and restored neuron cell death in motor neurons	[100]
In Vitro	PDC-induced mouse organotypic spinal cord	VPA (an inhibitor of GSK3β) treatment protected spinal motor neurons against glutamate toxicity	[88]
In Vitro	SOD1^{G93A}-transfected VSC 4.1 cells	EGCG (an inhibitor of GSK3β) treatment restored viability in oxidative stress-induced cell death via activation of PI3-K/Akt signaling	[103]
In Vitro	SOD1^{G93A}-transfected VSC 4.1 cells	2-thio(3-iodobenzyl)-5-(1-pyridyl)-[1,3,4]-oxadiazole (an inhibitor of GSK3β) treatment restored viability via activation of HSF-1 and reduction of cytochrome c release, caspase-3 activation, and PARP cleavage	[104]

CTE-ALS, chronic traumatic encephalopathy-amyotrophic lateral sclerosis; SOD1, superoxide dismutase 1; TDP-43, TAR DNA binding protein-43; FUS, fused in sarcoma; NMJ, neuromuscular junction; VPA, valproic acid; ER, endoplasmic reticulum; ALSci, amyotrophic lateral sclerosis cognitive inhibition; IGF-2, insulin-like growth factor-2; VEGF, vascular endothelial growth factor; EGCG, epigallocatechin gallate; HSF-1, heat shock transcription factor-1; PARP, poly (ADP-ribose) polymerase; VSC, ventral spinal cord.

Addressing these questions will improve our understanding of ALS pathology and may provide valuable information about potential targets that can be used in the development of novel therapeutic drugs for treating ALS.

Author Contributions: Literature review and writing—original draft preparation, H.-J.C., S.J.C., H.-J.K., J.-W.L., and K.K.; writing—review and editing, H.-J.K. and K.K.; supervision, K.K.; funding acquisition, K.K. All authors have read and agreed to the published version of the manuscript.

Funding: This work was supported by the Basic Science Research Program through the National Research Foundation of Korea (NRF), funded by the Ministry of Science and information & communication technology (ICT) (MSIT) (2019R1F1A1045639), and by the Soonchunhyang University Research Fund.

Conflicts of Interest: The authors declare no conflict of interest.

References

1. Cleveland, D.W.; Rothstein, J.D. From Charcot to Lou Gehrig: deciphering selective motor neuron death in ALS. *Nat. Rev. Neurosci.* **2001**. [CrossRef]
2. Rowland, L.P.; Shneider, N.A. Amyotrophic lateral sclerosis. *N. Engl. J. Med.* **2001**, *344*, 1688–1700. [CrossRef]
3. Alonso, A.; Logroscino, G.; Jick, S.S.; Hernán, M.A. Incidence and lifetime risk of motor neuron disease in the United Kingdom: A population-based study. *Eur. J. Neurol.* **2009**, *16*, 745–751. [CrossRef]
4. Rosen, D.R.; Siddique, T.; Patterson, D.; Figlewicz, D.A.; Sapp, P.; Hentati, A.; Donaldson, D.; Goto, J.; O'Regan, J.P.; Deng, H.-X. Mutations in Cu/Zn superoxide dismutase gene are associated with familial amyotrophic lateral sclerosis. *Nature* **1993**, *362*, 59–62. [CrossRef] [PubMed]
5. Kabashi, E.; Valdmanis, P.N.; Dion, P.; Spiegelman, D.; McConkey, B.J.; Velde, C.V.; Bouchard, J.-P.; Lacomblez, L.; Pochigaeva, K.; Salachas, F. TARDBP mutations in individuals with sporadic and familial amyotrophic lateral sclerosis. *Nat. Genet.* **2008**, *40*, 572–574. [CrossRef] [PubMed]
6. Rutherford, N.J.; Zhang, Y.-J.; Baker, M.; Gass, J.M.; Finch, N.A.; Xu, Y.-F.; Stewart, H.; Kelley, B.J.; Kuntz, K.; Crook, R.J. Novel mutations in TARDBP (TDP-43) in patients with familial amyotrophic lateral sclerosis. *PLoS Genet.* **2008**, *4*, e1000193. [CrossRef]
7. Kwiatkowski, T.J.; Bosco, D.; Leclerc, A.; Tamrazian, E.; Vanderburg, C.; Russ, C.; Davis, A.; Gilchrist, J.; Kasarskis, E.; Munsat, T. Mutations in the FUS/TLS gene on chromosome 16 cause familial amyotrophic lateral sclerosis. *Science* **2009**, *323*, 1205–1208. [CrossRef]
8. Vance, C.; Rogelj, B.; Hortobágyi, T.; De Vos, K.J.; Nishimura, A.L.; Sreedharan, J.; Hu, X.; Smith, B.; Ruddy, D.; Wright, P. Mutations in FUS, an RNA processing protein, cause familial amyotrophic lateral sclerosis type 6. *Science* **2009**, *323*, 1208–1211. [CrossRef] [PubMed]
9. Renton, A.E.; Majounie, E.; Waite, A.; Simón-Sánchez, J.; Rollinson, S.; Gibbs, J.R.; Schymick, J.C.; Laaksovirta, H.; Van Swieten, J.C.; Myllykangas, L. A hexanucleotide repeat expansion in C9ORF72 is the cause of chromosome 9p21-linked ALS-FTD. *Neuron* **2011**, *72*, 257–268. [CrossRef] [PubMed]
10. DeJesus-Hernandez, M.; Mackenzie, I.R.; Boeve, B.F.; Boxer, A.L.; Baker, M.; Rutherford, N.J.; Nicholson, A.M.; Finch, N.A.; Flynn, H.; Adamson, J. Expanded GGGGCC hexanucleotide repeat in noncoding region of C9ORF72 causes chromosome 9p-linked FTD and ALS. *Neuron* **2011**, *72*, 245–256. [CrossRef] [PubMed]
11. Embi, N.; Rylatt, D.B.; Cohen, P. Glycogen Synthase Kinase-3 from Rabbit Skeletal Muscle: Separation from Cyclic-AMP-Dependent Protein Kinase and Phosphorylase Kinase. *Eur. J. Biochem.* **1980**, *107*, 519–527. [CrossRef] [PubMed]
12. Rylatt, D.B.; Aitken, A.; Bilham, T.; Condon, G.D.; Embi, N.; Cohen, P. Glycogen synthase from rabbit skeletal muscle: Amino acid sequence at the sites phosphorylated by glycogen synthase kinase-3, and extension of the N-terminal sequence containing the site phosphorylated by phosphorylase kinase. *Eur. J. Biochem.* **1980**, *107*, 529–537. [CrossRef] [PubMed]
13. Somanath, P.R.; Jack, S.L.; Vijayaraghavan, S. Changes in sperm glycogen synthase kinase-3 serine phosphorylation and activity accompany motility initiation and stimulation. *J. Androl.* **2004**, *25*, 605–617. [CrossRef] [PubMed]
14. Bhattacharjee, R.; Goswami, S.; Dey, S.; Gangoda, M.; Brothag, C.; Eisa, A.; Woodgett, J.; Phiel, C.; Kline, D.; Vijayaraghavan, S. Isoform-specific requirement for GSK3α in sperm for male fertility. *Biol. Reprod.* **2018**, *99*, 384–394. [CrossRef]

15. Kaidanovich-Beilin, O.; Lipina, T.V.; Takao, K.; Van Eede, M.; Hattori, S.; Laliberté, C.; Khan, M.; Okamoto, K.; Chambers, J.W.; Fletcher, P.J. Abnormalities in brain structure and behavior in GSK-3alpha mutant mice. *Mol. Brain* **2009**, *2*, 35. [CrossRef]

16. Kockeritz, L.; Doble, B.; Patel, S.; Woodgett, J.R. Glycogen synthase kinase-3-an overview of an over-achieving protein kinase. *Curr. Drug Targets* **2006**, *7*, 1377–1388. [CrossRef]

17. Frame, S.; Cohen, P. GSK3 takes centre stage more than 20 years after its discovery. *Biochem. J.* **2001**, *359*, 1–16. [CrossRef]

18. Woodgett, J.R. Molecular cloning and expression of glycogen synthase kinase-3/factor A. *EMBO J.* **1990**, *9*, 2431–2438. [CrossRef]

19. Mukai, F.; Ishiguro, K.; Sano, Y.; Fujita, S.C. Alternative splicing isoform of tau protein kinase I/glycogen synthase kinase 3β. *J. Neurochem.* **2002**, *81*, 1073–1083. [CrossRef]

20. Wood-Kaczmar, A.; Kraus, M.; Ishiguro, K.; Philpott, K.L.; Gordon-Weeks, P.R. An alternatively spliced form of glycogen synthase kinase-3β is targeted to growing neurites and growth cones. *Mol. Cell. Neurosci.* **2009**, *42*, 184–194 [CrossRef]

21. Castaño, Z.; Gordon-Weeks, P.R.; Kypta, R.M. The neuron-specific isoform of glycogen synthase kinase-3β is required for axon growth. *J. Neurochem.* **2010**, *113*, 117–130. [CrossRef] [PubMed]

22. Goold, R.G.; Gordon-Weeks, P.R. Microtubule-associated protein 1B phosphorylation by glycogen synthase kinase 3β is induced during PC12 cell differentiation. *J. Cell Sci.* **2001**, *114*, 4273–4284. [PubMed]

23. Mao, Y.; Ge, X.; Frank, C.L.; Madison, J.M.; Koehler, A.N.; Doud, M.K.; Tassa, C.; Berry, E.M.; Soda, T.; Singh, K.K. Disrupted in schizophrenia 1 regulates neuronal progenitor proliferation via modulation of GSK3β/β-catenin signaling. *Cell* **2009**, *136*, 1017–1031. [CrossRef] [PubMed]

24. Kim, W.-Y.; Wang, X.; Wu, Y.; Doble, B.W.; Patel, S.; Woodgett, J.R.; Snider, W.D. GSK-3 is a master regulator of neural progenitor homeostasis. *Nat. Neurosci.* **2009**, *12*, 1390–1397. [CrossRef] [PubMed]

25. Bultje, R.S.; Castaneda-Castellanos, D.R.; Jan, L.Y.; Jan, Y.-N.; Kriegstein, A.R.; Shi, S.-H. Mammalian Par3 regulates progenitor cell asymmetric division via notch signaling in the developing neocortex. *Neuron* **2009**, *63*, 189–202. [CrossRef]

26. Kim, N.-G.; Xu, C.; Gumbiner, B.M. Identification of targets of the Wnt pathway destruction complex in addition to β-catenin. *Proc. Natl. Acad. Sci. USA* **2009**, *106*, 5165–5170. [CrossRef]

27. Xu, C.; Kim, N.-G.; Gumbiner, B.M. Regulation of protein stability by GSK3 mediated phosphorylation. *Cell Cycle* **2009**, *8*, 4032–4039. [CrossRef]

28. Gregory, M.A.; Qi, Y.; Hann, S.R. Phosphorylation by glycogen synthase kinase-3 controls c-myc proteolysis and subnuclear localization. *J. Biol. Chem.* **2003**, *278*, 51606–51612. [CrossRef]

29. Jiang, J. Regulation of Hh/Gli signaling by dual ubiquitin pathways. *Cell Cycle* **2006**, *5*, 2457–2463. [CrossRef]

30. Barth, A.I.; Caro-Gonzalez, H.Y.; Nelson, W.J. Role of adenomatous polyposis coli (APC) and microtubules in directional cell migration and neuronal polarization. *Semin. Cell Dev. Biol.* **2008**, *3*, 245–251. [CrossRef]

31. Jiang, H.; Guo, W.; Liang, X.; Rao, Y. Both the establishment and the maintenance of neuronal polarity require active mechanisms: Critical roles of GSK-3β and its upstream regulators. *Cell* **2005**, *120*, 123–135. [PubMed]

32. Etienne-Manneville, S.; Hall, A. Cdc42 regulates GSK-3β and adenomatous polyposis coli to control cell polarity. *Nature* **2003**, *421*, 753–756. [CrossRef] [PubMed]

33. Mimori-Kiyosue, Y.; Tsukita, S. Where is APC going? *J. Cell Biol.* **2001**, *154*, 1105. [CrossRef] [PubMed]

34. Yoshimura, T.; Kawano, Y.; Arimura, N.; Kawabata, S.; Kikuchi, A.; Kaibuchi, K. GSK-3β regulates phosphorylation of CRMP-2 and neuronal polarity. *Cell* **2005**, *120*, 137–149. [CrossRef]

35. Zhou, F.-Q.; Zhou, J.; Dedhar, S.; Wu, Y.-H.; Snider, W.D. NGF-induced axon growth is mediated by localized inactivation of GSK-3β and functions of the microtubule plus end binding protein APC. *Neuron* **2004**, *42*, 897–912. [CrossRef]

36. Akhmanova, A.; Hoogenraad, C.C.; Drabek, K.; Stepanova, T.; Dortland, B.; Verkerk, T.; Vermeulen, W.; Burgering, B.M.; De Zeeuw, C.I.; Grosveld, F. Clasps are CLIP-115 and-170 associating proteins involved in the regional regulation of microtubule dynamics in motile fibroblasts. *Cell* **2001**, *104*, 923–935. [CrossRef]

37. Trivedi, N.; Marsh, P.; Goold, R.G.; Wood-Kaczmar, A.; Gordon-Weeks, P.R. Glycogen synthase kinase-3β phosphorylation of MAP1B at Ser1260 and Thr1265 is spatially restricted to growing axons. *J. Cell Sci.* **2005**, *118*, 993–1005. [CrossRef]

38. Stoothoff, W.H.; Johnson, G.V. Tau phosphorylation: physiological and pathological consequences. *Biochim. Biophys. Acta (BBA) Mol. Basis Dis.* **2005**, *1739*, 280–297. [CrossRef]

39. Zumbrunn, J.; Kinoshita, K.; Hyman, A.A.; Näthke, I.S. Binding of the adenomatous polyposis coli protein to microtubules increases microtubule stability and is regulated by GSK3β phosphorylation. *Curr. Biol.* **2001**, *11*, 44–49. [CrossRef]

40. Owen, R.; Gordon-Weeks, P.R. Inhibition of glycogen synthase kinase 3β in sensory neurons in culture alters filopodia dynamics and microtubule distribution in growth cones. *Mol. Cell. Neurosci.* **2003**, *23*, 626–637. [CrossRef]

41. Garrido, J.J.; Simón, D.; Varea, O.; Wandosell, F. GSK3 alpha and GSK3 beta are necessary for axon formation. *FEBS Lett.* **2007**, *581*, 1579–1586. [CrossRef] [PubMed]

42. Gärtner, A.; Huang, X.; Hall, A. Neuronal polarity is regulated by glycogen synthase kinase-3 (GSK-3β) independently of Akt/PKB serine phosphorylation. *J. Cell Sci.* **2006**, *119*, 3927–3934. [CrossRef] [PubMed]

43. Rui, Y.; Myers, K.R.; Yu, K.; Wise, A.; De Blas, A.L.; Hartzell, H.C.; Zheng, J.Q. Activity-dependent regulation of dendritic growth and maintenance by glycogen synthase kinase 3β. *Nat. Commun.* **2013**, *4*, 1–13. [CrossRef]

44. Peineau, S.; Taghibiglou, C.; Bradley, C.; Wong, T.P.; Liu, L.; Lu, J.; Lo, E.; Wu, D.; Saule, E.; Bouschet, T. LTP inhibits LTD in the hippocampus via regulation of GSK3β. *Neuron* **2007**, *53*, 703–717. [CrossRef]

45. Wei, J.; Liu, W.; Yan, Z. Regulation of AMPA receptor trafficking and function by glycogen synthase kinase 3. *J. Biol. Chem.* **2010**, *285*, 26369–26376. [CrossRef]

46. Tyagarajan, S.K.; Ghosh, H.; Yévenes, G.E.; Nikonenko, I.; Ebeling, C.; Schwerdel, C.; Sidler, C.; Zeilhofer, H.U.; Gerrits, B.; Muller, D. Regulation of GABAergic synapse formation and plasticity by GSK3β-dependent phosphorylation of gephyrin. *Proc. Natl. Acad. Sci. USA* **2011**, *108*, 379–384. [CrossRef]

47. Steinberg, G.R.; Carling, D. AMP-activated protein kinase: the current landscape for drug development. *Nat. Rev. Drug Discov.* **2019**, *18*, 527–551. [CrossRef] [PubMed]

48. Suzuki, T.; Bridges, D.; Nakada, D.; Skiniotis, G.; Morrison, S.J.; Lin, J.D.; Saltiel, A.R.; Inoki, K. Inhibition of AMPK catabolic action by GSK3. *Mol. Cell* **2013**, *50*, 407–419. [CrossRef]

49. Hoshi, M.; Takashima, A.; Noguchi, K.; Murayama, M.; Sato, M.; Kondo, S.; Saitoh, Y.; Ishiguro, K.; Hoshino, T.; Imahori, K. Regulation of mitochondrial pyruvate dehydrogenase activity by tau protein kinase I/glycogen synthase kinase 3beta in brain. *Proc. Natl. Acad. Sci. USA* **1996**, *93*, 2719–2723. [CrossRef]

50. Youle, R.J.; Van Der Bliek, A.M. Mitochondrial fission, fusion, and stress. *Science* **2012**, *337*, 1062–1065. [CrossRef]

51. Seo, A.Y.; Joseph, A.-M.; Dutta, D.; Hwang, J.C.; Aris, J.P.; Leeuwenburgh, C. New insights into the role of mitochondria in aging: mitochondrial dynamics and more. *J. Cell Sci.* **2010**, *123*, 2533–2542. [CrossRef]

52. Chou, C.-H.; Lin, C.-C.; Yang, M.-C.; Wei, C.-C.; Liao, H.-D.; Lin, R.-C.; Tu, W.-Y.; Kao, T.-C.; Hsu, C.-M.; Cheng, J.-T. GSK3beta-mediated Drp1 phosphorylation induced elongated mitochondrial morphology against oxidative stress. *PLoS ONE* **2012**, *7*, e49112. [CrossRef] [PubMed]

53. Kandimalla, R.; Reddy, P.H. Multiple faces of dynamin-related protein 1 and its role in Alzheimer's disease pathogenesis. *Biochim. Biophys. Acta (BBA) Mol. Basis Dis.* **2016**, *1862*, 814–828. [CrossRef] [PubMed]

54. Olson, B.L.; Hock, M.B.; Ekholm-Reed, S.; Wohlschlegel, J.A.; Dev, K.K.; Kralli, A.; Reed, S.I. SCFCdc4 acts antagonistically to the PGC-1α transcriptional coactivator by targeting it for ubiquitin-mediated proteolysis. *Genes Dev.* **2008**, *22*, 252–264. [CrossRef]

55. Martin, S.A.; Souder, D.C.; Miller, K.N.; Clark, J.P.; Sagar, A.K.; Eliceiri, K.W.; Puglielli, L.; Beasley, T.M.; Anderson, R.M. GSK3β regulates brain energy metabolism. *Cell Rep.* **2018**, *23*, 1922.e1924–1931.e1924. [CrossRef]

56. Hemmings, B.A.; Yellowlees, D.; Kernohan, J.C.; Cohen, P. Purification of Glycogen Synthase Kinase 3 from Rabbit Skeletal Muscle. Copurification with the Activating Factor (FA) of the (Mg-ATP) Dependent Protein Phosphatase. *Eur. J. Biochem.* **1981**, *119*, 443–451. [CrossRef] [PubMed]

57. Grimes, C.A.; Jope, R.S. The multifaceted roles of glycogen synthase kinase 3β in cellular signaling. *Prog. Neurobiol.* **2001**, *65*, 391–426. [CrossRef]

58. Llorens-Marítin, M.; Jurado, J.; Hernández, F.; Ávila, J. GSK-3β, a pivotal kinase in Alzheimer disease. *Front. Mol. Neurosci.* **2014**, *7*, 46.

59. Li, D.W.; Liu, Z.Q.; Chen, W.; Yao, M.; Li, G.R. Association of glycogen synthase kinase-3β with Parkinson's disease. *Mol. Med. Rep.* **2014**, *9*, 2043–2050. [CrossRef]

60. Lei, P.; Ayton, S.; Bush, A.I.; Adlard, P.A. GSK-3 in neurodegenerative diseases. *Int. J. Alzheimer's Dis.* **2011**, *2011*. [CrossRef]

61. Hu, J.H.; Zhang, H.; Wagey, R.; Krieger, C.; Pelech, S. Protein kinase and protein phosphatase expression in amyotrophic lateral sclerosis spinal cord. *J. Neurochem.* **2003**, *85*, 432–442. [CrossRef] [PubMed]

62. Yang, W.; Leystra-Lantz, C.; Strong, M.J. Upregulation of GSK3β expression in frontal and temporal cortex in ALS with cognitive impairment (ALSci). *Brain Res.* **2008**, *1196*, 131–139. [CrossRef] [PubMed]

63. Sabatelli, M.; Conte, A.; Zollino, M. Clinical and genetic heterogeneity of amyotrophic lateral sclerosis. *Clin. Genet.* **2013**, *83*, 408–416. [CrossRef]

64. Dudek, H.; Datta, S.R.; Franke, T.F.; Birnbaum, M.J.; Yao, R.; Cooper, G.M.; Segal, R.A.; Kaplan, D.R.; Greenberg, M.E. Regulation of neuronal survival by the serine-threonine protein kinase Akt. *Science* **1997**, *275*, 661–665. [CrossRef] [PubMed]

65. Datta, S.R.; Dudek, H.; Tao, X.; Masters, S.; Fu, H.; Gotoh, Y.; Greenberg, M.E. Akt phosphorylation of BAD couples survival signals to the cell-intrinsic death machinery. *Cell* **1997**, *91*, 231–241. [CrossRef]

66. Philpott, K.L.; McCarthy, M.J.; Klippel, A.; Rubin, L.L. Activated phosphatidylinositol 3-kinase and Akt kinase promote survival of superior cervical neurons. *J. Cell Biol.* **1997**, *139*, 809–815. [CrossRef]

67. Crowder, R.J.; Freeman, R.S. Phosphatidylinositol 3-kinase and Akt protein kinase are necessary and sufficient for the survival of nerve growth factor-dependent sympathetic neurons. *J. Neurosci.* **1998**, *18*, 2933–2943. [CrossRef]

68. Warita, H.; Manabe, Y.; Murakami, T.; Shiro, Y.; Nagano, I.; Abe, K. Early decrease of survival signal-related proteins in spinal motor neurons of presymptomatic transgenic mice with a mutant SOD1 gene. *Apoptosis* **2001**, *6*, 345–352. [CrossRef]

69. Yazdani, U.; Terman, J.R. The semaphorins. *Genome Biol.* **2006**, *7*, 211. [CrossRef]

70. Dolma, K.; Iacobucci, G.J.; Hong Zheng, K.; Shandilya, J.; Toska, E.; White, J.A.; Spina, E.; Gunawardena, S. Presenilin influences glycogen synthase kinase-3 β (GSK-3β) for kinesin-1 and dynein function during axonal transport. *Hum. Mol. Genet.* **2014**, *23*, 1121–1133. [CrossRef]

71. Gao, F.J.; Hebbar, S.; Gao, X.A.; Alexander, M.; Pandey, J.P.; Walla, M.D.; Cotham, W.E.; King, S.J.; Smith, D.S. GSK-3β phosphorylation of cytoplasmic dynein reduces Ndel1 binding to intermediate chains and alters dynein motility. *Traffic* **2015**, *16*, 941–961. [CrossRef] [PubMed]

72. Hida, T.; Nakamura, F.; Usui, H.; Takeuchi, K.; Yamashita, N.; Goshima, Y. Semaphorin3A-induced axonal transport mediated through phosphorylation of Axin-1 by GSK3β. *Brain Res.* **2015**, *1598*, 46–56. [CrossRef] [PubMed]

73. Venkova, K.; Christov, A.; Kamaluddin, Z.; Kobalka, P.; Siddiqui, S.; Hensley, K. Semaphorin 3A Signaling Through Neuropilin-1 Is an Early Trigger for Distal Axonopathy in the SOD1G93A Mouse Model of Amyotrophic Lateral Sclerosis. *J. Neuropathol. Exp. Neurol.* **2014**, *73*, 702–713. [CrossRef] [PubMed]

74. Allodi, I.; Comley, L.; Nichterwitz, S.; Nizzardo, M.; Simone, C.; Benitez, J.A.; Cao, M.; Corti, S.; Hedlund, E. Differential neuronal vulnerability identifies IGF-2 as a protective factor in ALS. *Sci. Rep.* **2016**, *6*, 25960. [CrossRef]

75. Moszczynski, A.J.; Strong, W.; Xu, K.; McKee, A.; Brown, A.; Strong, M.J. Pathologic Thr175 tau phosphorylation in CTE and CTE with ALS. *Neurology* **2018**, *90*, e380–e387. [CrossRef]

76. Cairns, N.J.; Neumann, M.; Bigio, E.H.; Holm, I.E.; Troost, D.; Hatanpaa, K.J.; Foong, C.; White III, C.L.; Schneider, J.A.; Kretzschmar, H.A. TDP-43 in familial and sporadic frontotemporal lobar degeneration with ubiquitin inclusions. *Am. J. Pathol.* **2007**, *171*, 227–240. [CrossRef]

77. Lee, E.B.; Lee, V.M.-Y.; Trojanowski, J.Q. Gains or losses: molecular mechanisms of TDP43-mediated neurodegeneration. *Nat. Rev. Neurosci.* **2012**, *13*, 38–50. [CrossRef]

78. Li, Y.; Ray, P.; Rao, E.J.; Shi, C.; Guo, W.; Chen, X.; Woodruff, E.A.; Fushimi, K.; Wu, J.Y. A Drosophila model for TDP-43 proteinopathy. *Proc. Natl. Acad. Sci. USA* **2010**, *107*, 3169–3174. [CrossRef]

79. Hanson, K.A.; Kim, S.H.; Wassarman, D.A.; Tibbetts, R.S. Ubiquilin modifies TDP-43 toxicity in a Drosophila model of amyotrophic lateral sclerosis (ALS). *J. Biol. Chem.* **2010**, *285*, 11068–11072. [CrossRef]

80. Sreedharan, J.; Neukomm, L.J.; Brown Jr, R.H.; Freeman, M.R. Age-dependent TDP-43-mediated motor neuron degeneration requires GSK3, hat-trick, and xmas-2. *Curr. Biol.* **2015**, *25*, 2130–2136. [CrossRef]

81. Stambolic, V.; Ruel, L.; Woodgett, J.R. Lithium inhibits glycogen synthase kinase-3 activity and mimics wingless signalling in intact cells. *Curr. Biol.* **1996**, *6*, 1664–1669. [CrossRef]

82. Klein, P.S.; Melton, D.A. A molecular mechanism for the effect of lithium on development. *Proc. Natl. Acad. Sci. USA* **1996**, *93*, 8455–8459. [CrossRef] [PubMed]

83. Eldar-Finkelman, H.; Martinez, A. GSK-3 inhibitors: preclinical and clinical focus on CNS. *Front. Mol. Neurosci.* **2011**, *4*, 32. [CrossRef] [PubMed]

84. Chalecka-Franaszek, E.; Chuang, D.-M. Lithium activates the serine/threonine kinase Akt-1 and suppresses glutamate-induced inhibition of Akt-1 activity in neurons. *Proc. Natl. Acad. Sci. USA* **1999**, *96*, 8745–8750. [CrossRef] [PubMed]

85. Forlenza, O.V.; De-Paula, V.d.J.R.; Diniz, B. Neuroprotective effects of lithium: implications for the treatment of Alzheimer's disease and related neurodegenerative disorders. *ACS Chem. Neurosci.* **2014**, *5*, 443–450. [CrossRef] [PubMed]

86. Shin, J.H.; Cho, S.I.; Lim, H.R.; Lee, J.K.; Lee, Y.A.; Noh, J.S.; Joo, I.S.; Kim, K.-W.; Gwag, B.J. Concurrent administration of Neu2000 and lithium produces marked improvement of motor neuron survival, motor function, and mortality in a mouse model of amyotrophic lateral sclerosis. *Mol. Pharmacol.* **2007**, *71*, 965–975. [CrossRef] [PubMed]

87. Chen, G.; Huang, L.D.; Jiang, Y.M.; Manji, H.K. The mood-stabilizing agent valproate inhibits the activity of glycogen synthase kinase-3. *J. Neurochem.* **2000**, *72*, 1327–1330. [CrossRef] [PubMed]

88. Sugai, F.; Yamamoto, Y.; Miyaguchi, K.; Zhou, Z.; Sumi, H.; Hamasaki, T.; Goto, M.; Sakoda, S. Benefit of valproic acid in suppressing disease progression of ALS model mice. *Eur. J. Neurosci.* **2004**, *20*, 3179–3183. [CrossRef]

89. Feng, H.-L.; Leng, Y.; Ma, C.-H.; Zhang, J.; Ren, M.; Chuang, D.-M. Combined lithium and valproate treatment delays disease onset, reduces neurological deficits and prolongs survival in an amyotrophic lateral sclerosis mouse model. *Neuroscience* **2008**, *155*, 567–572. [CrossRef]

90. Fornai, F.; Longone, P.; Cafaro, L.; Kastsiuchenka, O.; Ferrucci, M.; Manca, M.L.; Lazzeri, G.; Spalloni, A.; Bellio, N.; Lenzi, P. Lithium delays progression of amyotrophic lateral sclerosis. *Proc. Natl. Acad. Sci. USA* **2008**, *105*, 2052–2057. [CrossRef]

91. Koh, S.-H.; Kim, Y.; Kim, H.Y.; Hwang, S.; Lee, C.H.; Kim, S.H. Inhibition of glycogen synthase kinase-3 suppresses the onset of symptoms and disease progression of G93A-SOD1 mouse model of ALS. *Exp. Neurol.* **2007**, *205*, 336–346. [CrossRef] [PubMed]

92. Stoica, R.; De Vos, K.J.; Paillusson, S.; Mueller, S.; Sancho, R.M.; Lau, K.-F.; Vizcay-Barrena, G.; Lin, W.-L.; Xu, Y.-F.; Lewis, J. ER–mitochondria associations are regulated by the VAPB–PTPIP51 interaction and are disrupted by ALS/FTD-associated TDP-43. *Nat. Commun.* **2014**, *5*, 1–12. [CrossRef] [PubMed]

93. Stoica, R.; Paillusson, S.; Gomez-Suaga, P.; Mitchell, J.C.; Lau, D.H.; Gray, E.H.; Sancho, R.M.; Vizcay-Barrena, G.; De Vos, K.J.; Shaw, C.E. ALS/FTD-associated FUS activates GSK-3β to disrupt the VAPB–PTPIP 51 interaction and ER–mitochondria associations. *EMBO Rep.* **2016**, *17*, 1326–1342. [CrossRef] [PubMed]

94. Gohar, M.; Yang, W.; Strong, W.; Volkening, K.; Leystra-Lantz, C.; Strong, M.J. Tau phosphorylation at threonine-175 leads to fibril formation and enhanced cell death: implications for amyotrophic lateral sclerosis with cognitive impairment. *J. Neurochem.* **2009**, *108*, 634–643. [CrossRef] [PubMed]

95. Yang, Y.M.; Gupta, S.K.; Kim, K.J.; Powers, B.E.; Cerqueira, A.; Wainger, B.J.; Ngo, H.D.; Rosowski, K.A.; Schein, P.A.; Ackeifi, C.A. A small molecule screen in stem-cell-derived motor neurons identifies a kinase inhibitor as a candidate therapeutic for ALS. *Cell Stem Cell* **2013**, *12*, 713–726. [CrossRef]

96. Matsuzaki, H.; Tamatani, M.; Yamaguchi, A.; Namikawa, K.; Kiyama, H.; Vitek, M.P.; Mitsuda, N.; Tohyama, M. Vascular endothelial growth factor rescues hippocampal neurons from glutamate-induced toxicity: signal transduction cascades. *FASEB J.* **2001**, *15*, 1218–1220. [CrossRef]

97. Jin, K.; Mao, X.; Batteur, S.; McEachron, E.; Leahy, A.; Greenberg, D. Caspase-3 and the regulation of hypoxic neuronal death by vascular endothelial growth factor. *Neuroscience* **2001**, *108*, 351–358. [CrossRef]

98. Lambrechts, D.; Carmeliet, P. VEGF at the neurovascular interface: Therapeutic implications for motor neuron disease. *Biochim. Biophys. Acta (BBA) Mol. Basis Dis.* **2006**, *1762*, 1109–1121. [CrossRef]

99. Nicoletti, J.; Shah, S.; McCloskey, D.; Goodman, J.; Elkady, A.; Atassi, H.; Hylton, D.; Rudge, J.; Scharfman, H.; Croll, S. Vascular endothelial growth factor is up-regulated after status epilepticus and protects against seizure-induced neuronal loss in hippocampus. *Neuroscience* **2008**, *151*, 232–241. [CrossRef]

100. Li, B.; Xu, W.; Luo, C.; Gozal, D.; Liu, R. VEGF-induced activation of the PI3-K/Akt pathway reduces mutant SOD1-mediated motor neuron cell death. *Mol. Brain Res.* **2003**, *111*, 155–164. [CrossRef]

101. Zhang, B.; Rusciano, D.; Osborne, N.N. Orally administered epigallocatechin gallate attenuates retinal neuronal death in vivo and light-induced apoptosis in vitro. *Brain Res.* **2008**, *1198*, 141–152. [CrossRef] [PubMed]

102. Koh, S.-H.; Kim, S.H.; Kwon, H.; Park, Y.; Kim, K.S.; Song, C.W.; Kim, J.; Kim, M.-H.; Yu, H.-J.; Henkel, J.S. Epigallocatechin gallate protects nerve growth factor differentiated PC12 cells from oxidative-radical-stress-induced apoptosis through its effect on phosphoinositide 3-kinase/Akt and glycogen synthase kinase-3. *Mol. Brain Res.* **2003**, *118*, 72–81. [CrossRef] [PubMed]

103. Koh, S.-H.; Kwon, H.; Kim, K.S.; Kim, J.; Kim, M.-H.; Yu, H.-J.; Kim, M.; Lee, K.-W.; Do, B.R.; Jung, H.K. Epigallocatechin gallate prevents oxidative-stress-induced death of mutant Cu/Zn-superoxide dismutase (G93A) motoneuron cells by alteration of cell survival and death signals. *Toxicology* **2004**, *202*, 213–225. [CrossRef] [PubMed]

104. Koh, S.H.; Lee, Y.B.; Kim, K.S.; Kim, H.J.; Kim, M.; Lee, Y.J.; Kim, J.; Lee, K.W.; Kim, S.H. Role of GSK-3β activity in motor neuronal cell death induced by G93A or A4V mutant hSOD1 gene. *Eur. J. Neurosci.* **2005**, *22*, 301–309. [CrossRef] [PubMed]

105. Ryves, W.J.; Harwood, A.J. Lithium inhibits glycogen synthase kinase-3 by competition for magnesium. *Biochem. Biophys. Res. Commun.* **2001**, *280*, 720–725. [CrossRef]

106. Meijer, L.; Skaltsounis, A.-L.; Magiatis, P.; Polychronopoulos, P.; Knockaert, M.; Leost, M.; Ryan, X.P.; Vonica, C.A.; Brivanlou, A.; Dajani, R. GSK-3-selective inhibitors derived from Tyrian purple indirubins. *Chem. Biol.* **2003**, *10*, 1255–1266. [CrossRef]

107. Gompel, M.; Leost, M.; Joffe, E.B.D.K.; Puricelli, L.; Franco, L.H.; Palermo, J.; Meijer, L. Meridianins, a new family of protein kinase inhibitors isolated from the ascidian Aplidium meridianum. *Bioorg. Med. Chem. Lett.* **2004**, *14*, 1703–1707. [CrossRef]

108. Martinez, A.; Alonso, M.; Castro, A.; Pérez, C.; Moreno, F.J. First non-ATP competitive glycogen synthase kinase 3 β (GSK-3β) inhibitors: thiadiazolidinones (TDZD) as potential drugs for the treatment of Alzheimer's disease. *J. Med. Chem.* **2002**, *45*, 1292–1299. [CrossRef]

109. Bebchuk, J.M.; Arfken, C.L.; Dolan-Manji, S.; Murphy, J.; Hasanat, K.; Manji, H.K. A preliminary investigation of a protein kinase C inhibitor in the treatment of acute mania. *Arch. Gen. Psychiatry* **2000**, *57*, 95–97. [CrossRef]

110. Cechinel-Recco, K.; Valvassori, S.S.; Varela, R.B.; Resende, W.R.; Arent, C.O.; Vitto, M.F.; Luz, G.; de Souza, C.T.; Quevedo, J. Lithium and tamoxifen modulate cellular plasticity cascades in animal model of mania. *J. Psychopharmacol.* **2012**, *26*, 1594–1604. [CrossRef]

111. Effect of Lithium and Divalproex in Alzheimer's Disease. Available online: https://ClinicalTrials.gov/show/NCT00088387 (accessed on 14 September 2020).

112. Forlenza, O.V.; Diniz, B.S.; Radanovic, M.; Santos, F.S.; Talib, L.L.; Gattaz, W.F. Disease-modifying properties of long-term lithium treatment for amnestic mild cognitive impairment: randomised controlled trial. *Br. J. Psychiatry* **2011**, *198*, 351–356. [CrossRef] [PubMed]

113. Hampel, H.; Ewers, M.; Burger, K.; Annas, P.; Mortberg, A.; Bogstedt, A.; Frolich, L.; Schroder, J.; Schonknecht, P.; Riepe, M.W. Lithium trial in Alzheimer's disease: a randomized, single-blind, placebo-controlled, multicenter 10-week study. *J. Clin. Psychiatry* **2009**, *70*, 922. [CrossRef]

114. Macdonald, A.; Briggs, K.; Poppe, M.; Higgins, A.; Velayudhan, L.; Lovestone, S. A feasibility and tolerability study of lithium in Alzheimer's disease. *Int. J. Geriatr. Psychiatry A J. Psychiatry Late Life Allied Sci.* **2008**, *23*, 704–711. [CrossRef] [PubMed]

115. Sereno, L.; Coma, M.; Rodriguez, M.; Sanchez-Ferrer, P.; Sanchez, M.; Gich, I.; Agullo, J.; Perez, M.; Avila, J.; Guardia-Laguarta, C. A novel GSK-3β inhibitor reduces Alzheimer's pathology and rescues neuronal loss in vivo. *Neurobiol. Dis.* **2009**, *35*, 359–367. [CrossRef] [PubMed]

116. Del Ser, T.; Steinwachs, K.C.; Gertz, H.J.; Andres, M.V.; Gomez-Carrillo, B.; Medina, M.; Vericat, J.A.; Redondo, P.; Fleet, D.; Leon, T. Treatment of Alzheimer's disease with the GSK-3 inhibitor tideglusib: A pilot study. *J. Alzheimer's Dis.* **2013**, *33*, 205–215. [CrossRef]

117. Lovestone, S.; Boada, M.; Dubois, B.; Hüll, M.; Rinne, J.O.; Huppertz, H.-J.; Calero, M.; Andres, M.V.; Gomez-Carrillo, B.; Leon, T. A phase II trial of tideglusib in Alzheimer's disease. *J. Alzheimer's Dis.* **2015**, *45*, 75–88. [CrossRef]

Article

Resveratrol Prevents GLUT3 Up-Regulation Induced by Middle Cerebral Artery Occlusion

Germán Fernando Gutiérrez Aguilar [1], Iván Alquisiras-Burgos [1], Javier Franco-Pérez [2], Narayana Pineda-Ramírez [1], Alma Ortiz-Plata [3], Ismael Torres [4], José Pedraza-Chaverri [5] and Penélope Aguilera [1,*]

[1] Laboratorio de Patología Vascular Cerebral, Instituto Nacional de Neurología y Neurocirugía "Manuel Velasco Suárez", Insurgentes Sur #3877, Mexico City 14269, Mexico; gutierrezref@gmail.com (G.F.G.A.); burgos_inc@hotmail.com (I.A.-B.); narayana_pinedar@yahoo.com.mx (N.P.-R.)

[2] Laboratorio de Formación Reticular, Instituto Nacional de Neurología y Neurocirugía "Manuel Velasco Suárez", Insurgentes Sur #3877, Mexico City 14269, Mexico; jfranco@innn.edu.mx

[3] Laboratorio de Neuropatología Experimental, Instituto Nacional de Neurología y Neurocirugía "Manuel Velasco Suárez", Insurgentes Sur #3877, Mexico City 14269, Mexico; aortizplata@yahoo.com.mx

[4] Unidad del Bioterio, Facultad de Medicina, Universidad Nacional Autónoma de México, Mexico City 04510, Mexico; ismael.torres10@hotmail.com

[5] Departamento de Biología, Facultad de Química, Universidad Nacional Autónoma de México, Mexico City 04510, Mexico; pedraza@unam.mx

* Correspondence: penelope.aguilera@innn.edu.mx

Received: 20 August 2020; Accepted: 18 September 2020; Published: 20 September 2020

Abstract: Glucose transporter (GLUT)3 up-regulation is an adaptive response activated to prevent cellular damage when brain metabolic energy is reduced. Resveratrol is a natural polyphenol with anti-oxidant and anti-inflammatory features that protects neurons against damage induced in cerebral ischemia. Since transcription factors sensitive to oxidative stress and inflammation modulate GLUT3 expression, the purpose of this work was to assess the effect of resveratrol on GLUT3 expression levels after ischemia. Male Wistar rats were subjected to 2 h of middle cerebral artery occlusion (MCAO) followed by different times of reperfusion. Resveratrol (1.9 mg/kg; i. p.) was administered at the onset of the restoration of the blood flow. Quantitative-PCR and Western blot showed that MCAO provoked a substantial increase in GLUT3 expression in the ipsilateral side to the lesion of the cerebral cortex. Immunofluorescence assays indicated that GLUT3 levels were upregulated in astrocytes. Additionally, an important increase in GLUT3 occurred in other cellular types (e.g., damaged neurons, microglia, or infiltrated macrophages). Immunodetection of the microtubule-associated protein 2 (MAP2) showed that MCAO induced severe damage to the neuronal population. However, the administration of resveratrol at the time of reperfusion resulted in injury reduction. Resveratrol also prevented the MCAO-induced increase of GLUT3 expression. In conclusion, resveratrol protects neurons from damage induced by ischemia and prevents GLUT3 upregulation in the damaged brain that might depend on AMPK activation.

Keywords: GLUT3; cerebral ischemia; MCAO; resveratrol; astrocytes; AMPK

---

1. Introduction

Cerebral ischemia occurs when the blood flow in an important cerebral artery is interrupted, depriving the tissue of oxygen and glucose. Consequently, the induction of an irreversible cascade of events, such as excitotoxicity, oxidative stress, apoptosis, and inflammation, provokes neuronal damage. Paradoxically, restoring blood flow (reperfusion) increases reactive oxygen species (ROS) production [1,2]. Oxidative stress stimulates the generation of inflammatory mediators and inflammation that in turn generates ROS. Therefore, it is considered that the interactions between oxidative stress and inflammatory pathways involve positive feedback mechanisms [3].

Accordingly, the use of antioxidants has been recommended as a protective therapy for stroke [4–6]. Resveratrol is a stilbene with neuroprotective properties proved in diverse experimental models of neurological diseases including cerebral ischemia [5,7]. Resveratrol's mechanism of protection is principally associated with its antioxidant capacity, which prevents direct damage to biomolecules [8]. Furthermore, resveratrol directly interacts with proteins, triggering cell signaling pathways that activate anti-oxidant, anti-apoptotic, and anti-inflammatory processes [9–11].

Interestingly, ROS increases inflammation by promoting certain stress-activated kinases that stimulate transcription factors such as the nuclear factor kappa B (NF-κB) and the activator protein 1 (AP-1) to induce the expression of pro-inflammatory cytokines [3], as well as the expression of genes involved in controlling the transport of glucose [12,13]. Since the adjustment in glucose transport has been attributed to the neuronal metabolic demand during the post-ischemic period, it is important to understand how the glucose transporters function is regulated [14].

The facilitative glucose transporters (GLUT) are key proteins involved in cellular metabolism because they incorporate glucose to the cerebral parenchyma [15]. There are fourteen members of the GLUT family encoded by the Slc2 genes, the expression of which is regulated during stress conditions [16]. GLUT1 and GLUT3 are highly expressed in the brain and have an essential role in cerebral glucose metabolism [17,18]. GLUT1 is principally expressed in astrocytes, while GLUT3 is known as the neuronal transporter par excellence with a higher affinity for glucose compared to other transporters in the brain [19,20].

Interestingly, GLUT3 expression is up-regulated in the immature and adult brain after global and focal ischemia [14,21–23]. Although signaling that activates GLUTs synthesis is unknown, evidence supports that astrocytes are involved in the response. In cultured astrocytes, the oxygen and glucose deprivation (OGD) induces an increase in GLUT3 expression which is mediated by the NF-κB [24]. Conversely, glutamate-mediated excitotoxicity increases glucose transport through trafficking GLUT3 to the neuronal surface [25]. Remarkably, reactive astrocytes, neurons, and microglia secrete cytokines that modify the metabolic phenotype of astrocytes [26–28]. In particular, pro-inflammatory cytokines increase glucose consumption and decrease the astrocytic glycogen stores. Unlike astrocytes, neurons are unresponsive to the metabolic effects of cytokines [28]. Therefore, it is probable that the inflammatory process activates a signaling pathway in astrocytes to overcome the energetic challenge induced by ischemia.

Since it has been described that resveratrol's anti-oxidant and anti-inflammatory properties are associated with protection in ischemia, we hypothesized that resveratrol treatment should not affect GLUT3 up-regulation induced in astrocytes after cerebral ischemia. We confirmed that resveratrol has a protective effect, and intriguingly, our results also indicated that resveratrol significantly diminishes post-ischemic rise in GLUT3 expression at the mRNA and protein level. Treatment prevented GLUT3 up-regulation in astrocytes which might depend on AMPK activation. However, the tremendous rise in GLUT3 expression induced by ischemia was observed in a different type of cell. Since a significant loss of the microtubule-associated protein 2 (MAP2) level, which is considered as indicative of neuronal death, was also observed, we were not able to evaluate the participation of neurons in the increase of GLUT3 expression and consequently, the effect of resveratrol in its prevention. As the presented data are based on a single time of evaluation (24 h of reperfusion), future studies are needed to analyze the involvement of neurons and other types of cells from the very early period after ischemia onset.

2. Materials and Methods

2.1. Animals

The use of animals was carried out following the protocol No. 23/12 approved by the Institutional Committee for the Care and Use of Animals of the National Institute of Neurology and Neurosurgery

"Manuel Velasco Suárez", under NOM-062-200 and the NIH for the Care and Use of Laboratory Animals. Likewise, the protocol for the use of animals in biomedical research is compatible with the Declaration of the Helsinki World Medical Association. All surgical procedures were performed aseptically, avoiding their suffering as much as possible.

2.2. Resveratrol Treatment

An intravenous dose administration was chosen due to low resveratrol bioavailability to allow a greater amount of resveratrol to reach the tissue in its active form at the onset of reperfusion. This dose induces neuroprotection in a similar range than other studies do [7,29,30]. The maximum soluble concentration of resveratrol in 100 µL of the vehicle was used: 1.9 mg resveratrol/kg (R5010, Sigma-Aldrich, St. Louis, MO, USA) diluted in ethanol. Ethanol was prepared at 132 mg/kg of body weight. Resveratrol was administered to animals at the onset of reperfusion by intravenous via (tail vein). For cultures, resveratrol was dissolved in ethanol (0.01%) at a final concentration of 10 µM. Cells were treated at the onset of recovery.

2.3. Model of Induction of Transitory Focal Cerebral Ischemia in the Rat

Occlusion of the middle cerebral artery (MCAO) described by Longa et al. [31] was used to induce focal cerebral ischemia. Briefly, male Wistar rats (280–350 g) were anesthetized with 2.5% isoflurane (PiSA, Guadalajara, JAL, Mexico), maintaining oxygen 2% and temperature at 37 °C. The surgery consisted of obstructing the cerebral blood flow to the middle cerebral artery with an intraluminal suture (nylon 3-0) introduced by the left internal carotid artery. After 2 h, animals were anesthetized again and the suture was removed to restore the blood flow. Reperfusion was allowed for different times according to the experiment from 1 to 24 h, and then, animals were sacrificed. The control group underwent the same surgical procedure as the group with MCAO without the insertion of the suture (sham). Physiological parameters were monitored throughout the procedure and animals were maintained at 37 °C after surgery until were completely recovery from anesthesia.

2.4. Infarct Area Identification with 2,3,5-Triphenyl Tetrazolium (TTC)

The TTC staining was used to identify the cerebral region damaged by ischemia. Coronal slices of 2.5 mm thick were incubated in 2% TTC solution in PBS (30 min) at 37 °C in the darkness. Subsequently, slices were incubated with 4% paraformaldehyde for 15 min before being photographed. The infarct volume was calculated by measuring the volume of infarct in the slice with respect to the total volume of the slice. Analysis of the images was performed with the ImageJ software 1.8.1 [32].

2.5. Ribonucleic Acid (RNA) Extraction

TRIzol® (ThermoFisher Scientific, Waltham, MA, USA) was used to obtain total RNA following manufacturer instructions. One mL Trizol was added to tissue, and combined with 200 µL of chloroform. After 15 min, the mixture was centrifuged (12,000× *g*) for 15 min at 4 °C. Then, the superior phase was mixed with 500 µL of isopropanol. After 10 min at room temperature, the solution was centrifuged (12,000× *g*) for 10 min at 4 °C. Then, the pellet was washed with 75% ethanol and centrifuged (7500× *g*) for 5 min. Lastly, the supernatant was discarded, and the pellet was allowed to desiccate at room temperature. The transparent pill was resuspended in 50 µL of 0.1% diethylpyrocarbonate (DEPC) treated sterile bidistilled water.

2.6. Complementary Deoxyribonucleic Acid (cDNA) Synthesis

Five µg of total RNA was used to synthesize the cDNA. The RNA was mixed with hexamers (2.5 µM), reverse transcriptase enzyme buffer (M531A, Promega Corporation, Madison, WI, USA), deoxyribonucleotide triphosphates (dNTPs) (500 µM), RNAsin (20 U), and M-MLV reverse transcriptase (200 U; M1708, Promega Corporation) and DEPC-water. The mixture was incubated 1 h at 37 °C and 5 min at 94 °C. Samples were stored at −20 °C until use.

2.7. Quantitative Real-Time Polymerase Chain Reaction (qPCR)

To evaluate the relative level of GLUT3 mRNA expression, the TaqMan®FAM probe (Rn01492963_m1, Applied Biosystems, Foster, CA, USA) was employed. The TaqMan®VIC probe (Rn01455646_m1, Applied Biosystem) was used to detect the TATA-binding protein and normalize GLUT3 expression. A 7500 Real-Time PCR System (Applied Biosystems) was used to perform the qPCR reactions using Universal PCR Master Mix (4304437, Applied Biosystem). Reactions in triplicate were run under the following conditions: holding step, 95 °C for 10 min, followed by 40 cycles of 92 °C for 15 s and 60 °C for 1 min. The threshold cycle (Ct) was calculated using SDS Software 1.3.1 [33].

2.8. Western Blot

To quantify the GLUT3 protein levels, samples were lysed in radioimmunoprecipitation assay buffer (RIPA, 150 mM NaCl, 1% NP40, 1% sodium deoxycholate, 5 mM EDTA, 50 mM HEPES, pH 7.5) supplemented with protease inhibitors (P8340, Sigma-Aldrich). Then, samples were centrifuged (16,000× g) for 10 min. The supernatant was recovered, and the protein concentration was calculated by the bicinchoninic acid determination kit (BCA1, Sigma-Aldrich). Electrophoresis was performed on 10% acrylamide gels at constant 110 V for 1 h. Proteins were transferred to polyvinylidene fluoride (PVDF) membranes (162-0115, Millipore, Burlington, MA, USA) during 1 h at 100 V at 4 °C. After, membranes were blocked 1 h with 5% low-fat milk in Tris-buffered saline-Tween (TBS-T) (10 mM Tris Base, 200 mM NaCl, 0.1% Tween 20, pH 7.5). Then, membranes were incubated with antibody anti GLUT3 (1:3000, sc-30107, Santa Cruz Biotechnology, Dallas, TX, USA) overnight at 4 °C, followed by 4 washes of TBS-T (5 mL/10 min) before being incubated with the antibody anti-rabbit IgG conjugated with horseradish peroxidase (1:5000; 115-035-144, Jackson ImmunoResearch, West Grove, PA, USA) for 2 h at room temperature, followed by washes. Signal was detected by chemiluminescence (WBLUF0100, Millipore) using the imaging system Fusion Solo S (Vilber Lourmat, Collégien, France). After that, striping buffer (15 g glycine, 1 g sodium dodecyl sulfate (SDS), 10 mL Tween 20, pH 2.2) was added to the PVDF membrane for 20 min, washed with TBS-T, and blocked again to incubated with α-tubulin (1:3000; T9026, Sigma-Aldrich), the protein used to normalize protein concentration, followed by the antibody anti-mouse IgG conjugated with horseradish peroxidase (1:10,000; 115-035-003, Jackson ImmunoResearch).

2.9. Immunofluorescence

Rats were anesthetized intraperitoneally with 100 mg/kg of pentobarbital (PiSA), and then perfused transcardially with 4% paraformaldehyde in phosphate buffered saline (PBS). The brains were fixed in 4% paraformaldehyde (24 h), and successively placed in augmented concentrations of sucrose (10%, 20%, and 30%) for three days at 4 °C. Coronal sections of 10 μm were obtained (freezing microtome Sartorius-Werke, model 27, Gottingen, Germany) and stored at −20 °C in 30% ethylene glycol and 20% glycerol in PBS, pH 7.4, until use. Coronal sections were permeabilized with 0.1% Triton X-100 in PBS (30 min) and blocked with 10% goat serum in PBS (60 min). Then, antibodies anti Microtubule-Associated Protein 2 (1:750; MAP-2, AB5392, Abcam, Cambridge, UK), anti-Glial Fibrillary Acid Protein (1:750; GFAP, Z0334, Dako, Carpinteria, CA, USA), and anti GLUT3 (1:500; sc-30107, Santa Cruz) were incubated in 1% bovine serum albumin in PBS overnight at 4 °C. Then, sections were washed and incubated with Alexa Fluor® 594-conjugated anti-chicken IgG (703-585-155), Alexa Fluor® 594-conjugated anti-rabbit IgG (711-585-152) or DyLight™ 488-conjugated anti-mouse IgG (715-485-150) from Jackson ImmunoResearch Laboratories Inc or Alexa Fluor®488-conjugated anti-rabbit IgG (1500-73; Abcam) as required. To identify nuclei, coronal sections were incubated with 1 μg/mL 4′, 6-diamidino-2-phenylindole (DAPI) to detect nuclei (Sigma-Aldrich) (15 min) and mounted with Mowiol (4.8% Mowiol 4.88, 0.1% p-phenylenediamine, 12% glycerol, 0.02% NaN$_3$, and 0.2 M Tris-HCl). The sections were obtained from Bregma 1.0 to −0.4 mm and images were taken in the ipsilateral region of the cerebral cortex. Four images were acquired in each slice ($n = 3$) using 20× dry objective lenses in an inverted microscope Olympus 1 × 71 (Olympus Corporation of the Americas, Center Valley, PA, USA). Image J software 1.8.0 was used to perform the imaging analysis [32]. Colocalization analysis was performed as follows: 1) Channels were separated with the channel split function; 2) Colocalization of the green (GLUT3) and red (GFAP or MAP2) channels was analyzed using the function for colocalization thresholds. Colocalization was reported as "R" total of MAP2 or GFAP/GLUT3 corresponding to Pearson's correlation coefficient. The coefficient ranges from −1 to 1, a result near +1 means perfect correlation, 0 indicates no correlation, and −1 for perfect anti-correlation.

2.10. Primary Culture of Neurons and Astrocytes from the Cerebral Cortex

Embryos (E17–18) were removed from the yolk sac and placed in cold Hank's Buffer Saline Solution (0.14 mM NaCl, 5.33 mM KCl, 0.44 mM KH_2PO_4, 5.56 mM glucose, and 0.34 mM Na_2HPO_4). The brain was removed and transferred to cold a DMEM™ culture medium (D1152-1L, Thermo Scientific). Cerebral cortices were dissected and transferred to Neurobasal™ medium (21103049, Thermo Scientific) supplemented with B27™ 1X (17504-044, Thermo Scientific) and GlutaMAX™ 1X (35050-061, ThermoFisher Scientific). Then, tissue was enzymatically disintegrated with 3 mL of a trypsin solution for 5 min at 37 °C, and the reaction was inhibited with 1 mL of fetal bovine serum (FBS; S1810, BIOWEST, Nuaillé, France). The homogenous solution was passed through a 70-μm filter, and seeded at a density of 25,000 cells/cm^2. After three days, 50% of the culture medium was replaced (it was not supplemented with cytosine arabinoside to allow the growth of astrocytes). Cells were maintained at 37 °C, 21% O_2, and 5% CO_2. Experiments were carried out after 7 days in vitro (DIV). The presence of neurons was established by immunofluorescence, using the specific neuronal tag, the MAP-2 and astrocytes by the presence of the GFAP. Hoechst 3342 (Sigma Aldrich) was used to detect nuclei.

To induce excitotoxicity, cultures were treated with L-glutamate (100 μM) and glycine (10 μM) for 10 min. Cultures were divided into: CT, cells to which the culture medium was changed by Krebs–Henseleit solution (KHB) solution followed by recovery; GLU, cells with induced excitotoxicity followed by recovery; GLU + RSV, cells exposed to excitotoxicity plus resveratrol (40 μM).

2.11. Immunofluorescence in Cultures

Cells seeded on glass coverslips in 24-well plates were fixed with formaldehyde (0.5% in PBS, 10 min), permeabilized with cooled methanol (−70 °C, 1 min), washed with ice-cooled PBS (5 min), and blocked with bovine serum albumin (5% in PBS, 1 h, at room temperature). Subsequently, cells were incubated with Hoechst 33,342 (10 μg/mL) and the antibody anti-phospho-Thr 172-adenosine monophosphate (AMP)-activated protein kinase (AMPK) α (2531, Cell Signaling Technology, Danvers, MA, USA) at 4 °C, overnight. Additionally, the anti-MAP-2 antibody was used to identify neurons. After incubation, cells were washed with ice-cooled PBS (3 times, 10 min) and then incubated with the secondary antibodies Alexa Fluor® 594-conjugated anti-chicken IgG or DyLight™ 488-conjugated anti-mouse IgG (2 h, at 37 °C). Followed by washes with ice-cooled PBS (3 times, 10 min). The coverslips were mounted with Mowiol. Fluorescence was detected with an inverted microscope Olympus model 1 × 71 using a 20× dry objective.

2.12. Statistical Analysis

Statistical data were acquired from at least 3 independent experiments. Data were presented as the mean ± standard deviation. The analysis of variance of one way (ANOVA) followed by Tukey was performed using the GraphPad Prism 5 software (Inc. San Diego, CA, USA).

3. Results

3.1. Resveratrol Reduces the Damage Induced by the MCAO

We measured the infarct area induced by 2 h of MCAO with the TTC staining (Figure 1A,B) and found that resveratrol has a protective effect as has previously demonstrated [29]. We also performed an immunofluorescence assay to detect MAP2 expression because its loss reflects cytoskeletal breakdown and represents a sensitive marker of ischemic damage [34]. We observed an evident reduction in the

MAP2 signal in the MCAO + VH (vehicle) group, an alteration that represents a loss of axons integrity. When animals were treated with resveratrol, integrity of the fibers was conserved (Figure 1C,D). These results evidence that resveratrol avoided damage to the tissue and suggested the preservation of neuronal viability.

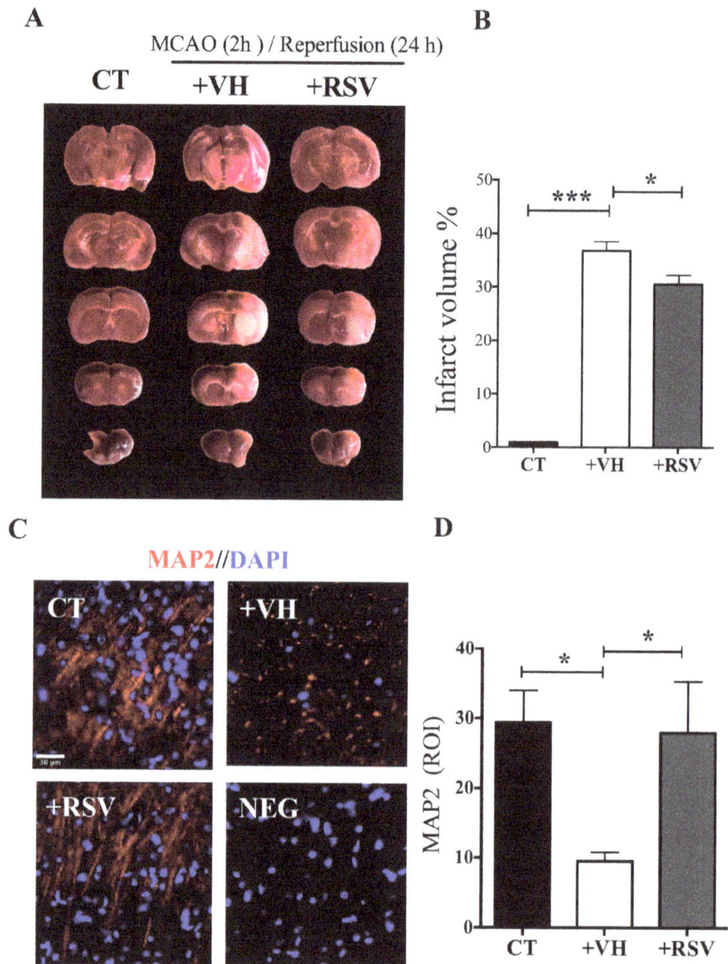

Figure 1. Resveratrol protects the brain from damage induced by ischemia. Rats were subjected to middle cerebral artery occlusion (MCAO) during 2 h followed by restoration of blood flow (reperfusion) for 24 h. Control animals (CT) were subjected to simulated MCAO. At the onset of reperfusion, rats received either vehicle, ethanol 50% (+VH) or resveratrol at 1.9 mg/kg, body weight (+RSV). Immunofluorescence was used to detect microtubule-associated protein 2 (MAP2) protein expression in neurons. 4′, 6-diamidino-2-phenylindole (DAPI) was used to detect nuclei. (**A**) Infarct area detected by 2,3,5-triphenyl tetrazolium (TTC) staining. (**B**) Quantification of the infarcted area. (**C**) Images show in red the signal associated to MAP2 and blue the nuclei. NEG, Negative control without primary antibody. (**D**) Quantification of MAP2 expression. Fluorescence was reported in % of the signal associated to MAP2 in the region of interest (ROI). Values were expressed as mean ± standard deviation. ANOVA, Tukey, * $p < 0.05$, *** $p < 0.001$.

3.2. The GLUT3 mRNA Up-Regulation Induced by MCAO is Prevented by Resveratrol Treatment

There are works showing that GLUT3 mRNA level is modified in different experimental models of ischemia [21]. We found that ischemia increased the level of expression of GLUT3 in the cerebral cortex at the ipsilateral side to the lesion. The highest level of expression was detected very early after reperfusion (2.5 ± 0.61-fold). Subsequently, GLUT3 mRNA returned to basal levels and was maintained from 3 to 8 h of reperfusion (Figure 2A).

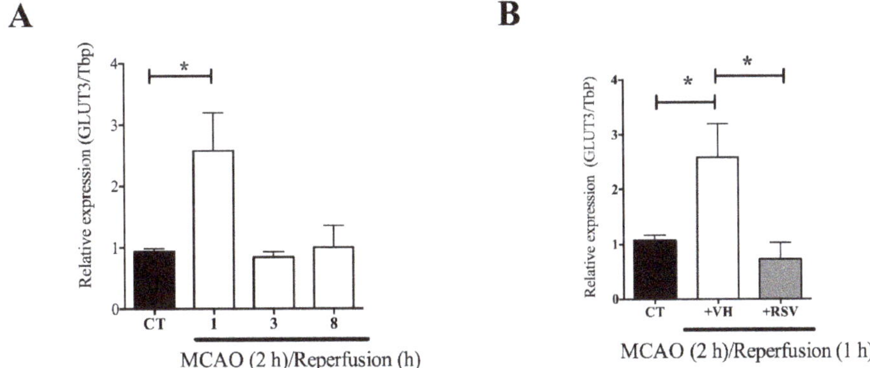

Figure 2. Resveratrol prevents the ischemia induced glucose transporter 3 (GLUT3) mRNA up-regulation in the brain. Rats were subjected to middle cerebral artery occlusion (MCAO) during 2 h followed by restoration of blood flow (reperfusion) for 24 h. Control animals (CT) were subjected to simulated MCAO. At the onset of restoration of blood flow, animals received either vehicle, ethanol 50% (+VH) or resveratrol at 1.9 mg/kg, body weight (+RSV). Relative mRNA expression was measured by quantitative PCR. (**A**) GLUT3 mRNA levels after MCAO and different times of reperfusion. (**B**) The effect of resveratrol on GLUT3 mRNA levels induced after MCAO and 1 h reperfusion. Values of expression were expressed as mean ± standard deviation. ANOVA, Tukey, * $p < 0.05$.

Because GLUT3 expression depends on transcription factors known to be regulated by resveratrol, we assessed its effect on the up-regulated GLUT3 mRNA levels induced by the MCAO. Experiments showed that resveratrol prevents the rise in the mRNA induced by 2 h of MCAO and 1 h of reperfusion (Figure 2B).

3.3. Resveratrol Prevents the Increase in the GLUT3 Protein Levels Induced by Ischemia and Reperfusion

The up-regulation in the mRNA levels is normally accompanied by a subsequent increase in protein concentration. Therefore, Western blot analysis was performed to corroborate that the effect of resveratrol on the GLUT3 mRNA levels correlated with GLUT3 protein expression. We found that 2 h of ischemia followed by 24 h of reperfusion stimulated a significant increase in GLUT3 protein level (3.4 ± 0.3-fold; Figure 3A,B). Consistent with the results found at the mRNA level, resveratrol administered at the onset of reperfusion also prevented the increase in the level of GLUT3 protein induced by the MCAO (Figure 3A,B).

3.4. GLUT3 Over Expression Induced after MCAO Does Not Overlap with MAP2 Signal

Immunofluorescence analysis was performed to assess whether MCAO and resveratrol alters expression of GLUT3 in a specific cell type. First, we confirm that GLUT3 is a protein with a high expression in neurons of the control brain while astrocytes are expressed at very low levels (Figure 4).

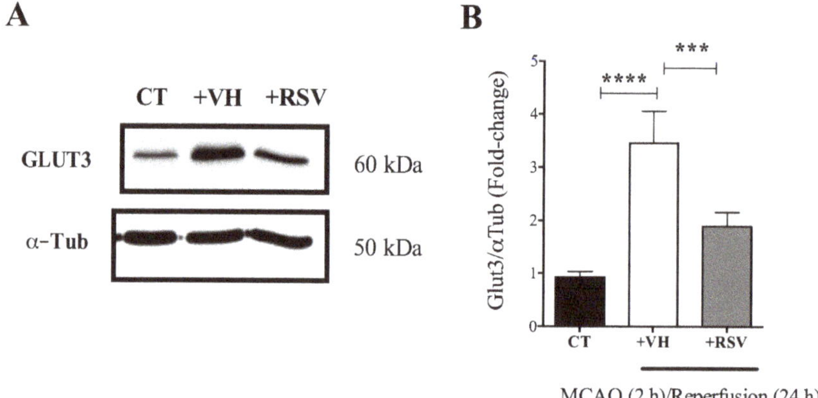

Figure 3. Resveratrol prevents glucose transporter 3 (GLUT3) up-regulation induced in the brain after ischemia. Rats were subjected to middle cerebral artery occlusion (MCAO) during 2 h followed by restoration of blood flow (reperfusion) for 24 h. Control animals (CT) were subjected to simulated MCAO. At the onset of restoration of blood flow, animals received either vehicle, ethanol 50% (+VH) or resveratrol at 1.9 mg/kg, body weight (+RSV). (**A**) GLUT3 protein expression measured by immunoblotting. (**B**) Quantification of the effect of resveratrol on GLUT3 protein after cerebral ischemia. Values were expressed as mean ± standard deviation. ANOVA, Tukey, *** $p < 0.001$, **** $p < 0.0001$.

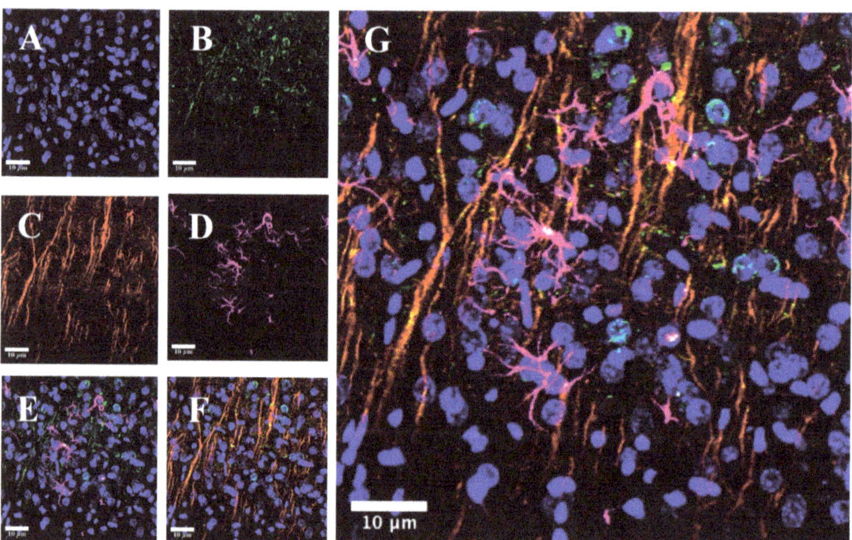

Figure 4. Glucose transporter 3 (GLUT3) expression in neurons and astrocytes in the control brain. Brain coronal sections were obtained from control rats and analyzed by immunofluorescence. Images were acquired on a Nikon Ti Eclipse inverted confocal microscope using a 20x objective lenses. (**A**) Nuclei (blue), 4′, 6-diamidino-2-phenylindole (DAPI) was used to detect nuclei, (**B**) GLUT3 (green), (**C**) Neuronal specific protein: the microtubule-associated protein 2 (MAP2, red), (**D**) Astrocyte specific protein: glial fibrillary acid protein (GFAP, violet), (**E**) Low expression of GLUT3 in GFAP positive cells, (**F**) High expression of GLUT3 in MAP2 positive cells, and (**G**) Merge of DAPI, GLUT3, MAP2 and GFAP signal, showing predominant expression of GLUT3 in neurons.

Then, we evaluated the effect of resveratrol in the MCAO groups. Because the rise induced by ischemia in the GLUT3 expression was huge, the intensity of the signal obtained in the microscope was downgraded in control groups to allow a proper quantification in Figure 5A,B (1.74 ± 0.69). After the insult, the increment on GLUT3 expression was significant, reaching 9.8 ± 2.4. As previously observed by Western blot, the increase induced by MCAO was prevented with resveratrol treatment (Figure 5A,B).

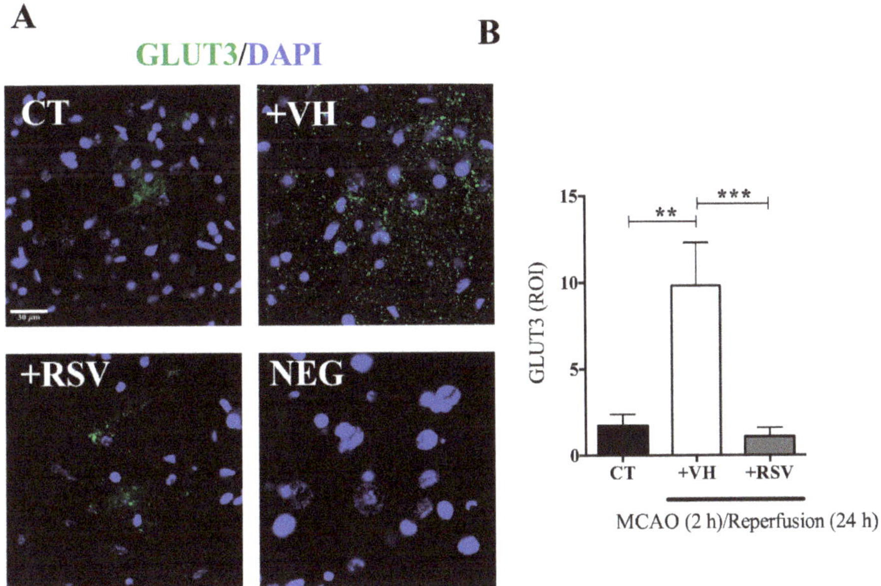

Figure 5. Resveratrol prevents glucose transporter 3 (GLUT3) up-regulation induced in the brain after ischemia. Rats were subjected to middle cerebral artery occlusion (MCAO) during 2 h followed by restoration of blood flow (reperfusion) for 24 h. Control animals (CT) were subjected to simulated MCAO. At the onset of restoration of blood flow, animals received either vehicle, ethanol 50% (+VH) or resveratrol at 1.9 mg/kg, body weight (+RSV). (**A**) Immunofluorescence of GLUT3 (green) protein expression. 4′, 6-diamidino-2-phenylindole (DAPI) was used to detect nuclei (blue). (**B**) Quantification of GLUT3 expression. Fluorescence was reported in % of the signal associated with GLUT3 in the region of interest (ROI). Values of ROI were expressed as mean ± standard deviation. ANOVA, Tukey, ** $p < 0.01$, *** $p < 0.001$.

After, the MAP2 level was detected to evaluate the abundance of GLUT3 in neurons. When colocalization of both proteins was quantified, we found that the GLUT3/MAP2 ratio did not increase after ischemia (Figure 6). However, MAP2 is a sensitive marker to damage, therefore the colocalization could be underestimated.

A **MAP2**/ **GLUT3**/**DAPI**

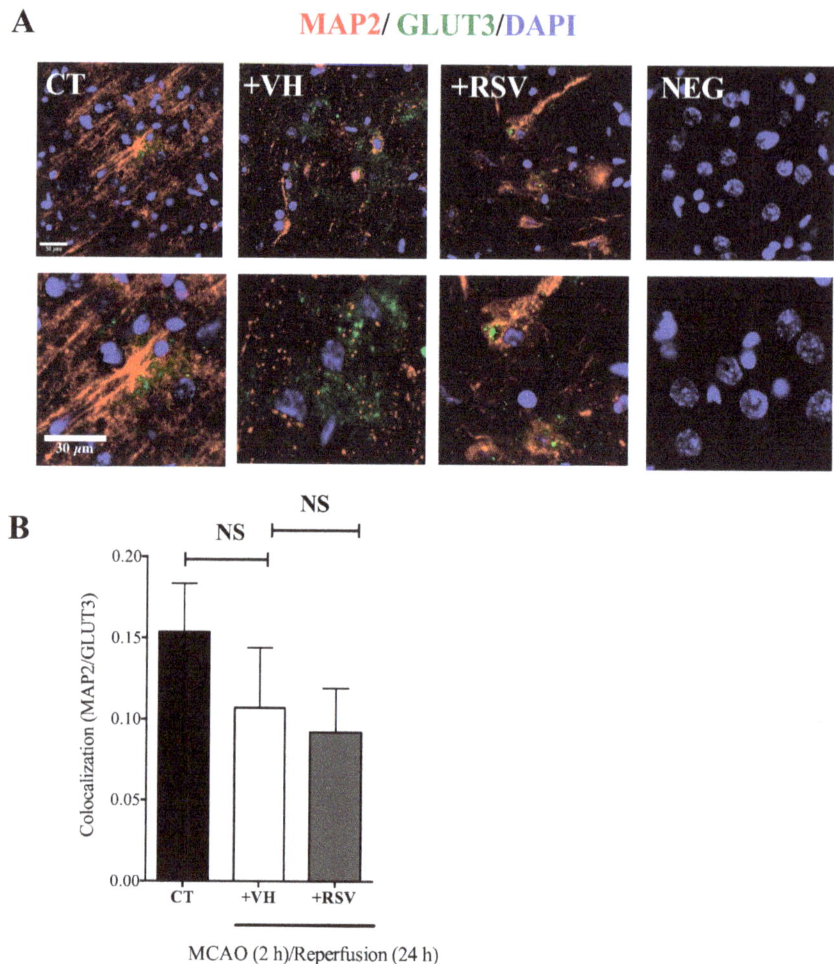

Figure 6. Glucose transporter 3 (GLUT3) expression is not increased in neurons after MCAO. Rats were subjected to middle cerebral artery occlusion (MCAO) during 2 h followed by restoration of blood flow (reperfusion) for 24 h. Control animals (CT) were subjected to simulated MCAO. At the onset of restoration of blood flow, animals received either vehicle, ethanol 50% (+VH) or resveratrol at 1.9 mg/kg, body weight (+RSV). Immunofluorescence was used to detect GLUT3 protein expression in neurons. 4′, 6-diamidino-2-phenylindole (DAPI) was used to detect nuclei. (**A**) Representative images show GLUT3 (green), microtubule-associated protein 2 (MAP2) (red), and nuclei (blue). NEG, Negative control without primary antibody. (**B**) Quantification of the fluorescence for GLUT3 expression in cells positive for the neuronal marker MAP2. Fluorescence was reported as R (Pearson's correlation coefficient) of the signal associated with the colocalization of MAP2 and GLUT3 proteins in the region of interest (ROI). Values were expressed as mean ± standard deviation. ANOVA, Tukey, NS, not significant.

3.5. MCAO Induces GLUT3 Up-Regulation in Astrocytes

Given the essential role of astrocytes in the transference of substrates to neurons [35], we next evaluated the expression of GLUT3 in cells positive to the astrocyte-specific mark GFAP. GFAP is a cytoskeleton protein, the levels of expression of which have been associated with inflammatory response to damage [36]. As previously showed, GLUT3 was up-regulated. Interestingly, GLUT3 overlapping with GFAP was significantly increased with MCAO and resveratrol prevented both the up-regulation and colocalization of GLUT3 with GFAP (Figure 7A,B).

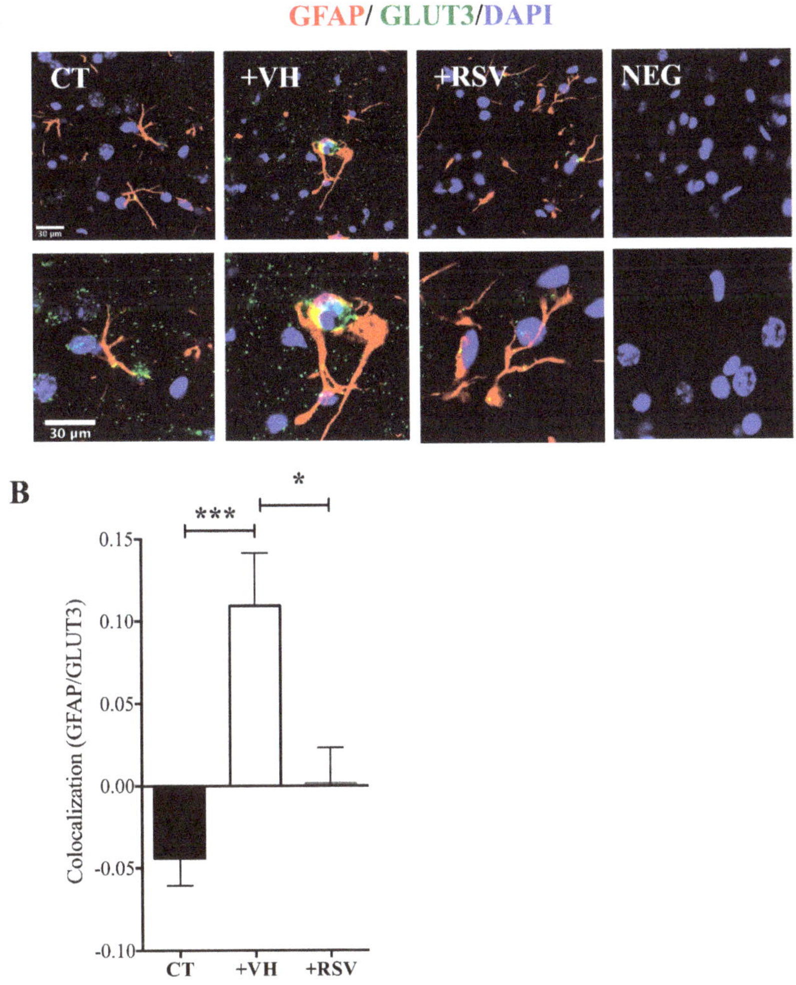

Figure 7. Resveratrol prevents the ischemia induced increase in glucose transporter 3 (GLUT3) expression. Rats were subjected to middle cerebral artery occlusion (MCAO) during 2 h followed by restoration of blood flow (reperfusion) for 24 h. Control animals (CT) were subjected to simulated MCAO. At the onset of restoration of blood flow, animals received either vehicle, ethanol 50% (+VH) or resveratrol at 1.9 mg/kg, body weight (+RSV). Immunofluorescence was used to detect GLUT3 protein expression in astrocytes. 4′, 6-diamidino-2-phenylindole (DAPI) was used to detect nuclei. (**A**) Representative images of GLUT3 (green), glial fibrillary acid protein (GFAP) (red), and nuclei (blue). (**B**) Quantification of GLUT3 expression astrocytes. Fluorescence quantification of the astrocyte (cells positive to GFAP) expressing GLUT3. Fluorescence was reported as R (Pearson's correlation coefficient) of the signal associated with the co-localization of GFAP and GLUT3 proteins in the region of interest (ROI). Values were expressed as mean ± standard deviation. ANOVA, Tukey, * $p < 0.05$, *** $p < 0.001$.

3.6. Resveratrol Induces Phosphorylation of AMPK

It is known that AMPK directly activates SIRT [37]. SIRT1 inhibits the transcriptional activity of p65-NF-κB, an important regulator of GLUT3 expression [24]. Therefore, we decided to evaluate the effect of resveratrol on the level of p-AMPK to identify a possible pathway of p65-NF-κB inactivation. We used mixed cultures of neurons and astrocytes exposed to glutamate-induced excitotoxicity, a model of ischemic stroke (Figure 8A). We observed that both excitotoxicity and resveratrol increased p-AMPK signal (Figure 8B,D). The increase in p-AMPK was found in neurons and also in non-MAP2 positive cells (i.e., astrocytes) (Figure 8C).

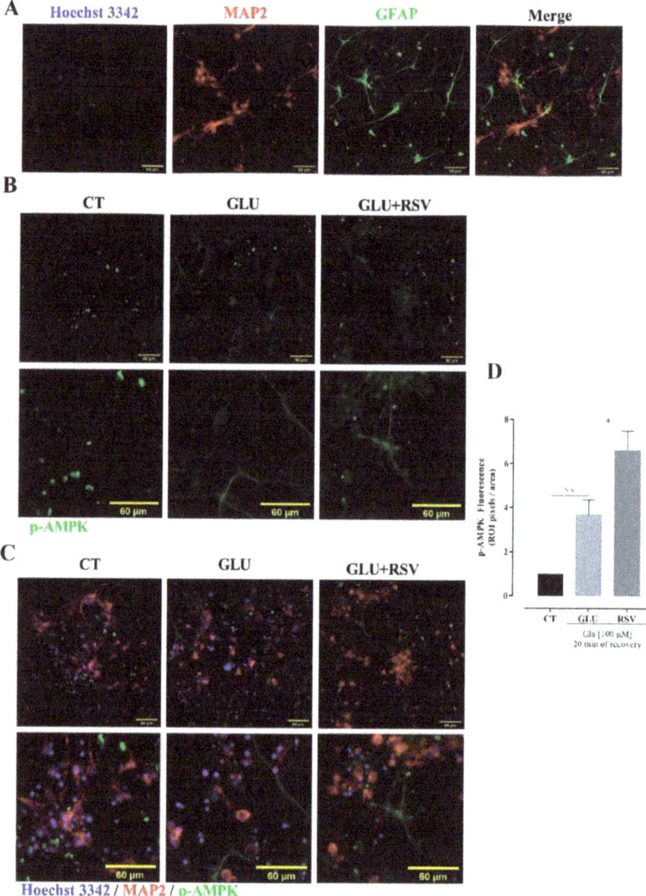

Figure 8. Resveratrol increases AMP-activated protein kinase (AMPK) phosphorylation. Primary mixed cultures of neurons and astrocytes of 7–8 days in vitro were stimulated with 100 µM glutamate (GLU) and 10 µM glycine to induce excitotoxicity and allowed to recover for 20 min. Cultures were divided into CT, control; GLU, exposed to excitotoxicity, GLU + RSV, plus resveratrol [40 µM]. Immunofluorescence. Cultures were stained with Hoechst 33,342 (blue) to identify nuclei, anti- microtubule-associated protein 2 (MAP2) followed by Alexa Fluor® 594 (red) to identify neurons, and anti- glial fibrillary acid protein (GFAP) or anti phospho AMPK(p-AMPK) followed by DyLight™ 488 (green) to identify astrocytes or phosphorylate AMPK. (**A**) Detection of the cellular types in the culture. (**B**) Detection of anti phospho AMPK (p-AMPK, green). (**C**) Co-localization of MAP2 and p-AMPK. (**D**) Quantification of the fluorescence for p-AMPK expression. Fluorescence was quantified and reported as "fold" of the intensity of the pixels in a region of interest (ROI). Values of ROI were expressed as mean ± standard deviation. ANOVA, * $p < 0.05$; N.S., not significant.

4. Discussion

The obstruction of blood flow during cerebral ischemia results in a critical depletion of cell energy. Consequently, the rapid restoration of nutrient supply is essential to maintain neuronal functions. Evidence suggests that cells integrating the neurovascular unit respond by up-regulating the expression of glucose transporters [21]. This response has been associated with a reduction of brain lesions in neurodegenerative diseases and traumatic brain injury [38,39].

In the present work, we observed that ischemia induced a substantial up-regulation of GLUT3 in the rat brain subjected to MCAO. We found an increase in GLUT3 expression within GFAP positive cells. Interestingly, in the healthy brain, GLUT3 gene expression is mainly limited to neurons [40,41]. Even so, in glioma cells, GLUT3 is the predominant glucose transporter [42]. Likewise, increased expression of GLUT3 occurs in astrocytes located in the center of chronic active lesions of multiple sclerosis disease, whereas axons showed a reduced expression [39]. Also, after 5 min of transient forebrain ischemia, a strong GLUT3 immunorcactivity is found in astrocytes in the hippocampus but not in differentiated neuroblasts [43]. Moreover, the specificity protein (SP), the hypoxia inducible factor-1α, and NF-κB modulate the expression of GLUT3 in astrocytes [24,42,44]. Therefore, the increase in GLUT3 synthesis observed after cerebral ischemia proposes the participation of astrocytes in the rescue of neurons to ensure sufficient nutrient transport under stress conditions and suggest the activation of the adaptative response to modulate glucose level [36].

We also found that MCAO induces an increase in GLUT3 expression in cells not expressing GFAP. Although neurons can detect glucose levels and adjust their transport, it is well known that fluctuating glucose levels modify the mitochondrial activity, activate apoptosis, and induce pro-inflammatory factors [45–47]. Therefore, the regulation of GLUT3 expression in neurons could represent a manner in which resveratrol induces protection. We used MAP2 as a neuronal marker to identify co-localization with GLUT3. However, MAP2 is a sensitive marker that begins to disappear as soon as 1 h after MCAO onset [48]. Because we performed the quantification 24 h after ischemia, the MAP2 signal was negligible and gave us little information. Therefore, the observed increased expression of GLUT3 could have occurred in neurons in which MAP2 was no longer detected because of the injury. Additionally, damage to the blood–brain barrier during ischemia could allow the infiltration of immune cells, which express high GLUT3 levels, and also in activated microglia [49–54]. Therefore, it will be essential to characterize the expression of GLUT3 in neurons at earlier times of reperfusion and in other types of cells.

The protective effect of resveratrol has been tested in many experimental models, where ROS provokes damage [52]. Previously, we demonstrated that resveratrol administered at the beginning of reperfusion reduces tissue damage and the neurological deficit induced by MCAO [29]. Here, we found that ischemia reduced the infarct area and the MAP2 level of expression, this effect represents an alteration in the microtubule structure of the cytoskeleton and signifies a reduction in the number of surviving neurons [34,48,53]. Importantly, the cytoskeleton alteration observed in neurons was reduced with resveratrol treatment. This situation might be associated with the presence of neurons that maintain viability and is in consensus to the resveratrol's protective effect.

When we investigated the effect of resveratrol in the up-regulation of GLUT3 expression induced by MCAO, we found that resveratrol significantly prevented it. Resveratrol modulates signaling pathways that depend on the redox state of the cell. For instance, SP, a transcription factor sensitive to oxidative stress has a relevant participation in GLUT3 mRNA synthesis [42,54]. The DNA binding activity of SP is blocked by resveratrol treatment after MCAO [29]. Consequently, it is possible that resveratrol blocks the SP binding to the promoter region of the GLUT3 gene, inhibiting its overexpression after MCAO.

In the current study, we found that resveratrol induces the phosphorylation of AMPK in mixed cultures of neurons and astrocytes subjected to excitotoxicity. It is relevant because glutamate excitotoxicity, cytokines, and high glucose induce NF-κB activation, a transcription factor involved in GLUT3 up-regulation in astrocytes [24,55,56]. Interestingly, resveratrol reverses high-glucose-induced inflammation by activating the AMPK/SIRT1/NF-κB pathway [55,57,58] and inhibits the transactivation of p65-NF-κB [59]. Additionally, resveratrol induces AMPK activation at early times of reperfusion in the in vivo model of MCAO, and compound C, an AMPK inhibitor prevents its protective effects [30]. Thus, it is also possible that resveratrol, through the activation of AMPK, down-regulates the GLUT3 expression induced after MCAO, which could occur in both neurons and astrocytes.

5. Conclusions

Resveratrol treatment resulted in a decrease in ischemic-induced damaged and also prevented the increase in GLUT3 expression. Our results indicated that astrocytes were partially responsible for the over-expression of GLUT3 after 2 h of MCAO and 24 h of reperfusion, and neurons could be also implicated at earlier times. It is known that GLUT3 up-regulation has a protective role against cerebral damage. However, we cannot exclude the possibility that the effect of resveratrol could be associated with an endogenous response that prevents glucose-induced toxicity in neurons and activates a compensatory mechanism in astrocytes. Nonetheless, our current study is initial, and additional research is required to better understand of the process.

Author Contributions: Conceptualization, G.F.G.A. and P.A.; Data curation, G.F.G.A. and I.A.-B.; Formal analysis, G.F.G.A., I.A.-B., J.F.-P., N.P.-R., A.O.-P., I.T., J.P.-C. and P.A.; Funding acquisition, J.P.-C. and P.A.; Investigation, G.F.G.A. and I.A.-B.; Methodology, G.F.G.A., I.A.-B., J.F.-P., N.P.-R., A.O.-P. and I.T.; Resources, J.P.-C. and P.A.; Supervision, P.A.; Writing—original draft, G.F.G.A. and I.A.-B.; Writing—review & editing, J.F.-P., N.P.-R., A.O.-P., I.T., J.P.-C. and P.A. All authors have read and agreed to the published version of the manuscript.

Funding: This research was funded by Consejo Nacional de Ciencia y Tecnología (CONACYT), grand number CB-2012-01-182266 to P.A.

Acknowledgments: Germán Fernando Gutiérrez Aguilar is a student from the Programa de Doctorado en Ciencias Biomédicas, Universidad Nacional Autónoma de México (UNAM), and beneficiary of scholarship No. 245537 from Consejo Nacional de Ciencia y Tecnología (CONACYT). Iván Mijail Alquisiras-Burgos and Narayana Pineda-Ramírez received the scholarship Nos. 275610 and 484304 from CONACYT.

Conflicts of Interest: The authors declare no conflict of interest.

References

1. Peters, O.; Back, T.; Lindauer, U.; Busch, C.; Megow, D.; Dreier, J.; Dirnagl, U. Increased formation of reactive oxygen species after permanent and reversible middle cerebral artery occlusion in the rat. *J. Cereb. Blood Flow Metab.* **1998**. [CrossRef]

2. Domínguez, C.; Delgado, P.; Vilches, A.; Martín-Gallán, P.; Ribó, M.; Santamarina, E.; Molina, C.; Corbeto, N.; Rodríguez-Sureda, V.; Rosell, A.; et al. Oxidative stress after thrombolysis-induced reperfusion in human stroke. *Stroke* **2010**, *41*, 653–660. [CrossRef] [PubMed]

3. Oguntibeju, O.O. Type 2 diabetes mellitus, oxidative stress and inflammation: Examining the links. *Int. J. Physiol. Pathophysiol. Pharmacol.* **2019**, *11*, 45–63. [PubMed]

4. Dirnagl, U.; Iadecola, C.; Moskowitz, M.A. Pathobiology of ischaemic stroke: An integrated view. *Trends Neurosci.* **1999**, *22*, 391–397. [CrossRef]

5. Lopez, M.S.; Dempsey, R.J.; Vemuganti, R. Resveratrol neuroprotection in stroke and traumatic CNS injury. *Neurochem. Int.* **2015**, *89*, 75–82. [CrossRef] [PubMed]

6. Zhang, F.; Liu, J.; Shi, J.S. Anti-inflammatory activities of resveratrol in the brain: Role of resveratrol in microglial activation. *Eur. J. Pharmacol.* **2010**, *636*, 1–7. [CrossRef]

7. Pineda-Ramírez, N.; Gutiérrez Aguilar, G.F.; Espinoza-Rojo, M.; Aguilera, P. Current evidence for AMPK activation involvement on resveratrol-induced neuroprotection in cerebral ischemia. *Nutr. Neurosci.* **2018**, *21*, 229–247. [CrossRef]

8. Shang, Y.J.; Qian, Y.P.; Liu, X.D.; Dai, F.; Shang, X.L.; Jia, W.Q.; Liu, Q.; Fang, J.G.; Zhou, B. Radical-scavenging activity and mechanism of resveratrol-oriented analogues: Influence of the solvent, radical, and substitution. *J. Org. Chem.* **2009**, *74*, 5025–5031. [CrossRef]

9. Borra, M.T.; Smith, B.C.; Denu, J.M. Mechanism of human SIRT1 activation by resveratrol. *J. Biol. Chem.* **2005**, *280*, 17187–17195. [CrossRef]

10. Chen, C.Y.; Jang, J.H.; Li, M.H.; Surh, Y.J. Resveratrol upregulates heme oxygenase-1 expression via activation of NF-E2-related factor 2 in PC12 cells. *Biochem. Biophys. Res. Commun.* **2005**, *331*, 993–1000. [CrossRef]
11. Lan, F.; Weikel, K.A.; Cacicedo, J.M.; Ido, Y. Resveratrol-induced AMP-activated protein kinase activation is cell-type dependent: Lessons from basic research for clinical application. *Nutrients* **2017**, *9*, 751. [CrossRef] [PubMed]
12. Zha, X.; Hu, Z.; Ji, S.; Jin, F.; Jiang, K.; Li, C.; Zhao, P.; Tu, Z.; Chen, X.; Di, L.; et al. NFκB up-regulation of glucose transporter 3 is essential for hyperactive mammalian target of rapamycin-induced aerobic glycolysis and tumor growth. *Cancer Lett.* **2015**, *359*, 97–106. [CrossRef]
13. Kao, Y.S.; Fong, J.C. Endothelin-1 induces glut1 transcription through enhanced interaction between Sp1 and NF-kappaB transcription factors. *Cell Signal.* **2008**, *20*, 771–778. [CrossRef] [PubMed]
14. Vannucci, S.J.; Seaman, L.B.; Vannucci, R.C. Effects of hypoxia-ischemia on GLUT1 and GLUT3 glucose transporters in the immature rat brain. *J. Cereb. Blood Flow Metab.* **1996**, *16*, 77–81. [CrossRef] [PubMed]
15. Patching, S.G. Glucose Transporters at the Blood-Brain Barrier: Function, Regulation and Gateways for Drug Delivery. *Mol. Neurobiol.* **2017**, *54*, 1046–1077. [CrossRef] [PubMed]
16. Mueckler, M.; Thorens, B. The SLC2 (GLUT) family of membrane transporters. *Mol. Aspects Med.* **2013**, *34*, 121–138. [CrossRef]
17. Maher, F.; Vannucci, S.J.; Simpson, I.A. Glucose transporter proteins in brain. *FASEB J.* **1994**, *8*, 1003–1011. [CrossRef]
18. Simpson, I.A.; Dwyer, D.; Malide, D.; Moley, K.H.; Travis, A.; Vannucci, S.J. The facilitative glucose transporter GLUT3: 20 Years of distinction. *Am. J. Physiol. Endocrinol. Metab.* **2008**, *295*, 242–253. [CrossRef]
19. Maher, F.; Davies-Hill, T.M.; Simpson, I.A. Substrate specificity and kinetic parameters of GLUT3 in rat cerebellar granule neurons. *Biochem. J.* **1996**, *315*, 827–831. [CrossRef]
20. Uldry, M.; Thorens, B. The SLC2 family of facilitated hexose and polyol transporters. *Pflugers Arch. Eur. J. Physiol.* **2004**, *447*, 480–489. [CrossRef]
21. Espinoza-Rojo, M.; Ivonne Iturralde-Rodriguez, K.; Elena Chanez-Cardenas, M.; Eugenia Ruiz-Tachiquin, M.; Aguilera, P. Glucose Transporters Regulation on Ischemic Brain: Possible Role as Therapeutic Target. *Cent. Nerv. Syst. Agents Med. Chem.* **2012**, *10*, 317–325. [CrossRef]
22. Gerhart, D.Z.; Leino, R.L.; Taylor, W.E.; Borson, N.D.; Drewes, L.R. GLUT1 and GLUT3 gene expression in gerbil brain following brief ischemia: an in situ hybridization study. *Mol. Brain Res.* **1994**, *25*, 313–322. [CrossRef]
23. Li, X.; Han, H.; Hou, R.; Wei, L.; Wang, G.; Li, C.; Li, D. Progesterone treatment before experimental hypoxia-ischemia enhances the expression of glucose transporter proteins GLUT1 and GLUT3 in neonatal rats. *Neurosci. Bull.* **2013**, *29*, 287–294. [CrossRef] [PubMed]
24. Iwabuchi, S.; Kawahara, K. Inducible astrocytic glucose transporter-3 contributes to the enhanced storage of intracellular glycogen during reperfusion after ischemia. *Neurochem. Int.* **2011**, *59*, 319–325. [CrossRef]
25. Weisová, P.; Concannon, C.G.; Devocelle, M.; Prehn, J.H.M.; Ward, M.W. Regulation of glucose transporter 3 surface expression by the AMP-activated protein kinase mediates tolerance to glutamate excitation in neurons. *J. Neurosci.* **2009**, *29*, 2997–3008. [CrossRef] [PubMed]
26. Du, Y.; Deng, W.; Wang, Z.; Ning, M.; Zhang, W.; Zhou, Y.; Lo, E.H.; Xing, C. Differential subnetwork of chemokines/cytokines in human, mouse, and rat brain cells after oxygen-glucose deprivation. *J. Cereb. Blood Flow Metab.* **2017**, *37*, 1425–1434. [CrossRef]
27. Gavillet, M.; Allaman, I.; Magistretti, P.J. Modulation of astrocytic metabolic phenotype by proinflammatory cytokines. *Glia* **2008**, *56*, 975–989. [CrossRef]
28. Bélanger, M.; Allaman, I.; Magistretti, P.J. Differential effects of pro- and anti-inflammatory cytokines alone or in combinations on the metabolic profile of astrocytes. *J. Neurochem.* **2011**, *116*, 564–576. [CrossRef]
29. Alquisiras-Burgos, I.; Ortiz-Plata, A.; Franco-Pérez, J.; Millán, A.; Aguilera, P. Resveratrol reduces cerebral edema through inhibition of de novo SUR1 expression induced after focal ischemia. *Exp. Neurol.* **2020**, *330*. [CrossRef]
30. Pineda-Ramírez, N.; Alquisiras-Burgos, I.; Ortiz-Plata, A.; Ruiz-Tachiquín, M.E.; Espinoza-Rojo, M.; Aguilera, P. Resveratrol Activates Neuronal Autophagy Through AMPK in the Ischemic Brain. *Mol. Neurobiol.* **2020**, *57*, 1055–1069. [CrossRef]
31. Longa, E.Z.; Weinstein, P.R.; Carlson, S.; Cummins, R. Reversible middle cerebral artery occlusion without craniectomy in rats. *Stroke* **1989**, *20*, 84–91. [CrossRef] [PubMed]

32. Rasband, W.S. *Image*; U.S. National Institutes of Health: Bethesda, MD, USA, 1997.

33. Livak, K.J.; Schmittgen, T.D. Analysis of relative gene expression data using real-time quantitative PCR and the 2-ΔΔCT method. *Methods* **2001**, *25*, 402–408. [CrossRef] [PubMed]

34. Härtig, W.; Krueger, M.; Hofmann, S.; Preißler, H.; Märkel, M.; Frydrychowicz, C.; Mueller, W.C.; Bechmann, I.; Michalski, D. Up-regulation of neurofilament light chains is associated with diminished immunoreactivities for MAP2 and tau after ischemic stroke in rodents and in a human case. *J. Chem. Neuroanat.* **2016**, *78*, 140–148. [CrossRef]

35. Descalzi, G.; Gao, V.; Steinman, M.Q.; Suzuki, A.; Alberini, C.M. Lactate from astrocytes fuels learning-induced mRNA translation in excitatory and inhibitory neurons. *Commun. Biol.* **2019**. [CrossRef] [PubMed]

36. Sims, N.R.; Yew, W.P. Reactive astrogliosis in stroke: Contributions of astrocytes to recovery of neurological function. *Neurochem. Int.* **2017**, *107*, 88–103. [CrossRef]

37. Chang, C.; Su, H.; Zhang, D.; Wang, Y.; Shen, Q.; Liu, B.; Huang, R.; Zhou, T.; Peng, C.; Wong, C.C.L.; et al. AMPK-Dependent Phosphorylation of GAPDH Triggers Sirt1 Activation and Is Necessary for Autophagy upon Glucose Starvation. *Mol. Cell.* **2015**, *60*, 930–940. [CrossRef]

38. Hamlin, G.P.; Cernak, I.; Wixey, J.A.; Vink, R. Increased expression of neuronal glucose transporter 3 but not glial glucose transporter 1 following severe diffuse traumatic brain injury in rats. *J. Neurotrauma* **2001**, *18*, 1011–1018. [CrossRef]

39. Nijland, P.G.; Michailidou, I.; Witte, M.E.; Mizee, M.R.; Van Der Pol, S.M.A.; Van Het Hof, B.; Reijerkerk, A.; Pellerin, L.; van der Valk, P.; de Vries, H.E.; et al. Cellular distribution of glucose and monocarboxylate transporters in human brain white matter and multiple sclerosis lesions. *Glia* **2014**, *62*, 1125–1141. [CrossRef]

40. Kong, L.; Zhao, Y.; Zhou, W.J.; Yu, H.; Teng, S.W.; Guo, Q.; Chen, Z.; Wang, Y. Direct neuronal glucose uptake is required for contextual fear acquisition in the dorsal hippocampus. *Front. Mol. Neurosci.* **2017**, *10*, 1–9. [CrossRef]

41. Lundgaard, I.; Li, B.; Xie, L.; Kang, H.; Sanggaard, S.; Haswell, J.D.R.; Sun, W.; Goldman, S.; Blekot, S.; Nielsen, M.; et al. Direct neuronal glucose uptake heralds activity-dependent increases in cerebral metabolism. *Nat. Commun.* **2015**. [CrossRef]

42. Zheng, C.; Yang, K.; Zhang, M.; Zou, M.; Bai, E.; Ma, Q.; Xu, R. Specific protein 1 depletion attenuates glucose uptake and proliferation of human glioma cells by regulating GLUT3 expression. *Oncol. Lett.* **2016**, *12*, 125–131. [CrossRef] [PubMed]

43. Yoo, D.Y.; Lee, K.Y.; Park, J.H.; Jung, H.Y.; Kim, J.W.; Yoon, Y.S.; Won, M.-H.; Choi, J.H.; Hwang, I.K. Glucose metabolism and neurogenesis in the gerbil hippocampus after transient forebrain ischemia. *Neural Regen Res.* **2016**, *11*, 1254–1259. [CrossRef] [PubMed]

44. Wang, P.; Li, L.; Zhang, Z.; Kan, Q.; Chen, S.; Gao, F. Time-dependent homeostasis between glucose uptake and consumption in astrocytes exposed to CoCl$_2$ treatment. *Mol. Med. Rep.* **2016**, *13*, 2909–2917. [CrossRef] [PubMed]

45. Russo, V.C.; Higgins, S.; Werther, G.A.; Cameron, F.J. Effects of fluctuating glucose levels on neuronal cells in vitro. *Neurochem. Res.* **2012**, *37*, 1768–1782. [CrossRef]

46. Zhang, X.C.; Gu, A.P.; Zheng, C.Y.; Li, Y.B.; Liang, H.F.; Wang, H.J.; Tang, X.L.; Bai, X.X.; Cai, J. YY1/LncRNA GAS5 complex aggravates cerebral ischemia/reperfusion injury through enhancing neuronal glycolysis. *Neuropharmacology* **2019**, *158*. [CrossRef]

47. Huang, X.Y.; Leng, T.D.; Inoue, K.; Yang, T.; Liu, M.; Horgen, F.D.; Fleig, A.; Li, J.; Xiong, Z.G. TRPM7 channels play a role in high glucose-induced endoplasmic reticulum stress and neuronal cell apoptosis. *J. Biol. Chem.* **2018**, *293*, 14393–14406. [CrossRef]

48. Dawson, D.A.; Hallenbeck, J.M. Acute focal ischemia-induced alterations in MAP2 immunostaining: description of temporal changes and utilization as a marker for volumetric assessment of acute brain injury. *J. Cereb. Blood Flow Metab.* **1996**, *16*, 170–174. [CrossRef]

49. Hernvann, A.; Aussel, C.; Cynober, L.; Moatti, N.; Ekindjian, O.G. IL-1β, a strong mediator for glucose uptake by rheumatoid and non-rheumatoid cultured human synoviocytes. *FEBS Lett.* **1992**. [CrossRef]

50. Jurcovicova, J. Glucose transport in brain—Effect of inflammation. *Endocr. Regul.* **2014**. [CrossRef]

51. Xiong, X.Y.; Liu, L.; Yang, Q.W. Functions and mechanisms of microglia/macrophages in neuroinflammation and neurogenesis after stroke. *Prog. Neurobiol.* **2016**, *142*, 23–44. [CrossRef]

52. Gambini, J.; Inglés, M.; Olaso, G.; Lopez-Grueso, R.; Bonet-Costa, V.; Gimeno-Mallench, L.; Mas-Bargues, C.; Abdelaziz, K.M.; Gomez-Cabrera, M.C.; Vina, J.; et al. Properties of Resveratrol: In Vitro and In Vivo Studies about Metabolism, Bioavailability, and Biological Effects in Animal Models and Humans. *Oxid. Med. Cell. Longev.* **2015**, *2015*. [CrossRef]
53. Atalay, B.; Caner, H.; Can, A.; Cekinmez, M. Attenuation of microtubule associated protein-2 degradation after mild head injury by mexiletine and calpain-2 inhibitor. *Br. J. Neurosurg.* **2007**, *21*, 281–287. [CrossRef]
54. Rajakumar, A.; Thamotharan, S.; Raychaudhuri, N.; Menon, R.K.; Devaskar, S.U. Trans-activators regulating neuronal glucose transporter isoform-3 gene expression in mammalian neurons. *J. Biol. Chem.* **2004**, *279*, 26768–26779. [CrossRef] [PubMed]
55. Jiang, T.; Gu, J.; Chen, W.; Chang, Q. Resveratrol inhibits high-glucose-induced inflammatory "metabolic memory" in human retinal vascular endothelial cells through SIRT1-dependent signaling. *Can. J. Physiol. Pharmacol.* **2019**, *97*, 1141–1151. [CrossRef] [PubMed]
56. Marchetti, P.; Bugliani, M.; Boggi, U.; Masini, M.; Marselli, L. The pancreatic beta cells in human type 2 diabetes. *Adv. Exp. Med. Biol.* **2012**, *771*, 288–309. [CrossRef] [PubMed]
57. Wang, W.; Bai, L.; Qiao, H.; Li, Y.; Yang, L.; Zhang, J.; Lin, R.; Ren, F.; Zhang, J.; Ji, M. The protective effect of fenofibrate against TNF-α-induced CD40 expression through SIRT1-mediated deacetylation of NF-κB in endothelial cells. *Inflammation* **2014**, *37*, 177–185. [CrossRef]
58. Zheng, Z.; Chen, H.; Li, J.; Li, T.; Zheng, B.; Zheng, Y.; Jin, H.; He, Y.; Gu, Q.; Xu, N. Sirtuin 1-mediated cellular metabolic memory of high glucose via the LKB1/AMPK/ROS pathway and therapeutic effects of metformin. *Diabetes* **2012**, *61*, 217–228. [CrossRef]
59. Yeung, F.; Hoberg, J.E.; Ramsey, C.S.; Keller, M.D.; Jones, D.R.; Frye, R.A.; Mayo, M.W. Modulation of NF-κB-dependent transcription and cell survival by the SIRT1 deacetylase. *EMBO J.* **2004**, *23*, 2369–2380. [CrossRef]

MDPI

St. Alban-Anlage 66

4052 Basel

Switzerland

Tel. +41 61 683 77 34

Fax +41 61 302 89 18

www.mdpi.com

Brain Sciences Editorial Office

E-mail: brainsci@mdpi.com

www.mdpi.com/journal/brainsci